化工过程强化关键技术丛书
CHEMICAL PROCESS INTENSIFICATION SERIES

中国化工学会组织编写

Ionic Liquid in Process Intensification

离子液体过程强化

张锁江（Suojiang Zhang） 等编著

ELSEVIER

化学工业出版社
Chemical Industry Press
·北京·

内容简介

Ionic Liquid in Process Intensification explores the applications of ionic liquids for process intensification research;focuses on computational simulation methods of ionic liquids, their structural design, regulation, and interaction; describes process intensification reactions, separation, photo/electrochemistry, and material synthesis related to ionic liquids.

Ionic Liquid in Process Intensification's combination of computational chemistry, physical chemistry, chemical engineering, materials science, and many other basic and applied disciplines makes it a key reference for scientists, technical personnel, teachers, and students in chemical engineering, chemistry, materials science, energy, and environment science.

图书在版编目（CIP）数据

离子液体过程强化 = Ionic Liquid in Process Intensification : 英文 / 中国化工学会组织编写 ; 张锁江等编著. -- 北京 : 化学工业出版社, 2025. 7.
ISBN 978-7-122-47837-5

Ⅰ. TQ15

中国国家版本馆CIP数据核字第20255BY197号

本书由化学工业出版社与爱思唯尔（Elsevier）出版公司合作出版。版权由化学工业出版社所有。本版本仅限在中华人民共和国境内（不包括中国台湾地区和中国香港、澳门特别行政区）销售。

责任编辑：成荣霞　任睿婷　杜进祥　　　　　装帧设计：张　辉
责任校对：边　涛

出版发行：化学工业出版社
　　　　　（北京市东城区青年湖南街13号　邮政编码100011）
印　　装：北京建宏印刷有限公司
710mm×1000mm　1/16　印张18　字数380千字
2025年8月北京第1版第1次印刷

购书咨询：010-64518888　　　　　售后服务：010-64518899
网　　址：http://www.cip.com.cn
凡购买本书，如有缺损质量问题，本社销售中心负责调换。

定　　价：498.00元　　　　　　　　　　版权所有　违者必究

About the author

Prof. Suojiang Zhang is a member of CAS. He is a researcher of the Institute of Process Engineering (IPE), CAS; the President of Henan University.

Prof. Zhang is mainly engaged in the research of green chemical engineering and new energy and materials, especially focusing on the fundamental and application studies of ionic liquids and green low-carbon process engineering. He has developed a number of green media-enhanced reaction/separation technologies, breaking through major challenges in ionic liquid design, process innovation, and system integration, and realizing the industrial application of more than 10 green technologies. He has also won several academic awards, including the Second Class Prize of the National Natural Science Award, Second Class Prize of the National Technological Invention Award, the TWAS-Award in Chemistry, the Ho Leung Ho Lee Foundation Science and Technology Progress Award, and the Hou Debang Chemical Science and Technology Outstanding Achievement Award.

Preface

Chemical industry is a significant industry that contributes to the national economic development and a typical high-energy consuming industry. Chemical process enhancement is an effective means to solve the contradiction between development and pollution brought by process industry. It is also an important measure for the sustainable development of chemical processes. At present, most chemical processes involve multiphase complex reactions and separation systems such as gas–liquid, liquid–liquid, and gas–liquid–solid. However, they are still limited to traditional static, offline, and macroscopic research modes, and cannot achieve in-depth analysis of reaction and transfer behavior at the micro level. There are many methods for intensifying chemical processes, among which medium strengthening is one of the most effective means. One of the important challenges facing scientists is to advance chemical research from the existing macro static level to the micro level, and achieve a fundamental breakthrough in traditional chemical processes.

Ionic liquid is a green medium composed of anions and cations, it appears as a liquid at room temperature. It can serve as a medium to replace traditional organic solvents for chemical reactions and separation, achieving the greening of chemical processes. After decades of development, ionic liquid has now reached the stage of industrial practice, attracting high attention from academia and industry, and becoming an international scientific frontier and research hotspot for chemical process enhancement. Especially in recent years, the understanding of the scientific essence of ionic liquid structures has been continuously deepened, forming the scientific foundation for enhancing the chemical process of ionic liquids.

This book focuses on the chemical process of intensifying the ionic liquid structure. Starting from the interaction and interface of ionic liquids, it elaborates on the processes of strengthening reactions and separation, photo and electrochemical processes with ionic liquids, synthesis of ionic liquid-based materials, and so on. On the basis of original, further introduces the latest research achievements at home and abroad, analyzes and discusses them with some typical examples. It is a comprehensive monograph that introduces the structure control and process intensification of ionic liquids.

This book was organized and written by academician Suojiang Zhang of the Institute of Process Engineering, Chinese Academy of Sciences (CAS), and proposed the writing idea and outline. This book is divided into seven chapters, and the authors who participated in this book are as follows:

Chapter 1 Introduction
Sections 1—3   Suojiang Zhang, Institute of Process Engineering, CAS
               Yao Li, Institute of Process Engineering, CAS

Chapter 2 Interaction and interface of ionic liquids
Section 1   Hongyan He, Institute of Process Engineering, CAS
Section 2   Zhiping Liu, Beijing University of Chemical Technology
Sections 3—4   Hongyan He, Institute of Process Engineering, CAS
               Yanlei Wang, Institute of Process Engineering, CAS
               Weilu Ding, Institute of Process Engineering, CAS
Section 5   Zhiping Liu, Beijing University of Chemical Technology

Chapter 3 Ionic liquids intensify reaction process
Section 1   Jiayu Xin, Institute of Process Engineering, CAS
Sections 2—3   Juan Zhang, Hebei University of Science and Technology
Section 4   Yuhong Huang, Institute of Process Engineering, CAS
               Xuan Zhang, Institute of Process Engineering, CAS
               Boxia Guo, Institute of Process Engineering, CAS
Section 5   Ying Kang, Liaoning University
               Dawei Fang, Liaoning University

Chapter 4 Ionic liquids intensify separation process
Section 1   Junfeng Wang, Institute of Process Engineering, CAS
Section 2   Shaojuan Zeng, Institute of Process Engineering, CAS
Section 3   Hui Wang, Institute of Process Engineering, CAS
               Jing Wang, Sino Danish Center, University of Chinese Academy of Sciences/
                   Institute of Process Engineering, CAS
               Bingtong Chen, Institute of Process Engineering, CAS
Section 4   Shiying Zheng, Dalian Institute of Chemical Physics, CAS
               Qun Zhao, Dalian Institute of Chemical Physics, CAS
               Baofeng Zhao, Dalian Institute of Chemical Physics, CAS
               Lihua Zhang, Dalian Institute of Chemical Physics, CAS
               Yukui Zhang, Dalian Institute of Chemical Physics, CAS
Section 5   Junfeng Wang, Institute of Process Engineering, CAS
               Shangqing Chen, Institute of Process Engineering, CAS

Chapter 5 Photo- and electrochemical process with ionic liquids
Section 1   Yongmei Chen, Beijing University of Chemical Technology
Section 2   Lijun Han, Institute of Process Engineering, CAS
Section 3   Qingqing Miao, Institute of Process Engineering, CAS
               Aamir Saeed, Institute of Process Engineering, CAS
Section 4   Yongmei Chen, Beijing University of Chemical Technology
               Haomin Jiang, Beijing University of Chemical Technology
Section 5   Weiwei Qian, Institute of Process Engineering, CAS
               Bing Xue, Institute of Process Engineering, CAS

Chapter 6 Synthesis of ionic liquid-based materials
Section 1   Yibo Wang, Beijing Technology and Business University
Section 2   Yibo Wang, Beijing Technology and Business University

	Xingwang Gao, Cancer Hospital Chinese Academy of Medical Sciences, Shenzhen Center
Section 3	Tao Zhang, Institute of Process Engineering, CAS
Section 4	Xingwang Gao, Cancer Hospital Chinese Academy of Medical Sciences, Shenzhen Center
	Yibo Wang, Beijing Technology and Business University
Section 5	Tao Dong, Institute of Process Engineering, CAS
	Xinyue Liu, Institute of Process Engineering, CAS

Chapter 7 Future outlook

Sections 1−3	Xiaoqian Yao, Institute of Process Engineering, CAS
	Chunyan Shi, Institute of Process Engineering, CAS
	Zongxu Wang, Huizhou Institute of Green Energy and Advanced Materials

This book provides a new perspective for in-depth research on the behavioral patterns and regulatory mechanisms of chemical process intensification, as well as new research ideas and methods for the correlation between different scales from the microscopic world of molecular engineering to the macroscopic world of chemical engineering. This book covers many fundamental and applied disciplines such as computational chemistry, physical chemistry, chemical engineering, materials science, and biology. It is a research achievement that crosses multiple disciplines. The research in this book is expected to greatly enhance the theoretical depth and international influence of chemical process intensification, and provide technological support for transforming traditional chemical processes.

We would like to express our gratitude to all those who participated in the compilation and proofreading of this book! Due to the limited level and knowledge of the authors, we kindly ask readers to give us your feedback!

Suojiang Zhang

Contents

1. Introduction   **1**

  1.1 Brief introduction of ionic liquids and the understanding of
       nanomicrostructures   1
  1.2 Nanostructure regulation and process intensification of ionic liquids 9
  1.3 Conclusion and perspective   13
  References   13

2. Interaction and interface of ionic liquids   **15**

  2.1 Overview   15
  2.2 Methods for simulating ionic liquid structures   16
  2.3 Simulation study of ionic liquid structures in the interface   30
  2.4 Simulation and regulation of two-dimensional ionic liquids   37
  2.5 Prediction and control of ionic liquid structures   49
  References   52

3. Ionic liquids intensify reaction process   **57**

  3.1 Overview   57
  3.2 Ionic liquids regulate homogeneous catalysis reaction   58
  3.3 The multiphase reaction based on ionic liquids   63
  3.4 Bioionic liquids: tunable microenvironment for biocatalysis   71
  3.5 Ionic liquids intensified biomass conversion   79
  References   90

4. Ionic liquids intensify separation process   **97**

  4.1 Overview   97
  4.2 Ionic liquids intensify gas separation   98
  4.3 Application of ionic liquids in liquid−liquid extraction   108
  4.4 Ionic liquids for protein and protein complex extraction   117
  4.5 Membrane separation process with ionic liquids   128
  References   135

5. Photo- and electrochemical process with ionic liquids   **143**

5.1 Overview   143

5.2 Photocatalytic and photoelectrocatalytic process with ionic liquids   144

5.3 Application of ionic liquids in solar cells   150

5.4 Ionic liquids intensify the electrochemistry process   159

5.5 Application of ionic liquids in new energy batteries   184

References   197

6. Synthesis of ionic liquid-based materials   **203**

6.1 Overview   203

6.2 Preparation of nanomaterials with ionic liquids   204

6.3 Ionic liquids intensify the synthesis of molecular sieve materials   225

6.4 Synthesis of ionic liquid-based metal organic complexes   229

6.5 Synthesis and application of polymerized ionic liquids   236

References   245

7. Future outlook   **257**

7.1 Structural characteristics and dynamic stability mechanism of
     ionic liquids   259

7.2 Action mechanism and regulation law of ionic liquids in process
     intensification   260

7.3 The development of the chemical process for ionic liquids   261

References   264

Index   265

CHAPTER 1

Introduction

Contents

1.1 Brief introduction of ionic liquids and the understanding of nanomicrostructures   1
    1.1.1 Brief introduction of ionic liquids   1
    1.1.2 Understanding the history of ionic liquid structures   3
    1.1.3 Nanostructure of ionic liquids at different scales   5
1.2 Nanostructure regulation and process intensification of ionic liquids   9
    1.2.1 Ionic liquid nanostructures intensifying the reaction process   10
    1.2.2 Ionic liquid structures intensifying the separation process   11
    1.2.3 Ionic liquid intensification in electrical and magnetic environments   12
1.3 Conclusion and perspective   13
References   13

1.1 Brief introduction of ionic liquids and the understanding of nanomicrostructures

1.1.1 Brief introduction of ionic liquids

Ionic liquids are a class of room-temperature liquid state substances composed entirely of positive and negative ions, which can be regarded as a special form of ions [1]. Traditional salts such as NaCl have a high melting point due to their good ion symmetry, small ionic radius, and ability to be firmly bonded together by electrostatic force, while the anions and cations of ionic liquid are larger in size and symmetry, and it is difficult to form an orderly arrangement, resulting in a lower melting point. Ionic liquids are usually composed of organic cations and inorganic or organic anions and can be divided into four main categories according to the different organic cation hosts, namely, imidazole salts, pyridine salts, quaternary ammonium salts, and quaternary phosphonium salts, and common anion and cation types are shown in Fig. 1.1.

Since ionic liquids are composed entirely of ions, they have special structural and internal properties. Traditionally, the essence of the special properties of ionic liquids, such as high viscosity and changes with temperature and size, is due to the electrostatic interaction between the cations and anions inside ionic liquids. However, with the progress of scientific research, it has been found that hydrogen bonding in ionic liquids is also very important after studying the structures and properties of ionic liquids by various experimental and calculation researches [2]. At present, it is generally believed that the interaction of anions and cations in ionic liquids is the result of a combination

Cations

Anions

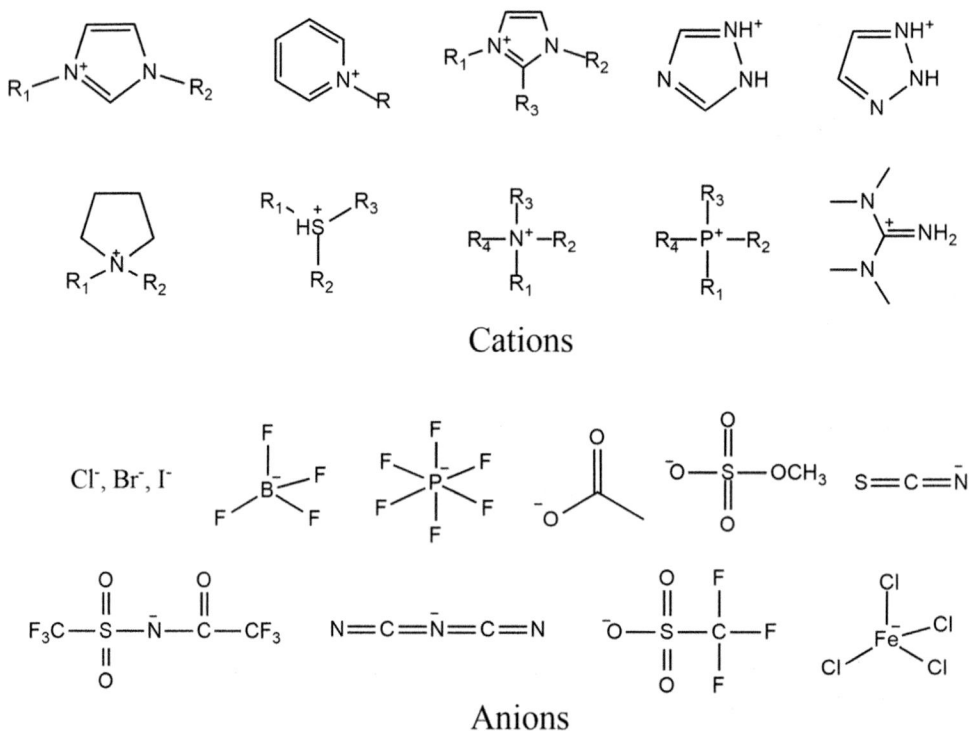

Figure 1.1 Chemical structures of 10 representative cations and anions commonly used in ionic liquids.

of forces, including electrostatic force, hydrogen bonding, van der Waals force, and polarization, among which van der Waals force (including induction force, dispersion force, and dipole−dipole interaction) and hydrogen bonding cannot be ignored [3].

Compared with conventional molecular solvents and high-temperature molten salts, ionic liquids show extremely low vapor pressure, wide liquid range, good solvent properties, moderate conductivity, and dielectric properties and have other characteristics such as catalytic performance, acidity, coordination, chirality, and so forth, resulting in great application potential in the fields of chemical industry, metallurgy, energy, environment, materials, biology, electrochemistry, and energy storage. Ionic liquids are expected to replace traditional heavy pollution media in traditional processes [4]. According to the market analysis, ionic liquids are expected to replace about 300 widely used organic solvents (170 million tons of global emissions annually, worth $6 billion), including benzene (carcinogenic), formaldehyde (disruptive immune function), and carbon disulfide (flammable and explosive). BASF has developed a deacidification process using ionic liquids, which improves the removal efficiency by a factor of 80000 [4]. Ionic liquids also exhibit good solubility of a variety of biomass materials, including cellulose, and are expected to replace the fiber-viscose process that has been used for more than a century (using more than 30 kinds of reagents, including a large number of acids and alkalis, heavy pollution, and high energy consumption). Professor Rogers, a leader in related research, won the U.S. President's Green Chemistry Challenge Award [5]. Ionic liquids are a new generation of green media/materials in the field of chemistry and chemical engineering and have the potential to form a new industrial technological revolution, but there is still a lack of research on the microscopic nature, structure−activity relationship, and reaction−amplification law of ionic liquids, which has become the bottleneck of large-scale industrialization of ionic liquids.

1.1.2 Understanding the history of ionic liquid structures

The development of ionic liquids has gone through a history of nearly 100 years. In the 1920s of the last century, Paul Walden reported the discovery of the first ionic liquid nitrate ethylamine ($C_2H_5NH_3NO_3$, melting point 13−14°C), which is produced by the neutralization reaction of amino ethane and concentrated nitric acid [6]. In 1934, Graenacher filed a patent for N-ethylpyridine chloride salt [7]. The structure of this salt contains the more common pyridine cation, which has a melting point of nearly 118 °C and is solid at room temperature. In 1951, Hurley and Wier [8] created the first generation of ionic liquids, chloro–aluminate ionic liquids, by heating N-alkyl pyridine and $AlCl_3$ together in order to find mild electrolytic conditions. In 1976, Osteryoung's group [9] solved the purification problem with the help of Bernard Gilbert [10], and they resynthesized [Bpy]Cl/$AlCl_3$ ionic liquid, and its physicochemical properties, especially its electrochemical properties, were studied in depth. In 1982, Wilkes and Hussey [11] calculated about 60 heterocyclic cations from semiempirical molecular orbitals, and with the help of computational predictions, they successfully synthesized 1,3-dialkyl imidazole aluminum chloride salt ([Emim]Cl/$AlCl_3$). The ionic liquid has a lower melting point and lower viscosity, which is more stable than pyridine ionic liquids. In 1992, Wilkes and Zaworotko [12] synthesized a series of new ionic liquids, whose cationic structure is mainly 1,3-dialkyrimidazole, and the anions are [CH_3CO_2]$^-$ and [NO_3], and [BF_4]$^-$. These ionic liquids do not absorb much water, are very stable, and have a wide range of anions. Based on this work, Bonhote et al. [13] and Gratzel et al. [14] used more anions, such as [CF_3SO_3]$^-$ and [N(CF_3SO_2)$_2$]$^-$, and a series of imidazole ionic liquids with stronger hydrophobicity were synthesized, which basically did not absorb water and had better electrochemical properties. Since then, more ionic liquids have been developed and expanded to a wider range of research fields, including organic synthesis, catalysis, dissolution, and separation, in addition to the earliest electrochemistry [15,16].

The attention of ionic liquids in the academic community began in the 1990s of the last century, and the number of related ionic liquid scientific research papers began to grow rapidly in recent 20 years. In 2003, Robin D. Rogers and Kenneth

R. Seddon, the world's leading experts on ionic liquids, pointed out in *Science* that ionic liquids represent the future development direction of solvents. Since then, the scientific research of ionic liquids has moved from preliminary exploration to in-depth research, and many important microscopic phenomena have been discovered and important theories have been proposed. Fig. 1.2 shows the development history of ionic liquids related papers in the past 30 years, and the progress of these studies in the deep insights of the structure of ionic liquids is marked in the box. When the earliest ionic liquids were synthesized, scientists thought that their structure was noncondensed ionic compounds. Subsequently, scientists further proposed that the hydrogen bonding network is an important structural factor, and at the same time, the ion pairs would produce oriented accumulation at the interface. In the 21st century, it was discovered that cations and anions are not always combined in the form of ion pairs, for example, ionic liquids have been found to have inhomogeneous structures such as clusters, and a large number of clusters have been widely studied. Moreover, for ionic liquid solutions, under the joint action of ion association and solvation effect, there are nano—microstructures such as ions, ion clusters, super ions, and micelles in the systems, which have an important impact on the processes of transport, separation, and catalysis. In 2011, the nanostructure of amphiphilic ionic liquids was extensively studied. Since then, ionic liquid nanostructures in three-dimensional space have become a hot topic. At present, in terms of the nanostructure of ionic liquids, the consensus of scientists is that ionic liquids exhibit a locally heterogeneous structure at the nanoscale and a continuous uniformity at a larger scale [17].

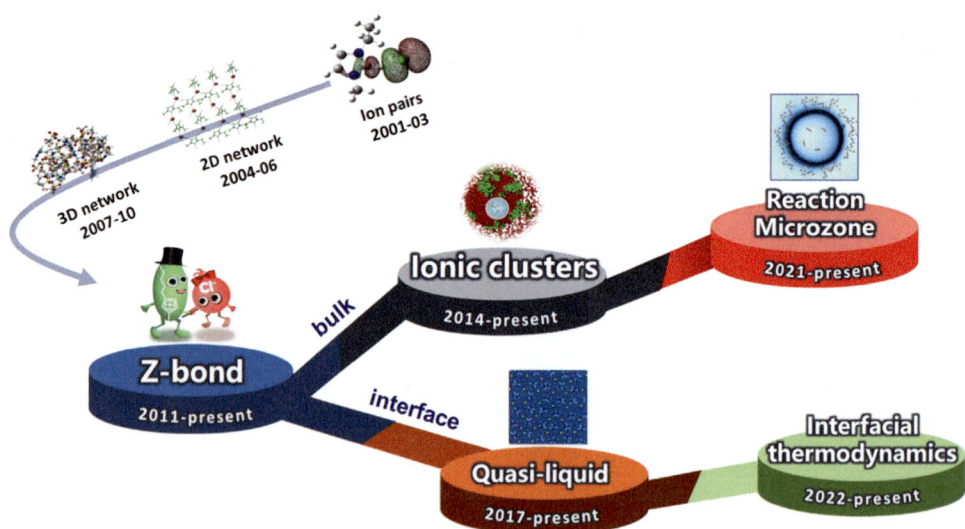

Figure 1.2 The brief evolution history of structural researches in ionic liquids [17].

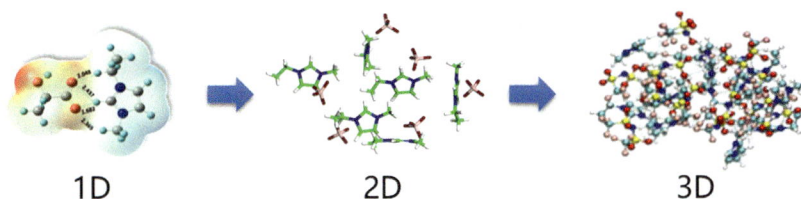

1D 2D 3D

Figure 1.3 1D→2D→3D models for ionic liquids and the hydrogen bonds network [20].

The macroscopic property depends on the microscopic nature. Through the study of the nanostructure of ionic liquids, the influence of the structure and interaction of cations and anions on their physical and chemical properties can be revealed in essence, so as to realize the design and optimization of the structure of ionic liquids according to the application. The conventional wisdom is that ionic liquids differ from high-temperature molten salts only by the temperature at which they appear in liquid state, and that ionic liquids are also completely ionized electrolytes, while there are other conventional beliefs that ionic liquids are also molecular liquids. This understanding cannot explain many experimental phenomena, such as ionic liquids dissolve cellulose [5] and can form a multiphase structure with water [18], or ionic liquids appear to be homogeneous on a macroscopic basis exhibiting heterogeneous structures at the nano-micro and flow field scales [19]. The unsystematic and incomplete understanding of ionic liquids has become a bottleneck in the development of ionic liquid theory and related technologies. Summarizing the research findings of the last 2 decades, ionic liquids are structural solvents ranging from molecular level (1D ion pairs, 2D hydrogen bond networks, and 3D ion clusters, as shown in Fig. 1.3) to mesoscopic (hydrogen bond networks, micelle clusters, and bicontinuous morphology) length scales, and understanding this is the key to revealing their mixed physicochemical properties and dynamic behavior [20]. Therefore, for the industrial application of ionic liquids, it is of great significance to establish the structure–activity relationship based on the hydrogen bonding of ionic liquids and the structure of nano-micro clusters and to realize the directional regulation of the properties of ionic liquid systems at the atomic/molecular level.

1.1.3 Nanostructure of ionic liquids at different scales

1.1.3.1 Ion pairs

An ion pair is the smallest unit in ionic liquids, so it is possible to describe a local ionic liquid microstructure based on the ion pair structure and the concentration of free individual anions and cations. There is conclusive evidence that ionic liquids can evaporate in the form of ion pairs, suggesting that the structure of ion pairs is also present in the bulk phase. In the beginning, this physicochemical concept, proposed at the beginning of the 20th century, has been significant for the research of phase transitions in aqueous electrolyte solutions as well as coulombic fluids, which operate as a single unit of a pair of opposite charged ions dispersed in H_2O. Because ionic liquids represent infinitely concentrated or solvent-free ionic solutions, tightly contacted ion pairs occur in this system. The concept of ion pairing has been widely used in computational studies of ionic liquids.

Scientists in China [3] studied [Bmim]Cl ion pairs and many other different kinds of ionic liquids using ab initio calculations, and the molecular orbital results showed that cations and anions can form special hydrogen bond structures, which are significantly different from other typical hydrogen bonds, with a mixed interaction of coulombic interactions and hydrogen bonds (Fig. 1.4). Due to hydrogen bonding, the anion distribution in certain positions in the cation ring plane, especially in front of C_2 carbon, is also a relatively stable structure. Interestingly, these hydrogen bonds are long enough (> 0.25 nm) and nonlinear (<170°), compared to the ideal hydrogen bond structure. The results show that both conformations are present because the reduced electrostatic attraction in the ion pairing allows forming other hydrogen bond-driven structures. The infrared spectroscopy data are similar to this, and the presence of hydrogen bond conformational isomers explains the vibrational redshift of the absorption peak of the anion—cation interaction.

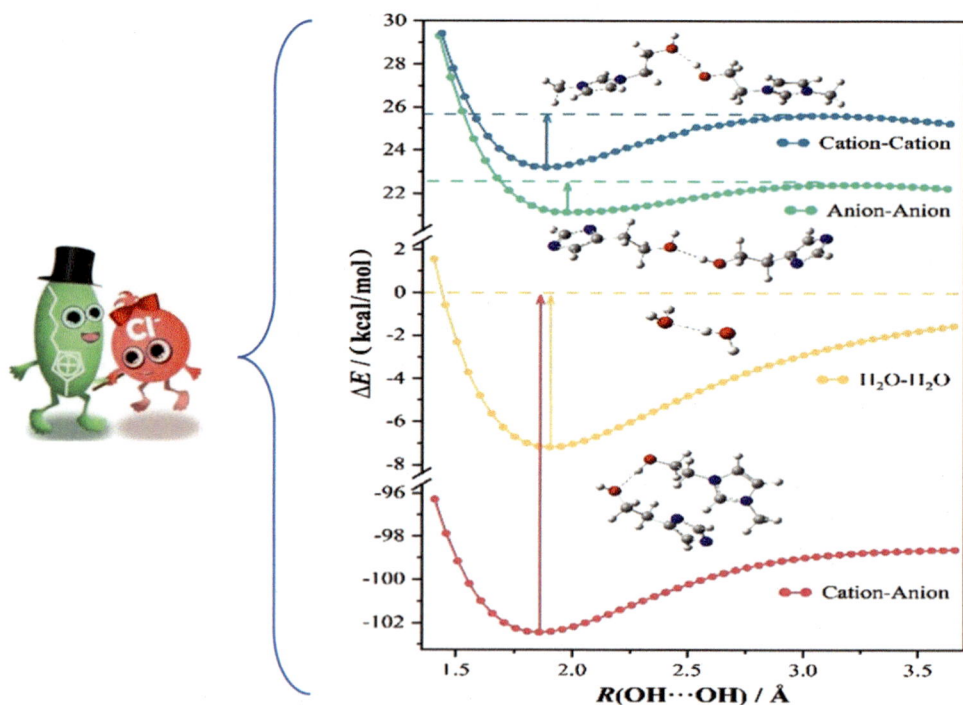

Figure 1.4 Z-bonds and the main types in ionic liquids [3].

1.1.3.2 Hydrogen bond network

Some researchers [21] first proposed that ionic liquids have a well-defined hydrogen bond structure. They studied the relationship between the solubility of the gas in EAN (ethylamine nitrate) as a function of temperature and found that the transfer of noble gases and methane from the cyclohexane phase to the EAN has enthalpy and entropy values below zero, similar to those in the H_2O environment. Although ionic liquids exhibit long-range disordered structures in the bulk phase, they always form staggered quasi-liquid structures at the interface induced by Z-bonds. This has also been confirmed by theoretical simulations and experiments (see Fig. 1.5). This hypothesis explains the peculiar structure between cationic and nonionic surfactant clusters previously detected in EAN, and this solvent-hydrogen bonding is thought to induce the necessary interaction to drive the amphiphilic assembly in dilute solvents.

In 2009, Ludwig et al. proved the existence of hydrogen bonds in protonic ionic liquids [22]. They measured the far-infrared spectra of EAN, PAN, and DMAN (dimethylammonium nitrate) in the region of $30-600$ cm^{-1} to excite hydrogen bond bending, stretching, and vibrational patterns in molecular liquids and performed complementary density functional theory (DFT) calculations to generalize the spectra of hydrogen bond interactions. In each protonic ionic liquid, the frequency difference between asymmetric and symmetrical stretch is about 65 cm^{-1}, indicating a certain hydrogen bond strength. The measured peak position and frequency difference are consistent with the far-infrared spectrum of pure water. This led the conclusion that there is a hydrogen bond network in proton-type ionic liquids, which is not like a tetrahedron but is still structurally similar to water. In a follow-up report, the authors continued to quantify the strength of hydrogen bonding interactions in protonic ionic

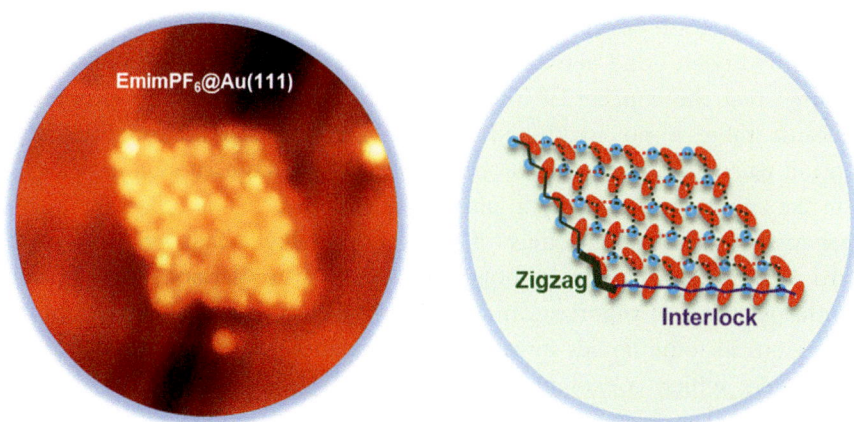

Figure 1.5 The structure of quasi-liquids formed by ionic liquids at interfaces; Z-bonds are shown as dashed lines [21].

liquids using DFT calculations. Trimethylammonium nitrate (a hydrogen bond donor) was found to have a higher energy per ion pair than tetramethylammonium nitrate (no hydrogen bond donor) by about 49 kJ/mol, which was attributed to the formation of a single cation-anion hydrogen bond. Notably, this value is more than twice that of hydrogen bonds in water (about 22 kJ/mol), indicating that they can be classified as moderate to strong hydrogen bonds in the established hydrogen bonding model.

Interestingly, recent spectroscopic studies have provided evidence of ionic hydrogen bonding in amino acid ionic liquids. This suggests that in ionic liquids of certain specific structures, hydrogen bonds can not only be formed between ions, but also stabilize the conformation of individual ions. This provides strong evidence that ionic liquids form hydrogen bonds. Both protonic ionic liquids and amino acid-based ionic liquids can form a network of hydrogen bonds that leads to typical fluid properties. In order to achieve this, the number of hydrogen bond donor and acceptor sites in the ionic liquid should be nearly equal and located in spatially accessible locations. However, there are some differences in hydrogen bonds of ionic liquids compared to molecular liquids. Unlike conventional hydrogen bonds, which make liquids more structured by increasing cohesion and inducing interactions, hydrogen bonds in ionic liquids appear to promote directed interactions because ionic liquids have similar electrostatic forces, which in solids can induce defect formation in the coulombic lattice, promoting the formation of larger-scale regular arrangements, such as the formation of ion clusters or amphiphilic structures.

1.1.3.3 Ionic clusters

In recent years, a lot of attempts have been made to describe the ionic liquid bulk phase structure as the shape of (net zero or net charged) ion clusters or aggregates. Most of the earlier work was summarized in Dupont's review [23], in which it was hypothesized that ionic liquids form a clustered large structure to form a three-dimensional hydrogen bond network. Recent studies on the structures of ionic liquids [20] reported that the concept of ion clusters in bulk phase must be expressed carefully, because there is no established standard for ion clusters, and the distinction between ion pairs and ion clusters is often arbitrary, although clusters have been confirmed in the amino acid ionic liquid phase [24].

Lopes et al. illustrated the effect of the cationic alkyl chain length on the bulk phase structure of ionic liquids by molecular dynamics simulations and calculations of molar enthalpy values of vaporization data [25]. Instead of being uniformly distributed, charged regions in ionic liquids form a continuous three-dimensional network similar to ion channels, which coexist with uncharged regions. For short alkyl chains (C_2), small globular alkane islands are formed within a continuous polar network. By increasing the alkyl chain length (C_6, C_8, and C_{12}), the cationic alkane groups can interconnect in a bicontinuous sponge-like nanostructure. Meanwhile, the ionic

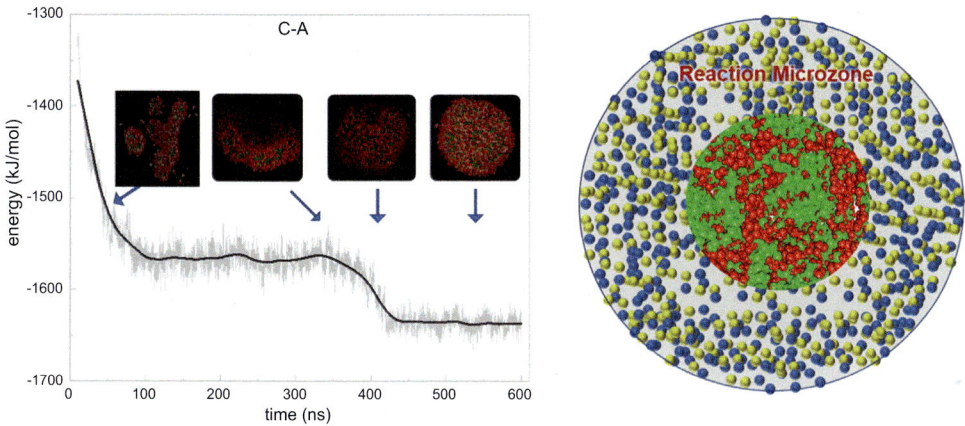

Figure 1.6 Evolution of ionic liquid clusters over time in molecular simulation and microscopic structures in typical reaction/separation processes.

liquids with long alkyl side chains (>6) are likely to form clusters in many kinds of solvents which can be further extended to micro-zone at micrometer scales, as shown in Fig. 1.6.

Hardacre et al. have published several papers on ionic liquid clusters, which have investigated structures in alkyl ionic liquids through EPSR simulations and neutron diffraction experiments [26,27]. EPSR fitting of diffraction spectra can elucidate local ion-ion distributions in imidazole ionic liquids. The arrangement of structures in these models shows a pronounced charge ordering, which resembles the structure of the crystalline state of some substances, and the outer charge distribution of cations and anions resembles an alternating "onion skin" structure. In these studies, ionic liquids with short alkyl chains such as $[C_1mim]^+$ cations are used, so they cannot be called amphiphilic cluster structures. Therefore, the bulk phase structure of ionic liquids is mainly determined by electrostatic interaction, and the cations and anions are not easily separated. However, the local ionic arrangement is also very important, because this structure persists with the increase of alkyl chain length while the characteristic peak of the bulk phase structure is present [28]. The structure in the polar charged region is largely unaffected by the increased cationic amphiphilicity [29].

1.2 Nanostructure regulation and process intensification of ionic liquids

The structure of ionic liquids plays a key role in many applications. Ionic liquids and their solvent mixtures are often used in chemical processes, and their properties can be adjusted by selecting different ions or functional groups. The properties and functions of solvents are greatly influenced by ion clusters in solution, which requires a deeper understanding of the structure—property relationship between ionic liquids and their clusters and how they can be controlled to achieve technological advances. In the following sections, we will focus on the impact and application of ionic liquid nanostructures in reaction and separation processes, and the understanding of the chemistry and cluster structure of ionic liquids can provide a scientific basis for realizing the application potential of ionic liquids.

1.2.1 Ionic liquid nanostructures intensifying the reaction process

A lot of studies have shown that the nano—microstructure of ionic liquids can enhance a wide range of organic chemical reactions. When ionic liquids participate in catalytic reactions, the role of ionic liquids can be roughly divided into two categories: one is as a green reaction solvent. Using their special solubility of eaction substrates and organometallic catalysts, the reaction is carried out in the ionic liquid phase, and at the same time, it is used to be immiscible with some organic solvents to make the product enter the organic solvent phase, which can not only realize the separation of the products well, but also realize the recovery and reuse of the catalyst in the ionic liquid phase simply through the method of physical phase separation. The other type is functionalized ionic liquids, that is, ionic liquids are used not only as green reaction media, but also as catalysts for reactions. For example, the inherent Lewis acidity of ionic liquids is used to catalyze esterification reactions, alkylation reactions, isomerization, etherification, and so forth or the purposeful synthesis of catalysts with special catalytic properties. In these reactions, not only are ionic liquids used as green reaction media or catalysts, but, due to the "designability" of their structures, the selection of appropriate ionic liquids can often play a synergistic catalytic role, so that the catalytic activity and selectivity are improved.

Among the existing reports on ionic liquid catalytic homogeneous reactions, the most noteworthy is the catalytic C_4 alkylation reaction by ionic liquids, and the related process has entered the pilot stage. Due to its acid strength, chloro-aluminic acid ionic liquid has been used by many researchers and companies to catalyze the C_4 alkylation reaction. Lu et al. [30] carried out a series of studies on the alkylation reaction of isobutane and 2-butene catalyzed by ionic liquids, which provides new ideas and basic data for the design and synthesis of catalysts for high-performance C_4 alkylation, the development of C_4 alkylation cleaning processes, and the cleaning of petrochemical, biochemical, and coal chemical processes catalyzed by other strong acids. The results showed that 1-methoxyethyl-3-methylimidazole bromo chloroaluminate ionic liquid ([MOEmim] $Br/AlCl_3$) was the most acidic when the molar ratio of $AlCl_3$ was 0.75, and the catalytic alkylation reaction had the best effect. Under the reaction temperature of 35°C and the volume ratio of isobutane and 2-butene at 10:1, the alkylation reaction with C_8 selectivity of 66.6% can be obtained, and its catalytic effect is much better than that of non-ether-functionalized chloro-aluminic acid ionic liquid, and the process can be recycled.

1.2.2 Ionic liquid structures intensifying the separation process

One of the most promising applications of ionic liquids is for biomass pretreatment or catalytic production of biofuels, where large biomass molecules including cellulose, wood, lignin, and so forth can be dissolved and regenerated under mild conditions by ionic liquids. Although much of the literature points to hydrogen bonding as the cause of cellulose dissolution in ionic liquids or molecular solvents, a recent review of experimental and simulation studies indicates the importance of hydrophobic interactions [31]. Cellulose itself has amphiphilic structural groups, and the hydrophobic domains in cellulose make the crystal structure stable in solution. In fact, every example of cellulose successfully dissolving in ionic liquids is achieved through the amphiphiles of one of the ions, usually cations. As described here, hydrophobic (soluble) interactions in ionic liquids are characterized by the aggregation of ion clusters, resulting in a well-defined arrangement of nanostructures within the bulk phase or at the interface. Therefore, a clear understanding of ionic liquid cluster aggregation can better illustrate the process of ionic liquid dissolution of cellulose/biomass, and ionic liquids can form typical micro-zone and have a good effect on the mild conversion of CO_2 (Fig. 1.7). Ionic liquids also produce distinct biological characteristics in these processes, as amphiphilic ionic liquids appear to inhibit microbial growth and reduce the efficiency of biofuel production. Ruegg et al. avoided this problem by genetically engineering microorganisms that are resistant to imidazole ionic liquids [32].

Ionic liquids also have a promising application in the capture of greenhouse gases such as carbon dioxide (CO_2). There are several major research efforts being carried out in this area aimed at reducing CO_2 emissions due to human activities and industries. Ionic liquids are now widely used in the development of technologies for separating CO_2 from flue gases from coal-fired power plants. In basic research, many physicochemical experiments and theoretical demonstrations have shown that it is feasible to separate CO_2 from ionic liquids, and the current challenge is to adjust the ionic reactivity, self-assembly, and carbon dioxide capture selectivity so that the technology can be scaled up and economically viable.

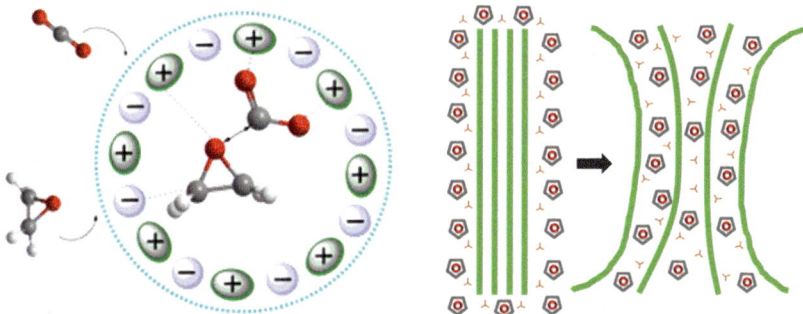

Figure 1.7 The great potential of ionic liquids has shown for application in CO_2 catalytic conversion and cellulose separation [32].

There are two main separation schemes: ionic liquid technology that reacts with CO_2 or physically dissolves CO_2 (at atmospheric pressure). Chemical capture of CO_2 is promising because the cations and anions of ionic liquids can be easily functionalized with amine groups, resulting in better stoichiometry of CO_2 reactions than single-molecule solvents such as monoethanolamide. This approach results in a large increase in solvent viscosity, but this can be overcome by disrupting the counter ions of the H-bond network, or by restricting the formation of H-bonds in highly unordered nanostructures. Moreover, an interesting approach is to use benzimidazole ionic liquids that change phase (solid $\rightarrow$ liquid) after dissolving CO_2. This allows the solvent melting heat to be used to maximize the release of CO_2 from the absorbent. Unlike chemical trapping, the physical separation of CO_2 is often difficult to achieve because it is often selectively mixed with other gas components, so that the solvent should have the ability to store large amounts of gas at partial pressure in the flue gas mixture.

1.2.3 Ionic liquid intensification in electrical and magnetic environments

Electrochemical processes involve electrodes/electrolytes/interfacial charge transfer between electrochemically active species, which is largely dependent on the properties of the electrolyte used. Ionic liquids have excellent characteristics such as electrical conductivity, low volatility, nonflammability, and wide electrochemical window, which can replace the original water and organic solvents as electrolytes, bringing electrochemistry into a new era. In recent years, important results have been achieved in the application of ionic liquids in supercapacitors, lithium-ion batteries, metal electrodeposition, biosensors, and electrochemical reduction of CO_2.

In addition, the effect of magnetic field on ionic liquids can be divided into two parts. One is the change in the properties of conventional ionic liquids under the magnetic field. For conventional ionic liquids, the viscosity under the magnetic field is reduced, and the greater the strength of the magnetic field, the smaller the viscosity. Moreover, the effect of the magnetic field on viscosity (δ value) is different for different ionic liquids. However, some studies have shown that the influence of magnetic field on viscosity changes regularly with the different structures of ionic liquids, and the specific mechanism is mainly because the magnetic field destroys the original hydrogen bond network structure of ionic liquids, so that the viscosity of ionic liquids decreases. So, the stronger the hydrogen bonds, the stronger the relative influence of the magnetic field. The second is the magnetic ionic liquid. It is a new type of functionalized ionic liquid, because its own cation or anion has a group that induces the magnetic field, so that it not only has excellent thermal stability, low volatility, good solubility, wide liquid range, and recoverability, but also can show a certain magnetization under the external magnetic field. Because of these properties, magnetic ionic liquids have a wide range of applications in the fields of extraction and separation, reaction catalysis, and carbon nanotube composites. The effect of magnetic field on the viscosity of magnetic ionic liquid can be understood as that magnetic ionic liquid is a small ball with weak magnetic field, and under the action of strong magnetic field, there is a magnetic attraction between the ball and the sphere, so the viscosity of the ionic liquid increases during the flow process.

1.3 Conclusion and perspective

There is no doubt that the unique nanostructure of ionic liquids offers a wide range of opportunities for them to innovate existing processes or develop entirely new engineering applications. Until now, most of the applications of ionic liquids have been limited to the laboratory or pilot phase. Ionic liquids have two constituent structures, anionic and cationic, consisting of charged and uncharged groups, while the ability of ionic liquids to self-assemble into ionic clusters gives the ionic arrangement regularity. On bulk phases above the nanoscale, ionic liquids have a homogeneous structure. Therefore, ionic liquids can be thought as fluids that are inhomogeneous on the nanoscale, but generally coherent overall. Recent advances in experimental and theoretical methods have demonstrated unprecedented results and discoveries in the study of ionic liquid structures, providing an opportunity for further study of ionic liquid nanostructures and their effects. It is believed that in the future, a cross-scale nanostructure map of ionic liquids will be established, which will promote the rapid development of ionic liquid related applications.

References

[1] Earle MJ, Seddon KR. Ionic liquids. Green solvents for the future. Pure and Applied Chemistry 2000;72(7):1391.
[2] Dong K, Zhang S, Wang J. Understanding the hydrogen bonds in ionic liquids and their roles in properties and reactions. Chemical Communications 2016;52(41):6744.
[3] Dong K, Liu X, Dong H, et al. Multiscale studies on ionic liquids. Chemical Reviews 2017;117 (10):6636.
[4] Rogers RD, Seddon KR. Ionic liquids-solvents of the future? Science 2003;302(5646):792.
[5] Swatloski RP, Spear SK, Holbrey JD, et al. Dissolution of cellose with ionic liquids. Journal of the American Chemical Society 2002;124(18):4974.
[6] Walden P. Molecular weights and electrical conductivity of several fused salts. Bulletin of the Imperial Academy of Sciences 1914;8:405–22.
[7] Charles G. Cellulose solution. US 1943176. 1934-01-09.
[8] Hurley FH, Wier TP. The electrodeposition of aluminum from nonaqueous solutions at room temperature. Journal of the Electrochemical Society 1951;98(5):207.
[9] Chum HL, Koch V, Miller L, et al. Electrochemical scrutiny of organometallic iron complexes and hexamethylbenzene in a room temperature molten salt. Journal of the American Chemical Society 1975;97(11):3264.
[10] Gale R, Gilbert B, Osteryoung R. Raman spectra of molten aluminum chloride: 1-butylpyridinium chloride systems at ambient temperatures. Inorganic Chemistry 1978;17(10):2728.

[11] Wilkes JS, Hussey CL. Selection of cations for ambient temperature chloroaluminate molten salts using MNDO molecular orbital calculations. Frank J Seiler Research Lab United States Air Force Academy Co, 1982.

[12] Wilkes JS, Zaworotko MJ. Air and water stable 1-ethyl-3-methylimidazolium based ionic liquids. Journal of the Chemical Society, Chemical Communications 1992;13:965.

[13] Bonhote P, Dias AP, Papageorgiou N, et al. Hydrophobic, highly conductive ambient-temperature molten salts. Inorganic Chemistry 1996;35(5):1168.

[14] Grätzel M. Dye-sensitized solar cells. Journal of Photochemistry and Photobiology C 2003;4(2):145.

[15] Wang J, Zhang S, Chen H, et al. Properties of ionic liquids and its applications in catalytic reactions. The Chinese Journal of Process Engineering 2003;3(2):177.

[16] Gu Y, Shi F, Deng Y, et al. Room temperature ionic liquids: a new type of soft media and functional materials. Chinese Science Bulletin 2004;49(6):515.

[17] Wang Y, He H, Wang C, et al. Insights into ionic liquids: from Z-bonds to quasi-liquids. JACS Au 2022;2(3):543−61.

[18] Liu X, Zhou G, Huo F, et al. Unilamellar vesicle formation and microscopic structure of ionic liquids in aqueous solutions. The Journal of Physical Chemistry C 2015;120(1):659−67.

[19] Wu X, Liu Z, Huang S, et al. Molecular dynamics simulation of room-temperature ionic liquid mixture of [Bmim][BF$_4$] and acetonitrile by a refined force field. Physical Chemistry Chemical Physics 2005;7(14):2771.

[20] Chen S, Zhang S, Liu X, et al. Ionic liquid clusters: structure, formation mechanism, and effect on the behavior of ionic liquids. Physical Chemistry Chemical Physics 2014;16(13):5893.

[21] Zhang S, Wang Y, He H, et al. A new era of precise liquid regulation: Quasi-liquid. Green Energy & Environment 2017;2(4):329−30.

[22] Fumino K, Wulf A, Ludwig R. Hydrogen bonding in protic ionic liquids: Reminiscent of water. Angewandte Chemie International Edition 2009;48(17):3184.

[23] Dupont J. On the solid, liquid and solution structural organization of imidazolium ionic liquids. Journal of the Brazilian Chemical Society 2004;15(3):341.

[24] Wakeham D, Niga P, Ridings C, et al. Surface structure of a "non-amphiphilic" protic ionic liquid. Physical Chemistry Chemical Physics 2012;14(15):5106.

[25] Lopes JN, Gomes MF. Pádua A. Nonpolar, polar, and associating solutes in ionic liquids. The Journal of Physical Chemistry B 2006;110(34):16816.

[26] Bradley AE, Hardacre C, Holbrey JD, et al. Small-angle X-ray scattering studies of liquid crystalline 1-alkyl-3-methylimidazolium salts. Chemistry of Materials 2002;14(2):629.

[27] Christopher H, Mcmath SEJ, Mark N, et al. Liquid structure of 1, 3-dimethylimidazolium salts. Journal of Physics: Condensed Matter 2003;15(1):S159.

[28] Triolo A, Russina O, Bleif HJ, et al. Nanoscale segregation in room temperature ionic liquids. Journal of Physical Chemistry B 2007;111(18):4641.

[29] Triolo A, Russina O, Fazio B, et al. Morphology of 1-alkyl-3-methylimidazolium hexafluorophosphate room temperature ionic liquids. Chemical Physics Letters 2008;457(4):362.

[30] Lu D, Zhao G, Ren B, et al. Isobutane alkylation catalyzed by ether functionalized ionic liquids. CIESC Journal 2015;66(7):2481.

[31] Medronho B, Romano A, Miguel MG, et al. Rationalizing cellulose (in)solubility: reviewing basic physicochemical aspects and role of hydrophobic interactions. Cellulose 2012;19(3):581.

[32] Zeng S, Zhang X, Bai L, et al. Ionic-liquid-based CO$_2$ capture systems: structure, interaction and process. Chemical Reviews 2017;117(14):9625−73.

CHAPTER 2

Interaction and interface of ionic liquids

Contents

2.1 Overview 15
2.2 Methods for simulating ionic liquid structures 16
    2.2.1 Force field 18
    2.2.2 Sampling method: Monte Carlo and molecular dynamics 28
    2.2.3 Software for molecular simulation 29
2.3 Simulation study of ionic liquid structures in the interface 30
    2.3.1 The structure and wetting behavior of ionic liquids at the solid surface 30
    2.3.2 The nanoconfined system 33
2.4 Simulation and regulation of two-dimensional ionic liquids 37
    2.4.1 Ionic liquid islands 37
    2.4.2 Two-dimensional ionic liquids with an anomalous stepwise melting process and ultrahigh CO_2 adsorption capacity 42
    2.4.3 Electron transfer and friction feature of two-dimensional ionic liquids 46
2.5 Prediction and control of ionic liquid structures 49
References 52

2.1 Overview

Ionic liquids (ILs) are unique because they can be made from various combinations of cations and anions that can be adapted to certain specific tasks by introducing fine-tuned functional groups. Therefore, it is crucial to understand the relationship between their properties and structures, which is determined by a delicate balance between the electrostatic interactions and other interactions, such as dispersion, repulsion, and hydrogen bonding.

The "ionic" nature of ILs, characterized by the strong long-range electrostatic interactions, results in cations/anions being surrounded by their counter-ions and responsible for very high cohesive energy, high viscosity, and low vapor pressure. Unlike conventional molten salts, ions in ILs are associated with functional groups, leading to diverse shapes and interactions. For instance, hydrophobic alkyl chain length influences nonpolar domain formation, while hydrogen bonding dictates specific cation—anion orientations between cations and anions. These unique ordered structures on different time and spatial scales are common and enable microenvironmental regulation of ion diffusion, cluster formation, and chemical reactions, diversifying the applications of ILs in process intensification.

Understanding the intricate structure of ILs is essential to optimize their performance in areas as diverse as catalysis, separation processes, and energy storage. Currently, the main experimental characterization techniques used include small-angle and wide-angle X-ray scattering and small-angle neutron scattering, nuclear magnetic resonance (NMR), Raman and infrared spectroscopy, electrospray ionization mass spectrometry, electron spin resonance spectroscopy, dielectric relaxation spectroscopy, dynamic light scattering, fluorescence spectroscopy, optical Kerr effect, and measurement techniques such as excess volume, excess enthalpy, heat capacity, viscosity, and conductivity. These experimental techniques have deepened our understanding of the structure of ionic liquids, but they also suffer from limitations such as low sample purity, difficulty in controlling conditions, and poor interpretation. On the other hand, molecular simulation has emerged as a powerful tool to complement experimental techniques, providing the ability to predict material properties and reveal atomic-scale mechanisms. It allows one to explore IL behavior from first principles, bridging the gap left by experimental limitations.

In this chapter, we give an introduction to the simulation of ILs, including how to describe ILs at the atomic level, predict their properties, and analyze their structure. The power of molecular simulations lies not only in the quantitative prediction of thermodynamic properties, but also in the understanding of mechanisms. We need to construct conceptual models more rationally to accomplish impossible tasks by controlling systems that are difficult to control or detect in experiments. The potential of molecular simulation is demonstrated in the interplay with theoretical models to build new and more effective hybrid solutions. If molecular simulation can only be used to fit decimal points behind experimental data or to passively explain experimental phenomena, it can hardly be the third pillar of scientific research besides experimental and theoretical approaches.

2.2 Methods for simulating ionic liquid structures

Molecular simulation, as the name suggests, is the numerical simulation of molecular systems governed by fundamental physical laws with the help of the power of modern computers. Taking molecules or atoms as the basic units, we can study their behavior under specific conditions, such as diffusion, aggregation, conformational or orientation transitions, and even chemical reactions, as well as calculate their macroscopic properties in the framework of statistical mechanics.

It is well known that the atom is the basic unit for constructing matter, which in turn consists of the nucleus and electrons. Generally speaking, physical properties do not involve any obvious changes in the electrons within the atom, whereas chemical properties are accompanied by significant transfers of electrons between atoms. Thus, there are roughly two kinds of simulation methods in terms of different ways to treat electrons. In the first, the basic units are atoms or molecules with a relatively stable structure (no bond generation or breakage, but there can be conformational changes), and the motion of molecules can be described by classical mechanics, so it is called molecular mechanics (MM), and simulations carried out based on this are generally referred to as atomistic molecular simulation. The second involves the motion or probability distribution of electrons, which must be described by quantum mechanics (QM). The starting point of its calculations is just the fundamental physical constants and the type and number of atoms contained in the system, hence the term "ab initio" and "first-principles calculations". Despite the inherent predictive nature of QM methods, in practice, they are heavily limited in time and spatial scales due to high computational costs and the sharp scaling to the system size and can currently only be applied to very small systems in good accuracy.

In order to more reliably simulate real systems, molecular simulations need to address two major challenges: first, how to describe interatomic interactions, that is, force fields (FFs, see Section 2.1); and second, how to mimic macroscopic systems (molar scales) on very limited spatial (<100 nm) and time scales (<100 ns). At the spatial scale, the simulation of macroscopic systems requires the use of a periodic boundary condition, that is, an infinite replication of the simulation box over the entire space (or over some dimensions in confined/interface systems) to eliminate interface effects caused by the limited size of the simulation box. Although the real macroscopic system does not consist of countless identical subsystems, as long as the size of the simulation box is much larger than the intermolecular correlation length (usually in nm scale), the results obtained can match the macroscopic system, and the size effect can be further investigated by gradually increasing the size of the simulation box. Regarding the time scales, since the correlation times of different microscopic quantities vary considerably, it is necessary to ensure that the simulation times cover the correlation times of the corresponding microscopic quantities when calculating thermodynamic properties, which are usually an order of magnitude higher. It is worth noting that since the molecular simulations start from a certain point in phase space (initial configuration), the simulation gradually converges to the thermodynamic equilibrium state, that is, the minimum of the free energy corresponding to the conditions, and thus if the system has multiple minima, enhanced sampling techniques are required to probe the free energy gap between the different states. Another technique in molecular simulations is the use of interaction truncation to reduce the computational intensity of the most time-consuming interparticle energy/force calculations. Errors from truncation require tail corrections, but for long-range electrostatic interactions, the truncation part requires specialized algorithms such as Ewald summation or PPPM. In virtue of modern algorithm optimizations, the computational complexity of molecular simulations can now be almost close to the linear scale.

2.2.1 Force field

Most of the published works on IL simulations have been carried out at the MM level, since the formation of IL structures results from intermolecular noncovalent interactions, such as hydrogen bonding and dispersion. These intermolecular interactions are often described by a set of mathematical functions that depend on the positions of the atoms, which can be regarded as acceptable approximations of the potential energy surface, that is, the FF. Typical FFs for molecular systems have a similar functional form as follows:

$$E = \sum_{\text{bonds}} K_r (r - r_{\text{eq}})^2 + \sum_{\text{angles}} K_\theta (\theta - \theta_{\text{eq}})^2 + \sum_{\text{dihedrals}} \sum_n \frac{K_\phi}{2} [1 + \cos(n\phi - \gamma)]$$
$$+ \sum_{i<j} \left\{ 4\epsilon_{ij} \left[\left(\frac{\sigma_{ij}}{r_{ij}} \right)^{12} - \left(\frac{\sigma_{ij}}{r_{ij}} \right)^{6} \right] + \frac{q_i q_j}{r_{ij}} \right\}$$

$$(2.1)$$

where E is the total potential energy of the system, the first line represents the intramolecular interactions, including the bond stretching, angle bending, and dihedral twisting, K_r, K_θ, and K_ϕ denote the force constants of bonds, angles, and dihedrals, respectively, and r_{eq} and θ_{eq} represent the equilibrium bond length and angle. The rotation of the dihedral is usually approximated by using one to six cosine functions. The second line represents the intermolecular interactions, expressed in terms of two-body pair potentials, including the electrostatic part in the coulomb form and dispersion part described in terms of Lennard-Jones (LJ), with ε_{ij} denoting the well depth, σ_{ij} denoting the distance between two atoms i and j when the interaction is zero, r_{ij} denoting the interatomic distance, and q denoting the partial charge on atom.

As mentioned above, a suitable FF and adequate sampling are two necessary prerequisites to obtain accurate and reliable molecular simulation results. The development of molecular FFs requires consideration of three issues, that is, accuracy, computational cost, and transferability, as it is often challenging to balance the former with the latter two. For ILs, classical FFs are still widely used, such as CLaP [1] developed by Padua et al., TEAM [2] developed by Sun et al., OPLS-2019-IL [3] developed by Acevedo et al., and LHW [4] developed by Liu et al., which has good compatibility with the AMBER/OPLS FFs, and is also supported by most simulation software.

The parameters of different terms in the FF represent different kinds of behaviors, which have different impacts on the final properties, for example, the LJ term has little impact on the vibration frequency. Therefore, it is not necessary to optimize the parameters of all objective functions at once, but it is possible to optimize them step by step in a certain order to reduce the difficulty. The objective function commonly used in FF optimization is the absolute or relative deviation between the calculated value and the "true value". The "true value" has two sources:

(1) *Experimental data*

They typically include bond lengths and angles in molecules (from X-ray diffraction, neutron scattering, or microwave spectroscopy), vibrational frequencies (from infrared or Raman spectroscopy), energy barriers to molecular torsion (from microwave spectroscopy), and condensed matter thermodynamic data such as densities, compressibility and thermal expansivity, enthalpies of vaporization or sublimation, heat capacities, and enthalpies of dissolution in aqueous solutions.

(2) *Ab initio calculations*

Ab initio calculations are attractive because they are based only on the basic physical constants and on the types of atoms and number of electrons in a system and do not require any experiments. However, ab initio calculations also have significant limitations. First of all, the results are always approximate, and to obtain sufficiently reliable results, larger basis sets and more complex methods must be used, which will be an endless challenge to the computational power. At present, the systems that can be handled with Gaussian software on an ordinary PC only reach tens of atoms and hundreds of electrons. Secondly, some properties calculated from scratch need to be scaled, for example, the vibrational frequencies calculated with HF/6–31G need to be multiplied by a scaling factor of 0.9 to be more consistent with the experimental results, and this scaling is purely empirical. The results of ab initio calculations commonly used in constructing FFs include the optimized geometry, interaction energies of molecules or molecular complexes, vibrational frequencies, and torsional energy profiles.

Although the mainstream molecular simulations still use the classical nonpolarized FF, the polarized FF of ILs has become increasingly mature in recent years with the increasing concern for interface studies and further improvement of algorithms and development of the corresponding software. On the other hand, coarse graining is also a significant trend to achieve simulations on larger spatial and time scales.

In the following, the construction of the classical FF is explained in more detail, and the polarized and coarse-grained (CG) FFs are briefly introduced.

2.2.1.1 Classical force field

Taking the imidazolium cation as an example, the atom type is first determined according to the AMBER FF, as shown in Fig. 2.1A. There are four types of hydrogen atoms: HC is the hydrogen on the ordinary alkyl chain, H_1 is the hydrogen on the methyl group neighboring the imidazolium ring (which is aromatic), and H_4 and H_5 are the two types of hydrogen on the imidazolium ring. We distinguish these types of hydrogens because of their relatively weak ability to bind electrons, and their repulsion and dispersion are strongly influenced by neighboring atoms or even sub-

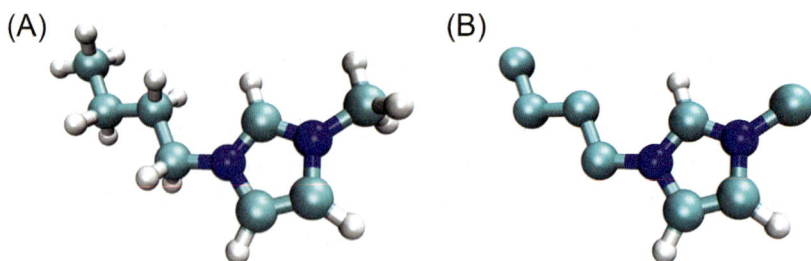

Figure 2.1 Atom types in imidazolium cation. (A) All atom; (B) united atom.

neighboring atoms, thus making them less transferable in different molecules. In fact, the chemical shift results of NMR also confirm the presence of several different hydrogens in the imidazolium cation. On the other hand, some groups like methyl and methylene have good transferability and can be treated as a whole, that is, united atoms (UA), as shown in Fig. 2.1B. This lightweight coarse-graining strategy is easy to implement, but will significantly reduce the number of atoms and degrees of freedom in the simulation, thus reducing the computational cost.

The second step is determining the equilibrium bond lengths, angles, and their force constants, which can generally be obtained through QM calculations on individual ions, with possible adjustments using experimental data such as crystal XRD and infrared Raman vibrations. In fact, the simple harmonic functions used in Eq. (2.1) make it difficult to reproduce the actual vibrational frequencies, but molecular simulations are mainly concerned with molecular translation, rotation, and distortion of intramolecular dihedrals, with energies in the level of thermal motion (kT) and an order of magnitude smaller than molecular vibrations. Therefore, there is no need to choose a more complex functional form for further optimization. Generally, the torsion energy barrier is the most essential intramolecular interaction in molecular simulations. If the parameters are not appropriate, it can cause a misleading distribution between different conformers and significantly impact the IL structure.

The third step is to optimize the intermolecular interaction parameters. Except for hydrogen atoms, the LJ parameters of other atoms have good transferability and can be basically used from existing FFs. It is worth noting that the LJ parameters of AMBER and OPLS FFs are very similar and can be generally used interchangeably, but they are noticeably different from the CHARMM FF, so it is not suggested to mix them.

The electrostatic interactions between cations and anions are described in coulomb terms using point charges located at the centers of atoms. Early FFs used the Mulliken population to determine charges, but the results were unsatisfactory. Currently, the most common method is to fit the electrostatic potential (ESP) around the molecule. This method generates grid points within a certain region around the molecular surface, and QM calculations are used to obtain the ESP at these grid points. The values of point charges on atoms are optimized to fit these ESPs. The method used in AMBER is called the restrained ESP (RESP) method. It adds a hyperbolic penalty function to non-hydrogen atoms in the optimization, which avoids some shortcomings of the unrestricted ESP method. Calculations on a large number of systems have shown that the RESP method is very successful.

When applying the RESP method, there are several points to note:

(1) *The basis set selection*

The results of any QM calculations are related to the basis set, and RESP is no exception. Calculating the molecular dipole moment using the $6-31G^*$ basis set combined with RESP generally gives values higher than experimental values. However, it is usually believed that this compensates for the lack of consideration of polarization.

(2) *Molecular conformation*

Starting from different initial conformations, multiple conformations can be optimized for the same molecule, corresponding to varying minima on the molecular energy surface. The charges obtained from different conformations are also different. In this case, multiple conformations can be used to fit the charges simultaneously or fit them separately and average the results. However, there is not enough evidence to prove that these methods are superior to fitting charges using a single conformation because it cannot resolve the fundamental problem of the absence of polarization term in the FF. In most cases, it is adequate to fit the charges based on the lowest energy conformation.

(3) *Penalty function factor*

When this factor is zero, RESP becomes an unrestricted ESP method. The selection of this value has a significant impact on the charges. The general approach is to start from zero and gradually increase the factor. In this case, the accuracy of the fit will definitely decrease. When the factor reaches a certain value, the accuracy will drop significantly. Therefore, a value slightly smaller than this can be chosen as the reasonable factor.

However, the value of atomic charge should vary with the microenvironment around the atom due to the polarization or charge transfer. For ILs, these effects are more significant than those in molecular fluids. In fact, early FFs predict the self-diffusion coefficients in ILs to be slower by one to two orders of magnitude compared to experimental values, indicating an overestimation of the interactions between cations and anions. From a physical point of view, polarization effects should be introduced explicitly (see section "Ab initio calculations" for a brief introduction), but this would significantly increase the computational cost, and the mathematical expression and parameters of the polarization effects would also require more sophisticated optimization.

An alternative approach is to use ESP for ion pairs rather than single ions to fit atomic charges. The absolute values of the total charges on each type of ion obtained in this way are less than 1, which can be regarded as the average effect of polarization. Early work by the Maginn group used configurations of ion pairs, but ion pairs generally have multiple configurations, and the differences in charges obtained from different configurations are significant. It is apparent that fitting charges based on configurations that are closer to the condensed phase can better reflect this average effect, for example, the use of double ion pair configurations (Fig. 2.2) proposed by the author, in which two cations share each anion, and vice versa. Recently, some researchers have achieved good results by fitting charges using ab initio molecular dynamics (AIMD) simulations of crystals or liquid phases. However, due to the complexity of the computational process, many studies have adopted simplified approaches, such as directly multiplying the charges obtained from fitting single ions by a scaling factor of 0.7–0.9, and have also obtained satisfactory simulation results.

It is worth mentioning that using more complex fitting strategies can better reproduce the ESPs, but there are still obvious shortcomings. In Fig. 2.3, we compared the ESPs obtained from the fitting of single cation and double ion pairs with the results of the QM calculations. It is found that at the center of the imidazolium ring, the former clearly overestimates the electrostatic effect, while in the latter, although it dramatically improves the results of the ESPs, some regions such as near the center of the imidazolium ring are underestimated.

In addition to electrostatic interactions, LJ parameters also significantly impact the structure of ILs, especially hydrogen atoms, which will be strongly affected by their nearby atoms. We proposed to optimize the LJ parameters of H_5 using the lowest energy configuration of the [Bmim][PF$_6$] ion pair, as shown in Fig. 2.4.

The fourth step is to optimize the dihedral angle parameters, which are very important for the distribution of the molecular conformation and need to be fitted by scanning the dihedral angles to obtain energy barrier profiles through high-precision QM calculations. A torsion refers to a rotation around a covalent bond, for example, C-A-B-D

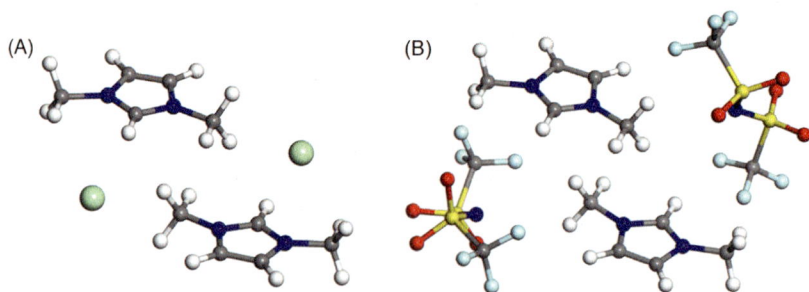

Figure 2.2 Fitting charges via double ion pairs. (A) [C$_1$mim]Cl; (B) [C$_1$mim][NTf$_2$].

ESP iso-surface of ion pair

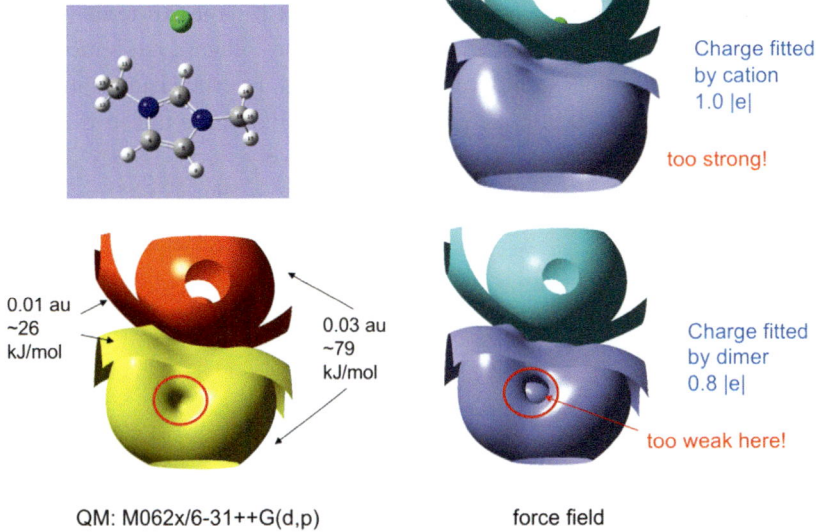

Charge fitted
by cation
1.0 |e|

too strong!

0.01 au
~26
kJ/mol

0.03 au
~79
kJ/mol

Charge fitted
by dimer
0.8 |e|

too weak here!

QM: M062x/6-31++G(d,p)   force field

Figure 2.3 Electrostatic potentials around [C₁mim]$^+$ by different fitting strategies.

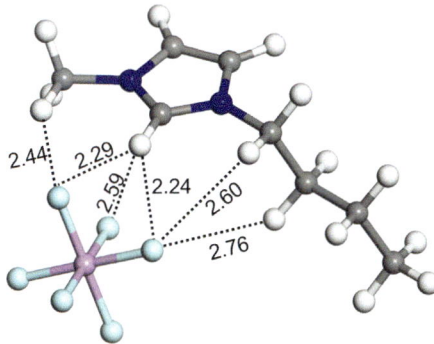

2.44   2.29   2.59   2.24   2.60   2.76

Figure 2.4 Optimized conformation of the [Bmim][PF₆] ion pair to be used in fitting LJ parameters.

represents the torsion of C and D atoms around an A-B bond, and the torsion angle [i.e., φ in Eq. (2.1)] is the dihedral angle between the two planes C-A-B and A-B-D. The torsion energy profiles describe the change in molecular energy relative to φ.

Unlike other terms in FF that have a clear physical meaning, torsion is essentially a correction term added artificially to improve the accuracy or remedy the deficiency of various simplifications in FF. Because of this, the torsion parameters in different FFs are significantly different and completely nontransferable. However, it directly affects the chain flexibility and tends to have a greater impact on the simulated structure and thermodynamic properties than the other bonding terms. It is worth mentioning that the torsion energy profile is not the sum of cos term in Eq. (2.1). As shown in Fig. 2.5 for the torsion of O-S-N-S of anion $[NTf_2]^-$, the cos term (in green) just contributes part of the torsion energy profile, and other contributions, including charges, LJ, and even angle/bond, cannot be ignored. For the charge and LJ contributions, special 1−4 scaling factors are usually used in FF because the atoms separated by three bonds are often too close, resulting in very high energy contributions.

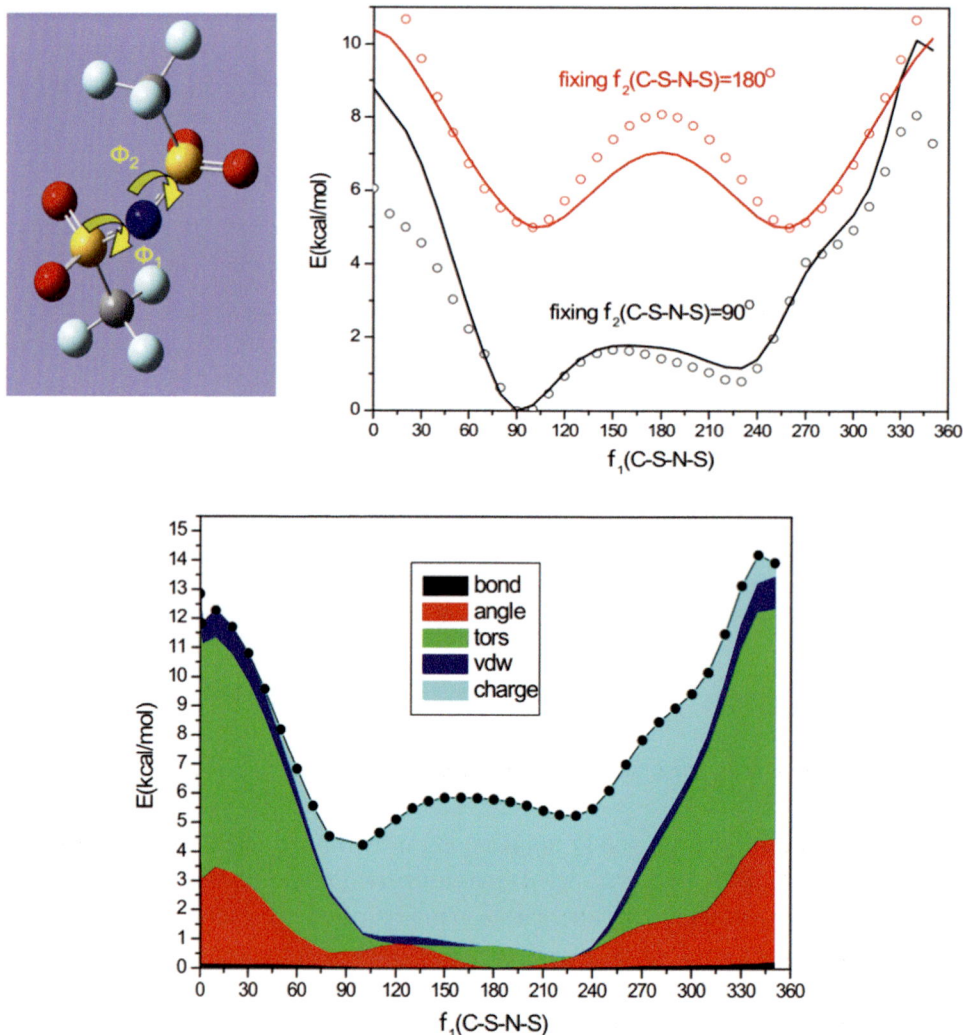

Figure 2.5 Torsion energy profiles of the anion of $[NTf_2]^-$. Top: QM (point) and force field (line); Down: breakdown of each contribution.

An empirical scaling, such as 0.5/0.5 in OPLS, can make the FF more robust. Therefore, after adjusting the atomic charges, LJ parameters, or the coefficients of the 1−4 scaling factors, it is, in principle, necessary to re-fit the dihedral parameters.

Although many out-of-box FFs were published for ILs, many parameters are still missing due to the massive number of cations and anions, especially those for functionalized ILs. The basic steps are introduced here for developing a nonpolarized molecular FF, which is now well-developed and implemented automatically in some software, such as ForceBalance [5]. It is convenient to develop these parameters independently using the methods described above.

2.2.1.2 Polarized force field

Although the nonpolarizable FF using charge scaling can significantly improve the simulation of dynamical properties such as diffusion coefficients of ILs, it is not satisfactory for simulating close-range structures. In addition, when ILs are mixed with other small molecular solvents, the scaling factor may be concentration-dependent, which is technically difficult to deal with. Therefore, polarizable FFs are the trend in future FF development because of their ability to explicitly deal with the changes in the atomic charges occurring in different microenvironments, and the addition of new parameters, such as polarizability, theoretically leads to better simulation results than the nonpolarizable FF can be achieved.

There are three typical schemes for implementing polarization in molecular FFs [6], namely, the fluctuating charge model (FCM), the Drude oscillator (DRD), and the induced point dipole (IPD).

The FCM follows the concept of atom partial charges in the classical FF, but these charges will change with the microenvironment around the atom during the simulation. The most popular algorithm is electronegativity equalization, which requires the introduction of atom-related properties: electronegativity χ° and hardness J°. Although both of them can be rigorously defined in QM theory for the same atom, their values in different molecules vary greatly and can be used as tunable parameters in practical applications. The DRD splits the charge on a polarizable atom into two parts, the partial charge located at the center of the nucleus as well as the Drude particles oscillating around the nucleus, which simulate the polarizable electron cloud by carrying a negative charge $-q_D$. The Drude particle is treated in MD as a real particle involved in the equations of motion, connected to its central atom by a simple harmonic spring potential. When there is no external electric field, the Drude particle coincides with the position of the central atom and its induced dipole moment is zero. An induced dipole is produced by an external electric field caused by its nearby atoms, which is determined by the atom polarizability. In fact, the DRD is essentially the same as the IPD model. The former has the convenience of avoiding the very time-consuming iteration of the self-consistent field computation instead of the algorithm of double heat bath, but there are some technical difficulties in dealing with the hydrogen atom due to its small mass. The IPD, on the other hand, introduces massless point dipoles on the atom, the size of which is proportional to the atom polarizability and varies with the change of the external electric field generated by other atoms in the microenvironment and usually needs to be derived by self-consistent field iteration based on the principle of the minimization of the energy of polarization.

Although the above three polarization models are implemented in popular MD software packages, such as AMBER/CHARMM/LAMMPS/GROMACS/OpenMM, each has its own advantages regarding implementation details. Not only does the addition of a polarization term significantly increase the computational cost of the simulation, but also some technical issues need to be solved, such as the algorithm instability, which may lead to the so-called "polarization catastrophe".

For ILs, most of the polarizable FFs are in the DRD framework, and relatively few FCMs are reported. Padua's group proposed the CL&Pol polarization FF by adding DRD terms to the CLaP of the aforementioned nonpolarized FF [7], which covers more than a dozen common anions and cations and provides plugins for LAMMPS and OpenMM, which are relatively easy to use. McDaniel et al. developed the SAPT-FF [8], splitting the total energy with the help of symmetry-adapted perturbation theory (SAPT) in order to optimize the parameters of each energy contribution separately. They also used the DRD model and implemented it in GROMACS. It is worth noting that all FF parameters are optimized using ab initio data computed by QM. A better fit is obtained by adopting a Born-Mayer repulsion term and describing the attraction through four power function expansion terms C_n/r^n ($n = 6,8,10,12$) instead of the commonly used LJ-6−12 terms. The SAPT-FF seems quite promising in its powerful prediction capability, with the disadvantage that it is difficult to develop and has fewer implementations in popular simulation software. Thus, it is not currently in the mainstream of IL research. Wang et al. were the first to report on the improvement of the IPD polarizable FF on the simulation results of [Emim][NO$_3$] [9], but did not extend their simulations to more ILs. Borodin et al. [10] also published an IPD-based polarizable FF named APPLE&P (Atomistic Polarizable Potential for Liquids, Electrolytes & Polymers), but their parameters and software are relatively less available.

2.2.1.3 Coarse-grained force field

As mentioned above, using UA can effectively reduce the number of degrees of freedom and lower the computational intensity. In general, nonpolar methyl/methylene groups can be treated as UA, but this idea can be further expanded, such as the whole imidazolium ring, or smaller rigid anions such as [BF$_4$]$^-$ can also be CG as a basic unit. In fact, the CG strategy is very flexible, thus significantly reducing the complexity and expanding the space and time scales of simulation. However, because of the reduced degrees of freedom, CG models generally suffer from overestimating transport properties such as diffusion coefficients.

There are two kinds of CG methods: bottom-up and top-down.

The former generally starts from the molecular FF at the atomic level and obtains the CG parameters by matching the simulation properties. Three kinds of methods have been proposed according to the properties to be matched. The most natural one is based on the potential of mean force, which can be regarded as matching the averaged free energy between CG groups, and can be realized by iterative Boltzmann inversion (IBI) or inverse Monte Carlo (MC). The other two approaches are force matching (FM) and relative entropy minimization (REM). For ionic liquids, Wang and Voth et al. established multi-scale CG based on FM [11]. Laaksonen's and Aluru's groups have also developed CG models based on IBI [12] and REM [13] to simulate larger systems involving ILs. The CG units are explicitly charged in their models to describe the electrostatic interactions. Recently, polarization was also introduced similarly to all-atom models in CG FFs [14]. However, the CG FFs are currently limited to a few ILs involving all imidazolium cations, and they are all nearly spherical anions, so large-scale applications have not been realized.

The most popular top-down CG FF is the MARTINI, which is more portable due to the deep optimization of the nonbonding interaction parameters using experimental thermodynamic data such as partition coefficients that are closely related to the free energy. While its earliest use was in biofilm simulation, a larger extension was realized in MARTINI version 3.0 for more organic small molecules and soft materials [15]. A new CG strategy of imidazolium was recently tested in its framework and successfully used to study the extraction of fine chemicals by ILs, but there is an order of magnitude overestimation of simulated diffusion coefficients [16].

2.2.1.4 Ab initio molecular dynamics

Unlike the FFs that require parameters to calculate intermolecular interactions as described in the previous section, AIMD does not require any predefined FF parameters and can, therefore, be called an FF-free simulation. The forces on the atoms are calculated in situ during the simulation by means of density functional theory (DFT) and are therefore referred to as ab initio or first-principles (FPMD). In addition, because Car and Parrinello established their theoretical foundation and developed the most efficient algorithm, which is implemented in the open-source software package CP2K, it is also referred to as CPMD. These acronyms are very similar in meaning in the literature.

It is worth noting that, although AIMD does not need to preset the FF, it needs to specify the necessary information for DFT calculation, such as the density functional and basis sets, so it is actually not parameter-free. On the other hand, DFT itself is an approximation, and the selection of a suitable density functional approximation (DFA) significantly impacts the simulation results. Recently, Grimme et al. reported [17] a detailed study on the accuracy of different DFA in the static calculations on IL clusters. The advantages of AIMD are that both the DFA and basis set are universal and can be applied to any atomic system, far superior to the FF parameters, which can only be applied to a limited range of systems.

The biggest bottleneck of AIMD is the computational cost (about three orders of magnitude higher than the classical FF), which significantly restricts its practical application. Therefore, the machine learning FF has recently been trained based on big data sampled by AIMD using modern machine learning methods such as deep neural networks. It is promising to significantly reduce the computational cost while maintaining the accuracy of the DFT-based AIMD.

In the field of ionic liquid simulation, due to the complexity of the system, limited by the computational cost, AIMD has not been fully applied, and its simulation results are mainly used to optimize the FF parameters.

2.2.2 Sampling method: Monte Carlo and molecular dynamics

In the framework of statistical mechanics, the macroscopic properties of a system are statistical averages of various microscopic states in the phase space. The core of statistical mechanics is the partition function, which describes the distribution probability of each microscopic state that satisfies certain macroscopic conditions (e.g., NVE or NPT). The macroscopic properties (e.g., enthalpy, temperature, pressure, density, viscosity, etc.) can be obtained as long as the partition function is known. Theoretically, a bridge is established in statistical mechanics between the partition function and the interaction of particles, but an analytical solution can only be derived for very simple systems such as hardsphere fluids through sophisticated mathematical techniques and the introduction of various approximations. Molecular simulation avoids these mathematical difficulties by obtaining an adequate sampling of the phase space through numerical methods, and then statistical averaging can be performed. Depending on the sampling method, there are two types of simulation methods, that is, MD and MC.

MD is a deterministic method that calculates the potential energy of the system and the force on each atom for a given molecular FF and solves the equations of motion based on classical mechanics to obtain the evolution of the momentum and position of each atom with respect to time, obtains a sampling of the phase space, and then calculates the properties of interest after equilibrium has been reached. MC is a stochastic method that calculates statistical averages from high-dimensional integrals employing importance sampling. The method explores the important area in phase space by continuously generating random configurations and using the accept-or-reject criterion that satisfies the detailed equilibrium. MD has the advantage of being able to compute time-dependent properties such as diffusion coefficient, thermal conductivity, and so forth. MD is also easy to parallelize and is therefore more rapidly developed and more widely used in recent years. In some cases, such as phase equilibrium, MC is more efficient because the change of particles is driven by random numbers rather than the physical forces in MD, making it easier to overcome the higher energy barriers.

It is not easy to get adequate sampling for either MD or MC. Typically, each MD/ MC run can get a sampling around the local minima of the system's free energy, and in many cases, multiple minima exist. If the energy difference between these local minima is small (comparable to the thermal motion kT), it is easier to obtain sufficient sampling by MD and thus obtain the system averages such as the internal energy and the density more accurately. However, when the complexity of the system increases, the number of local minima and the energy barrier between them will also increase accordingly. Thus, the simulation will be trapped in some minima and blind to others because the energy barrier is too high to cross, which requires enhanced sampling techniques to overcome the difficulty.

According to the Boltzmann distribution, the higher the energy barrier is, the more difficult it is to sample, that is, the so-called rare event. An effective enhanced sampling technique is to add a bias potential during simulation, forcing it to be sampled in the specified region, which requires predefined reaction coordinates or more generalized collective variables (CVs), then constructing and optimizing a suitable bias potential, and finally using maximum likelihood to reproduce the free energy profile without the bias potential. There have been great advances in efficiency and flexibility from the original umbrella sampling to the more advanced metadynamics. Another method that does not require predefined CVs is the so-called replica exchange MD, that is, multiple replica MDs are run independently simultaneously, and then their trajectories are exchanged between the replicas according to specific rules to enhance the sampling.

At present, the above algorithms have been implemented in mainstream MD packages, and they can also act as plug-ins through interfacing with MD packages, such as PLUMD and WESPA. However, there is still a high learning curve in practical applications. It is highly dependent on the researcher's experience to set reasonable parameters to control the simulation process and enhance the sampling effectively.

2.2.3 Software for molecular simulation

Early molecular simulations are primarily based on the home-made code developed by researchers, which has great limitations in running efficiency, ease of use, and bug-free and applicable systems. With the expansion of molecular simulation applications, a variety of efficient and open-source packages have become more and more mature. Here we only give a brief introduction for a few of them.

MCCCS Towhee (Monte Carlo for Complex Chemical Systems) is a Monte Carlo molecular simulation program that supports the classical FF and implements the configuration-bias sampling technique that can be used in Gibbs ensemble MC for predicting the phase equilibrium of fluids, as well as the grand canonical MC for simulating the adsorption of porous materials.

LAMMPS (Large-scale Atomic/Molecular Massively Parallel Simulator) is a classical molecular dynamics code focusing on materials modeling, with potential for solid materials (metals and semiconductors) and soft matter (biomolecules, polymers), as well as CG or mesoscopic systems. Its impressive high parallelization efficiency and easily extensible code framework attract many researchers to contribute their new algorithms and FFs.

AMBER (Assisted Model Building with Energy Refinement) is a suite of biomolecular simulation programs started in the late 1970s, including its built-in force-field and MD engines and many related modeling and analysis tools.

GROMACS (GROningen MAchine for Chemical Simulations) is mainly used to simulate proteins, lipids, and nucleic acids. It was initially developed at the University of Groningen in the Netherlands and is now maintained by contributors from universities and research centers worldwide. It is currently the most popular MD package with the aim of being fast, flexible, free.

OpenMM is a recently emerged high-performance toolkit for molecular simulation, with a more advanced code architecture, interfaces to a wide range of other codes that make it easier to use for multi-scale simulations, and modules optimized specifically for GPUs. The programming environment is also more flexible, for example, it can be called directly from Python.

Unlike the above-mentioned software, CP2K can carry out AIMD simulations of solid, liquid, molecular, periodic, material, crystalline, and biological systems in a computational framework that mainly uses DFT, with a mixed Gaussian and plane-wave method, and thus without molecular FF parameters. Since the states of electrons are directly considered, it is well suited for simulating systems in chemical reactions (with enhanced sampling). Currently, its biggest drawback is that the computational cost is much higher than that of molecular FF-based simulations, and in practical applications, it is often necessary to flexibly design simulation schemes to be used in combination with each other.

2.3 Simulation study of ionic liquid structures in the interface

2.3.1 The structure and wetting behavior of ionic liquids at the solid surface

The interfacial energy can be very effective in regulating the solidified ionic layer and then bringing about the abnormal IL wetting behavior, which will be invaluable and significant for the new IL-based coolant liquid, energy storage and conversion devices, and other chemical separation techniques [18]. The dynamical evolution and wetting behaviors of water, [Emim]Cl, and [Emim][PF$_6$] droplets on solid surfaces were explored by classical MD simulations, as shown in Fig. 2.6. Interestingly, two different equilibrium configurations, nonspreading structure and partial spreading structure, were identified for ILs when the solid interface property changes from hydrophobic to hydrophilic.

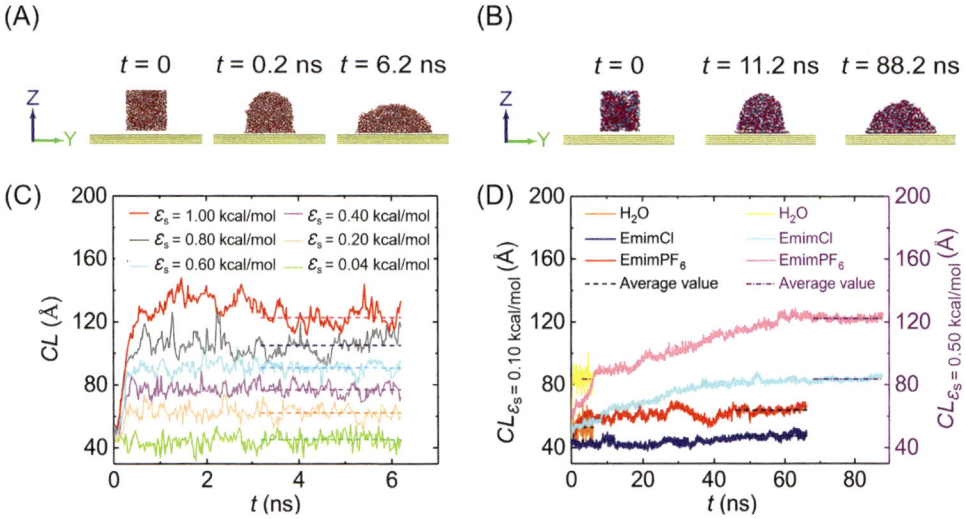

Figure 2.6 (A, B) Atomic structures of the water–solid and [Emim]Cl–solid systems with $\varepsilon_s = 0.60$ kcal/mol at different t, where red, white, magenta, cyan, and yellow colors represent an oxygen atom, hydrogen atom, cation, anion, and substrate, respectively. (C, D) Contact length (CL) of a water or the IL droplet on the surface with different ε_s changes with t.

The quantitative relation between the contact angle (CA) and the solid surface energy (ε_s) suggests that both of [Emim]Cl and [Emim][PF$_6$] show an abnormal wetting behavior compared with water, where the CA-ε_s relation for ILs deviates the theoretical linear model when ε_s is beyond the critical surface energy ε_c. The CA for water shows a linear dependence, while that for ILs shows a two-linear dependence with the depth of the potential well for the solid (ε_s) increasing from 0.04 to 1.0 kcal/mol, decreasing sharply first and then slowly. However, the IL droplets can form nonspreading and partial spreading structures, while the water droplets will always form nonspreading structures. For the partial spreading case, there exists a convergent CA for [Emim]Cl and [Emim][PF$_6$], reflecting the self-hating characteristics, which are 53.7° and 30.7°, respectively.

Moreover, the reduced density and orientation distribution show that ILs could form a denser solidified ionic layer near the solid surface, which will not happen to water. The reduced density and the orientation distribution of the adjacent liquid layer to the solid surface indicate that IL droplets on the solid surface with $\varepsilon_s > 0.4$ kcal/mol will form a dense solidified liquid layer, which will not happen to water. Furthermore, the dynamical properties, including the evolution of the center of mass, retention rate, and vibration displacement for both ions and water molecules in the adjacent liquid layer, show that the dense solidified ionic layer almost loses the fluid nature, while the adjacent water layer still possesses relatively high fluidity, especially when $\varepsilon_s > \varepsilon_c$, producing such an abnormal wetting behavior of ILs. All the abovementioned relations demonstrate that there exists a critical value of ε_s ($\varepsilon_c = 0.4$ kcal/mol); when $\varepsilon_s > \varepsilon_c$, the ions in the adjacent IL layer will be solidified, leading to the abnormal IL wetting behavior. Besides, the monolayer IL structure of the partial spreading droplet increases significantly with ε_s, and the arrangement of cations and anions shows a relatively ordered checkerboard structure as well, indicating that one can regulate the liquid structure via simply controlling the solid surface. These factors can serve as critical indicators in characterizing the mechanism of the structural transition of ILs wetting the solid surface and facilitate the rational design of the surface modification or strain engineering of the solid substrates.

The sensible design of IL-based electrolytes and other energy storage and conversion devices relies on a thorough knowledge of the structure and characteristics of electrolyte—electrode interfaces [19]. The wetting processes of the Li^+-doped IL droplets on the TiO_2-B(100) surface were investigated by MD simulations (Fig. 2.7). A series of

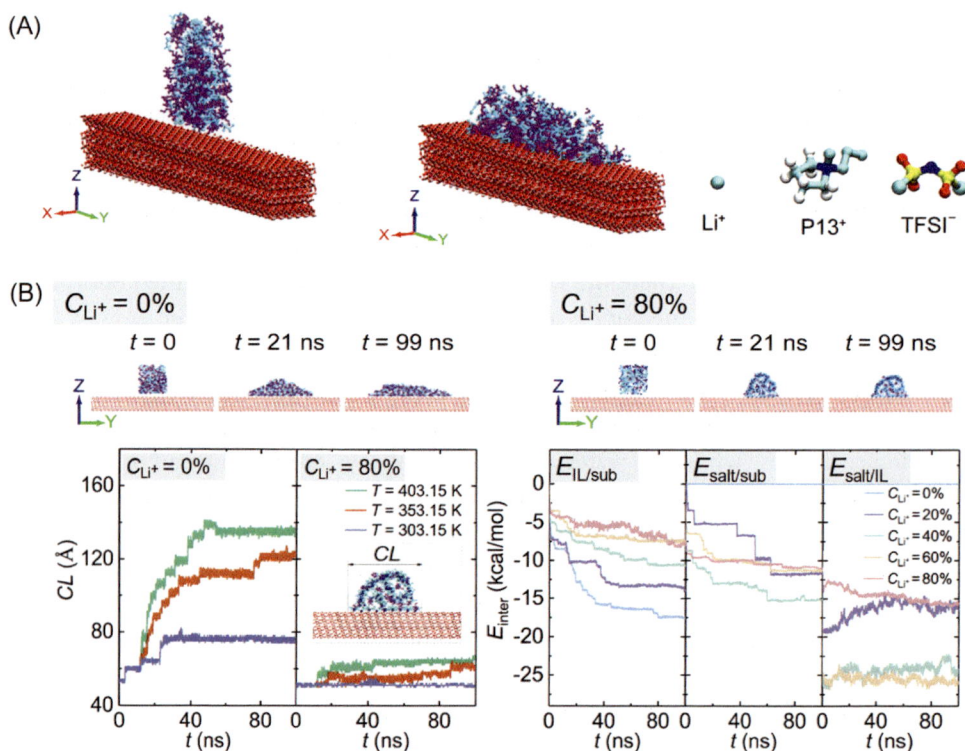

Figure 2.7 (A) The illustration of IL droplet on the TiO_2-B(100) surface and the atomic structures of Li^+, $P13^+$, and $TFSI^-$. (B) Snapshots of the evolution structure for the pure IL droplet (the left) and the Li^+-doped IL droplet (the right) on the TiO_2-B(100) surface at $T = 403.15K$, the evolution of contact length (CL) with simulated time (t), at $T = 303.15, 353.15,$ and $403.15K$ (the left of the bottom), as well as the interaction energy (E_{inter}) of P13 TFSI-TiO_2 ($E_{IL/sub}$), LiTFSI-TiO_2 ($E_{salt/sub}$) and LiTFSI-P13TFSI ($E_{salt/IL}$) in the composite system with different concentrations of Li^+ (C_{Li^+}) evolving with t, at $T = 353.15K$.

models of Li^+-doped ILs with the Li^+ concentration (C_{Li^+}) varied from 0% to 80% at temperatures of 303.15, 353.15, and 403.15K and were fabricated to better comprehend the wetting behavior and controlling mechanism of Li^+-doped IL droplets on the TiO_2-B(100) surface. As the C_{Li^+} grows from 0% to 80%, regardless of the temperature, the interaction energy between Li^+ and IL ($E_{salt/IL}$) and the interaction energy between Li^+ and substrate ($E_{salt/sub}$) of Li^+-doped ILs decrease first and then increase, reaching the minimum value at $C_{Li^+} = 40\%$, indicating that the suitable doping fraction of Li^+ can significantly enhance the interaction between salt and ILs or substrate. On the other hand, with the increase of C_{Li^+}, Li^+ in the adjacent layer will occupy the original positions of $P13^+$ cations near the TiO_2-B(100) surface, resulting in a monotonic rise in the $E_{IL/sub}$, demonstrating that the Li^+ will substitute for the ILs in the interfacial region and further weaken the ILs—substrate interaction. According to the spatial distributions of components, doped Li^+ prefers to substitute the ILs and adsorb to the substrate, causing the orientation changes of the ILs, weakening the ILs—substrate interaction, and slowing down the wetting process significantly. As Li^+ concentration rises from 0% to 80%, the CA increases from 86.97° to 131.18°, inducing the hydrophilic to hydrophobic transition. In detail, the $P13^+$ is gradually squeezed out of the adjacent layer, causing a change in the orientation of $TFSI^-$ and $P13^+$ and further increasing the CA, which means the wetting behavior of the Li^+-doped IL droplet entirely transforms from extremely hydrophilic to hydrophobic.

Furthermore, heating up would reduce the CA by extending the contact length and enhancing the maximum density of Li^+-doped ILs at the interface; the response of CA, $E_{salt/IL}$, $E_{salt/sub}$, and $E_{IL/sub}$ to temperature is monotonous compared to that of C_{Li^+}. The quantitative relationships between structures, wetting behaviors, interaction energies, and C_{Li^+} demonstrate that doping Li^+ can slow down the dynamical wetting process of Li^+-doped IL droplets on the TiO_2-B(100) surface, where the interface-induced dense adjacent ionic layer is the leading factor, which has implications for the new ILs-based electrolytes and the application.

2.3.2 The nanoconfined system

Several researchers have studied the interfacial structure of water in graphene (Gra) and graphene oxide (GO) nanochannels, showing an ordered water layer near the surfaces [20]. To investigate the distribution of ILs in nanochannels, we chose the most typical one, 1-butyl-3-methylimidazolium tetrafluoroborate, [Bmim][BF$_4$], as the research subject [21] (Fig. 2.8). We only consider the hydroxyl groups in the structure of GO herein, and the fraction of hydroxylation c is defined as $c = n_{OH}/n_C$, where n_{OH} and n_C are the numbers of hydroxyl groups and carbon atoms, respectively. The spatial density distribution in GO nanochannels with the interlayer distance (d) ranging from 0.75 to 2.00 nm was calculated. The ILs can form stable monolayer, bilayer, and

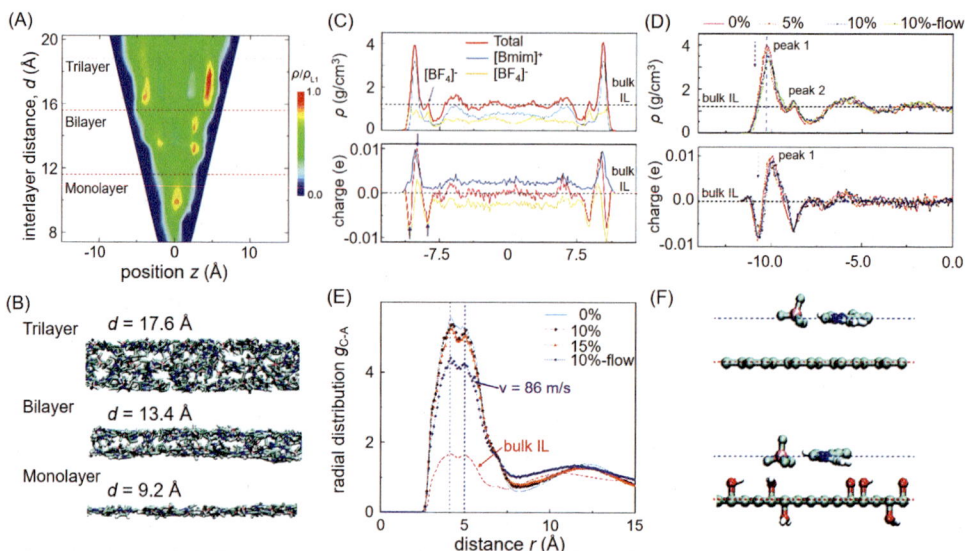

Figure 2.8 (A) Contour plot of the IL density distribution $\rho(z, d)$ at $c = 10\%$ with reference to the density of the first IL layer near GO sheet ρ_{L1}. (B) The structures of confined ILs with different values of d. (C) The density and charge distribution of confined ILs at $c = 0$, where the contributions from $[Bmim]^+$ and $[BF_4]^-$ are also shown. (D) The density and charge distribution of the IL across the GO nanochannel with different values of c. (E) RDF of $[Bmim][BF_4]$ in different GO nanochannels. (F) The interfacial structure of $[Bmim]^+$, $[BF_4]^-$, Gra and GO.

trilayer structures in the GO nanochannel. When d is less than 1.16 nm, an IL monolayer structure can be clearly observed. As d increases, the IL will form a bilayer structure when d is less than 1.57 nm and a trilayer structure if d is less than 1.88 nm. With d continuously increasing, the IL cannot retain the trilayer structure and will change to a structure like the bulk IL case, which is like the case of water in GO nanochannels. The structure of ILs in nanochannels is primarily determined by the balance between van der Waals (vdW) interactions, hydrogen bonds (HBs), electrostatic forces, and the geometrical properties of both ILs and hydroxyls in GO. Taking the interlayer distance of 2.8 nm as an example, it is indeed large enough to ensure that the properties of the center IL are close to those of the bulk IL. The density distribution as a function of z-position in the Gra nanochannel is plotted. The contributions from $[Bmim]^+$ and $[BF_4]^-$ to the total density are also shown. From the density distribution, two peaks near the GO sheets obviously existed, called peaks 1 and 2, possessing a higher density compared with the bulk one, while the center part ($Z > 0.75$ nm) has a density like that of the bulk one. Compared with $[Bmim]^+$, $[BF_4]^-$ is closer to the GO sheets, indicating that $[BF_4]^-$-GO interaction is stronger than $[Bmim]^+$-GO. The density distribution of the IL in nanochannels with different values of c was also summarized. The fact is that the intensity of peak 1 will decrease as c increases, while that of peak 2 increases, indicating that some ions will depart from GO sheets due to the interaction of IL-hydroxyls.

Generally, bulk IL is electrically neutral due to the high coulombic ordering. However, the nanoconfinement will partially break the coulombic ordering and induce the charge layer. The charge distribution of the IL across the nanochannel is also summarized, showing that obvious charge layers exist near the GO sheets. A negatively charged layer appears at 0.33 nm from GO, following a positively charged layer at 0.40 nm and a negatively charged layer at a distance of 0.52 nm. These three charged layers are mainly due to $[BF_4]^-$, the imidazole ring in $[Bmim]^+$, and $[BF_4]^-$, respectively. Hence, distributions of density and charge across the nanochannel are primarily determined by the GO—IL interactions, which can be tuned by changing the fraction of hydroxylation. Surprisingly, the peak positions of density and charge distribution hardly change as c increases. That means intercalating hydroxyls between Gra and the IL does not change the equilibrium distance between the IL and GO sheets due to the strong interactions between cations, anions, hydroxyls, and vdW interactions, resulting in a similar distance.

The first IL layer is defined as a distance between the IL and GO less than 0.6 nm. To capture the molecular insights of the first IL layer, the radial distribution function (RDF) between $[Bmim]^+$ and $[BF_4]^-$ was plotted. The results show that RDFs of confined ILs with $c = 0\%$, 10%, and 15% almost overlapped with each other, indicating that hydroxyls have a negligible effect on the first IL layer. Taking the RDF of bulk ILs as a comparison, it can be found that the position of the first IL peak is almost the same as that of the bulk one, suggesting that the average cation—anion interatomic distance of the neighboring ions is unaffected in GO nanochannels. From the snapshots of IL—GO interfaces, the imidazole ring in $[Bmim]^+$ prefers to lie on the GO sheet in parallel, showing strong π-π convergent interactions and leading to the arrangement of more IL ions in the neighboring spaces of GO sheets. Hence, nanoconfinement predominantly results in a higher degree of compaction of IL ions in the first liquid layer compared to the bulk scenario, while maintaining the same interionic distance.

To study the structure of confined ILs in the nanochannel with a larger range of interlayer distance (d), we changed the d from 0.75 to 5 nm [22] (Fig. 2.9). The IL considered herein is 1-ethyl-3-methylimidazolium bis(trifluoromethyl sulfonyl) imide, [Emim][NTf$_2$], which has been used widely as the electrolyte of the supercapacitor and other applications [23—25]. In order to provide a clear representation of the structure, we plotted the spatial mass density and charge distribution of the confined IL. Taking $d = 5$ nm as an example, the mass density distribution will result in the formation of a stable sandwich-shaped structure, characterized by two dense layers near the Gra wall, while the central region exhibits a density comparable to that of the liquid bulk IL. The IL adopts a sandwich-like structure, resulting in the

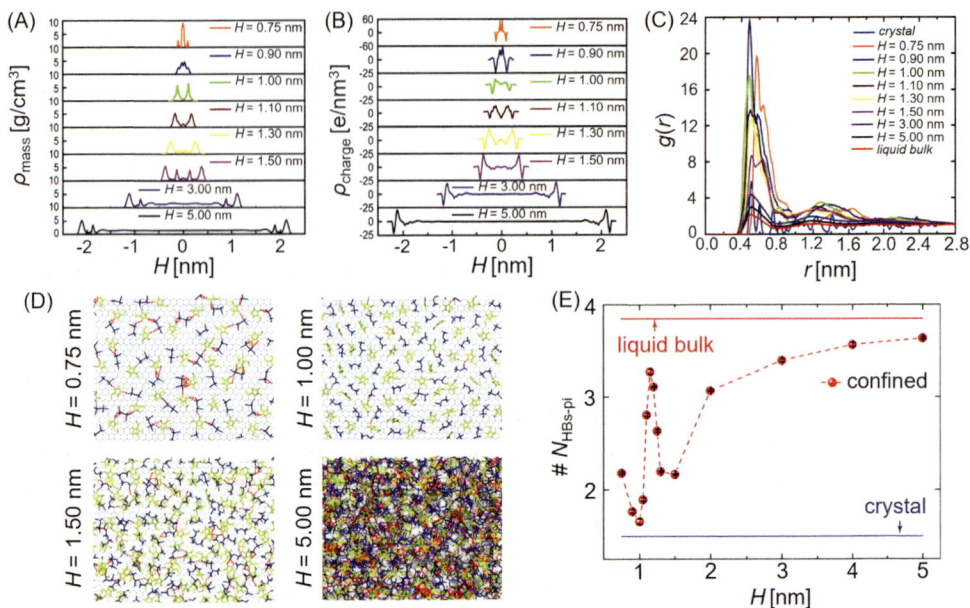

Figure 2.9 (A, B) Mass density profiles and charge density profiles of confined IL. (C) RDF for the [Emim][NTf$_2$] pair as a function of the distance r, where RDFs for crystal ILs and liquid bulk ILs are also plotted as comparisons. (D) The snapshots of the HB network of confined ILs with $d = 0.75$, 1.00, 1.50, and 5.00 nm, where the dash lines represent the HBs. (E) The average number of HBs per ion (N_{HBs-pi}) within confined IL as a function of H.

formation of multiple charge layers adjacent to the wall. The first charge layer, located 0.22 nm away from the Gra, consists of a positive charge. Following this, there is a negative charge layer at 0.30 nm from the Gra. Subsequently, a positive charge layer is found at 0.37 nm from the Gra, followed by a negative charge layer at a distance of 0.51 nm from the Gra. Beyond approximately 0.90 nm from the Gra, the charge distribution becomes similar to that observed in the bulk IL, where the charge density approaches zero. The confined IL with d greater than 1.5 nm exhibits nearly identical mass density and charge distribution as the confined IL with a d of 5.0 nm, despite having varying proportions of the liquid-bulk-like region. However, when $d < 1.5$ nm, the similar distribution of IL near the walls overlaps with each other, causing the liquid-bulk-like region to vanish. Instead, a partial bilayer IL forms, with a completely layer-by-layered charge distribution.

Moreover, the RDF for [Emim][NTf$_2$] pair is computed to examine the alterations in cation–anion interactions. The RDFs can be categorized into three regions as follows: When $d > 1.5$ nm, the RDF displays a single peak at the same position as the RDF for the bulk liquid, albeit with weaker intensity. For $d \leq 1.5$ nm and $d \geq 1.0$ nm, the RDF exhibits twin peaks, with the first peak aligning with that of the bulk liquid. As d continues to decrease, the peak position gradually shifts toward that of the solid crystal case. Notably, the variation in RDF peaks distinctly indicates the transition of confined IL from a liquid to a solid state as d diminishes from 5.00 to 0.75 nm. Additionally, the HB network of confined IL reveals a significant reduction in the number of HBs in the two-dimensional (2D) nanoconfinement due to the absence of out-of-plane interactions. To quantitatively describe the evolution of the confined IL, we calculate the average number of hydrogen bonds per ion (N_{HBs-pi}) within the confined IL. Interestingly, N_{HBs-pi} exhibits oscillatory behavior with respect to the d. Initially, N_{HBs-pi} decreases and then increases as the d varies from 0.75 to 1.1 nm. The lowest value of N_{HBs-pi} is observed at $d = 1.0$ nm. With further increase in d, N_{HBs-pi} reaches its maximum value at $d = 1.15$ nm and then decreases again. When d is beyond 1.5 nm, N_{HBs-pi} reaches a second extremely low value and gradually increases thereafter. Physically, the decrease in HBs arises from the formation of a partial bilayer structure, which reduces the average constraint on ions. This reduced constraint allows ions to exist in more complex structures.

2.4 Simulation and regulation of two-dimensional ionic liquids

2.4.1 Ionic liquid islands

Due to their green and environmentally friendly characteristics [26,27], ionic liquids (ILs) have become highly attractive as green solvents and catalysts in the fields of chemistry and chemical engineering. In practical applications, solid−surface−supported IL (SSIL) thin films are consistently present and exhibit various advantages compared to their bulk counterparts [28,29], such as higher stability, less usage of ILs, and faster adsorption kinetics. Consequently, SSIL thin films have garnered extensive attention from both the academic and industrial communities [30,31]. A quantitative understanding of the structure−function relationship of SSIL thin films is of paramount importance for the rational design and management of chemistry based on ILs.

Recently, a significant amount of experimentation and simulation has been carried out to elucidate the correlation between the structure and performance of SSIL thin films [32−34]. Additionally, owing to their customizable microenvironments and functionalities, thin IL films or islands hold vast application prospects, particularly in carbon dioxide capture and sequestration [35−38]. The outstanding performance of these SSIL thin films suggests that two-dimensional ionic liquid islands (2DIIs) should be the preferred candidates for IL-based applications [39]. Nevertheless, up to this point, besides qualitative understanding of 2DIIs, quantitative characterization remains a major challenge, such as detailed substructure, electronic performance, and gas adsorption capabilities. These limitations hinder the theoretical comprehension, rational design, and implementation of ILs in high-efficiency, low-cost gas separation and capture processes.

In our current work, our team utilizes image charge-augmented QM/MM (IC-QM/MM) and full-atomistic MD simulations to uncover the structure and properties of 2DIIs on graphite [40]. By exploring the potential substructures of ordered 2DIIs, we theoretically propose four distinct substructures and demonstrate the critical size (N_C) for various 2DIIs, beyond which thermodynamic stability converges. Simultaneously, we further reveal the mechanism by which substructures and size jointly control the melting behavior and electronic structure of 2DIIs, highlighting the tunable nature of these structures. Lastly, we investigate the dynamic gas adsorption process on 2DIIs, revealing that the edges of the 2DIIs serve as the primary adsorption sites for CO_2 and exhibit high selectivity for CO_2/CO, CO_2/CH_4, and CO_2/N_2 interactions.

Firstly, considering the strong correlation of special HBs in ILs [41], during the physical vapor deposition (PVD) process, gaseous particles will form cation–anion pairs (Fig. 2.10A). These pairs can deposit on the solid surface and readily assemble

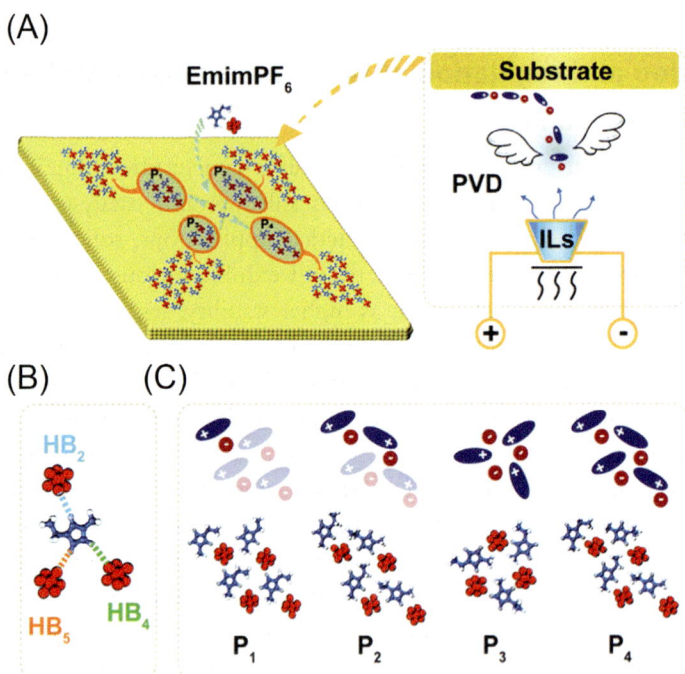

Figure 2.10 (A) A schematic diagram of 2DIIs on the graphite surface via PVD, where red, blue, yellow, white, and silver colors represent $[PF_6]^-$, $[Emim]^+$, C atoms in graphite, $H_{2/4/5}$ atoms, and H atoms in the alkyl chain, respectively. (B) Three main HBs in the ILs, where HB_2, HB_4, and HB_5 represent the HB between F and H in different sites of the imidazole ring. (C) The illustrated diagrams and atomic structures of the P_1-based 2DII with four subunits, the P_2-based 2DII with two subunits, the P_3-based 2DII with one subunit, and the P_4-based 2DII with one subunit, respectively.

into various island structures, particularly when the number of ionic liquid pairs is limited. As depicted in Fig. 2.10B, the HBs in 2DIIs mainly include three types: $C_2-H_2 \cdots F$, $C_4-H_4 \cdots F$, and $C_5-H_5 \cdots F$, denoted as HB_2, HB_4, and HB_5, respectively. Due to the unique directional features of HBs, cation−anion pairs can form multiple substructures (P_N), as shown in Fig. 2.10C, where $N = 1, 2, 3,$ and 4 represent the number of cation−anion pairs. For P_1, P_2, and P_4, the differences lie primarily in the combination of HBs and the relative orientations within neighboring cations. However, in the case of P_3, the substructure consists of three pairs of ILs, forming a circular arrangement. These distinct subunit structures can further dictate the arrangement of 2DIIs and meanwhile form feature edges.

To further quantify the impact of size on structural stability, E_f for 2DIIs with different N_{pairs} is summarized in Fig. 2.11A, demonstrating that E_f decreases as N_{pairs} increase for all 2DIIs. When N_{pairs} exceeds the critical value ($N_C = 18$), E_f of various 2DIIs only fluctuates within a small range, indicating that the thermodynamic stability of 2DIIs will continue to increase and converge with the increase of N_{pairs}. Interestingly, the E_f values for 2DIIs based

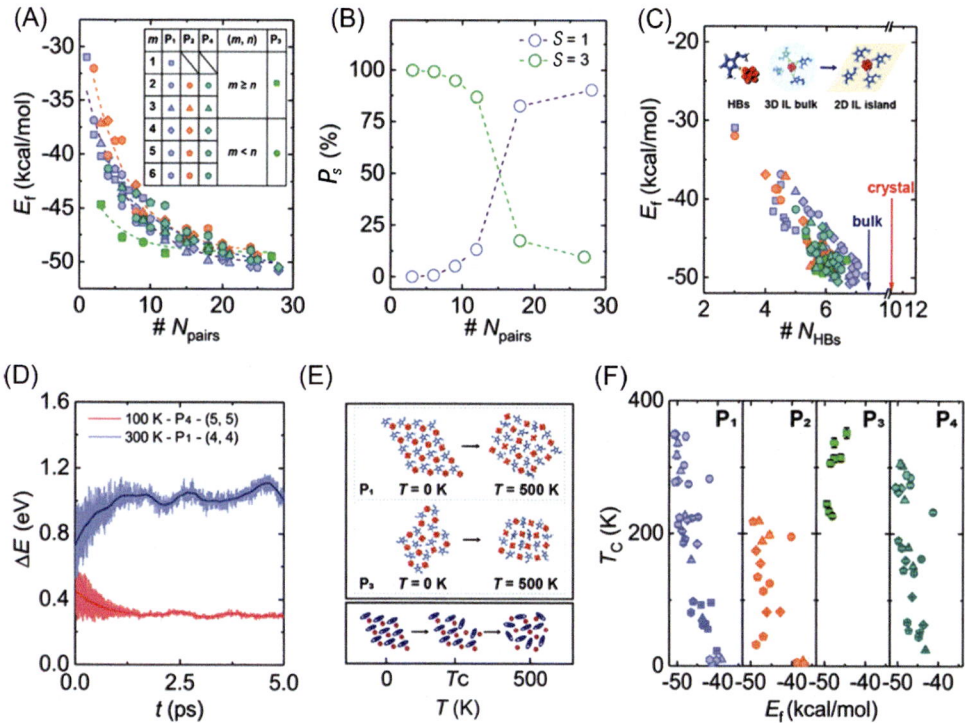

Figure 2.11 (A) E_f as a function of N_{pairs}, where the dashed lines represent exponential decay fitting. (B) P_S for different 2DIIs changes with N_{pairs}. (C) The correlation between E_f and N_{HBs}, where the inset shows the atomic structures of HBs in the bulk and a 2DII. (D) ΔE of two typical 2DIIs from the AIMD simulations. (E, F) T_C in the melting process of various 2DIIs as a function of E_f.

on $P_{1/2/4}$ are nearly the same, suggesting that the relative orientations within neighboring cations have little influence on the thermodynamic stability. Additionally, the tendencies of E_f-N_{pairs} for 2DIIs based on $P_{1/2/4}$ are entirely different from those based on P_3: when the size of 2DIIs is relatively small ($N_{pairs} < N_C$), the E_f of 2DIIs based on P_3 is the most negative; in contrast, once the size exceeds N_C, the E_f of 2DIIs based on $P_{1/2/4}$ becomes the lowest.

Assuming that N_{pairs} is fixed, the Boltzmann factor can be used to describe the distribution of $P_{1/3}$-based 2DIIs, that is,

$$P_s = \frac{e^{-E_f, P_s/k_B T}}{\sum_i e^{-E_f, P_i/k_B T}} \tag{2.2}$$

where $s = 1$ and 3, $i = 1$ versus 3, k_B is the Boltzmann constant, and T is set as 300K [42]. As displayed in Fig. 2.11B, there exist two stages in P_S-N_{pairs}. For instance, when $N_{pairs} = 12$, P_1 and P_3 are respectively 13.18% and 86.82%, while when $N_{pairs} = 18$, P_1 and P_3 are 82.51% and 17.49%, respectively. Hence, P_3 will be the likeliest subunit if $N_{pairs} < 18$, which is the opposite when $N_{pairs} \geq 18$. The turnover of P_S implies that the structural transition from P_3 to P_1 may occur during the assembly process.

The angles and energies of HBs in 2DII are larger and lower, respectively, compared to bulk and crystalline ILs, indicating that graphite enhances HBs in 2DII and further strengthens the structural stability of 2DII. As shown in Fig. 2.11C, E_f decreases almost linearly with the number of hydrogen bonds per cation−anion pair (N_{HBs}), suggesting that HBs play a dominant role in constructing 2DII. When $N_{pairs} > N_C$, N_{HBs} reaches the value of bulk ILs, aligning well with the converged E_f. Furthermore, representative 2DII structures were subjected to 5 ps long AIMD simulations at temperatures of 300 and 100K to validate their structural stability.

To further confirm the dynamic stability of 2DII, the melting process was elucidated through classical MD simulations. As the temperature (T) increases, 2DII initially retains its initial structure and then undergoes an in-plane transition at the critical temperature (T_C), simultaneously losing its ordered internal arrangement (Fig. 2.11D and E). Fig. 2.11F displays the evolution of T_C and E_f, indicating a negative correlation in 2DII based on $P_{1/2/4}$, while a positive correlation is observed in 2DII based on P_3. Hence, for some small P_3-based 2DIIs and large $P_{1/2/4}$-based 2DIIs, T_C is above room temperature, indicating their excellent dynamic stability. The different relationships between $P_{1/2/4}$-based and P_3-based 2DIIs coincide well with two stages of P_S-N_{pairs} (Fig. 2.11B). Combining E_f-N_{pairs}, E_f-N_{HBs}, T_C-E_f, and T_C-N_{pairs}, it can be concluded that HB networks can determine the stability of 2DIIs.

Indeed, SSIL thin films have found widespread applications in gas capture and conversion processes, where the interactions between gases and 2DII play a pivotal role. Understanding the gas adsorption process, especially the adsorption of CO_2, is of paramount importance for the rational design of corresponding catalysts. To evaluate the gas adsorption stability, we defined four different adsorption sites of 2DIIs (vertex site-R_{vertex}, edge sites-$R_{Z4}/R_{Z5}/R_{E1}/R_{E2}$, above site-$R_{above}$, and inside site-$R_{in}$), as shown in Fig. 2.12A. The adsorption energy (E_{gas}) of different sites is also calculated, as shown in Fig. 2.12B. For instance, E_{gas} of CO_2 at R_{vertex}, R_{Z4}, R_{Z5}, and R_{above} in the P_4-based 2DII with (4,4) is -9.88, -10.24, -12.08, and -5.99 kcal/mol, respectively, reflecting that the edge site (R_{Z5}) provides the strongest adsorption ability. The lower E_{gas} should originate from the anisotropic C_{image} distribution of graphite, providing the additional adsorption force to the target gas molecules. However, for the P_3-based 2DII, E_{gas} of CO_2 at R_{in}, R_{E1}, and R_{E2} is close, agreeing well with the isotropic C_{image} distribution. In addition, since the position of the CO_2 molecule at R_{above} is so far away from the graphite substrate, this leads to weaker substrate contribution, further resulting in a higher E_{gas} compared with that of other sites. Besides, E_{gas} of CO, CH_4, and N_2 at various sites of the P_4-based 2DII also shows the same order as that of CO_2.

To quantitatively explore the dynamic stability of gas adsorption on 2DII, further MD simulations were conducted at $T = 300K$ (Fig. 2.12C). With gas pressures (P) ranging from 0.44 to 1.76 bar, nearly all gas molecules adsorbed at the edges of 2DII based on P_4, while some gas molecules were adsorbed at internal positions of 2DII based on P_3. Excitingly, as shown in Fig. 2.12C, the Boltzmann factors of E_{gas} are in

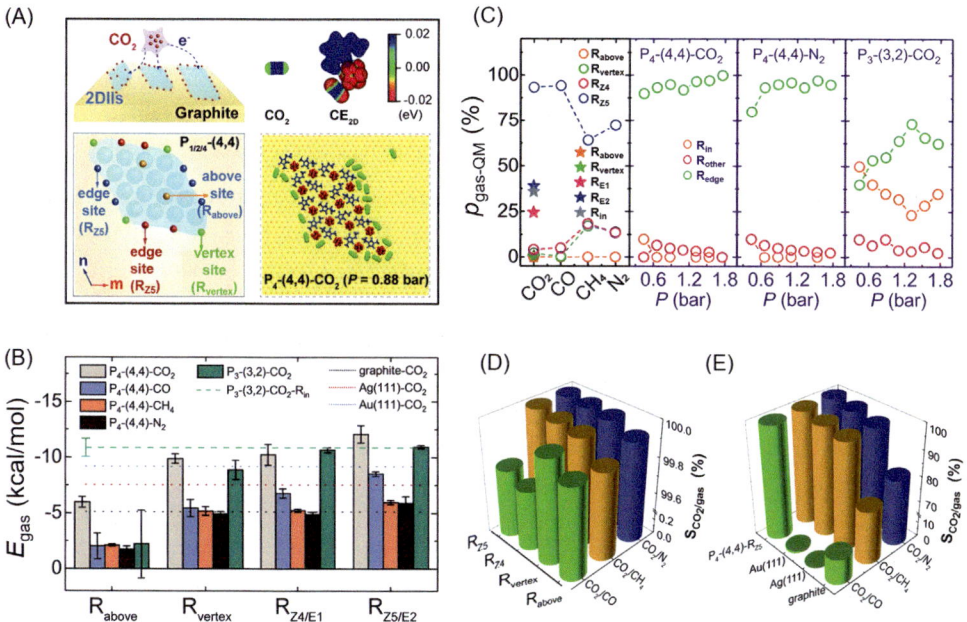

Figure 2.12 (A) Schematic diagrams of CO_2 adsorbed on different sites and ESP mapped to the vdW surface of CO_2 before (left) and after (right) adsorbing on the 2DII. (B) E_{gas} for different systems. (C) The P_{gas-QM} for different systems. (D, E) The CO_2 adsorption selectivity to other gases.

excellent agreement with the proportions of different gases adsorbed at different sites, confirming that the edges of 2DII are the main adsorption sites for different gases.

Furthermore, compared with many pure solids, E_{gas} of different gases on edge sites of 2DIIs is even stronger (Fig. 2.12B). Taking CO_2 as an example, E_{gas} of R_{Z5} in the P_4-based 2DII with (4,4), Au(111), Ag(111), and graphite is -12.08, -9.20, -7.54, and -5.13 kcal/mol, respectively. Considering that E_{gas} of CO_2 is always lower than that of CO, CH_4, and N_2 when they adsorb on the P_4-based 2DII, we further calculated the CO_2 adsorption selectivity to other gases ($S_{CO2/gas2}$) on different solid surfaces. Fantastically, $S_{CO2/gas2}$ for the P_4-based 2DII is bigger than 99.70% (Fig. 2.12D and E), indicating that all adsorption sites of 2DIIs show extremely high CO_2 selectivity.

Moreover, the $S_{CO2/CH4}$, $S_{CO2/CO}$, and $S_{CO2/N2}$ ratios on P_4-based 2DII are 1.31, 1.52, and 1.23 times higher than those on pure graphite surfaces. In addition, the $S_{CO2/CH4}$, $S_{CO2/CO}$, and $S_{CO2/N2}$ ratios on Au(111) are lower than 2DII by 98.59%, 1.26%, and 98.91%, respectively. The high CO_2 adsorption selectivity indicates that 2DII can be considered a promising adsorbent or reaction promoter in carbon capture and utilization processes, aligning well with recent experiments on stable single–atom catalysts supported by ILs [43]. Considering that some 2DIIs can maintain stable structures above room temperature, achieving higher gas adsorption and selectivity could be feasible by extending the edges of 2DII or tuning the ionic microenvironment around the catalyst.

2.4.2 Two-dimensional ionic liquids with an anomalous stepwise melting process and ultrahigh CO_2 adsorption capacity

The interfaces between ILs and solids are crucial for practical applications like high-efficiency catalysts, clean energy storage, and gas treatment [44–46]. Interfacial ILs enhance device performance compared to bulk ILs [47]. Upon contact with a surface, ILs transform into ultrathin films with complex structures, exhibiting various interactions like Coulombic, solvophobic, π-π, van der Waals, and hydrogen bonding [26]. These microenvironments enable functionalities such as CO_2 capture, and transistors [48]. Understanding ultrathin IL films, especially self-assembled HB networks [49], is vital for their chemistry, structures, catalytic properties, and thermodynamics [50].

Recent efforts employ experimental methods like STM and AFM to observe 2D monoionic layers on metal or graphite surfaces [51,52]. External potential and temperature precisely tune their structures and properties [53]. Molecular simulations analyze ionic arrangements in these films [54]. However, the melting behaviors of 2D ILs remain less clear due to their complexity, special HBs, and low-dimensionality. Multistage melting processes involve localized ion rotation, cation flipping, and interfacial free diffusion. Understanding these processes is essential for future applications in diverse fields.

The researchers conducted a comprehensive investigation into the dynamic melting behaviors and functionalities of 2D ILs using a combination of atomistic MD simulations and high-resolution scanning tunneling microscopy observations [55]. Their primary focus was to understand how the solid surface properties and chemical structures of ILs influence the unique HB networks within 2D IL films, ultimately governing their melting behaviors and structural transitions.

The researchers began by depositing six different ILs onto an Ag (111) surface, enabling them to investigate the interfacial structures and behaviors of the ILs (Fig. 2.13A). Interestingly, all the ILs formed 2D monoionic ordered structures at low temperatures, adopting a checkerboard arrangement (Fig. 2.13). The cations and anions were arranged in an interlocked form, forming highly ordered structures with transverse density profiles confirming the monoionic layer nature. High-resolution STM experiments validated the simulated checkerboard structures of 2D ILs, further confirming the monoionic features of these materials.

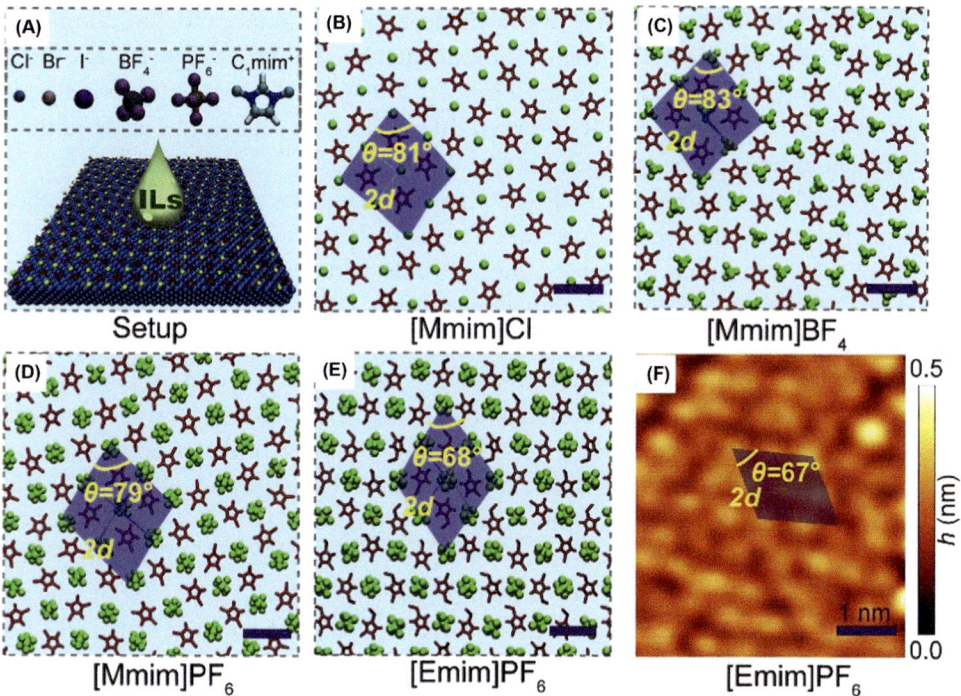

Figure 2.13 Structure feature of 2D IL. (A) Anion structure and simulation setup in the present work.(B−E) Ordered checkerboard structures of 2D ILs [Mmim]Cl, [Mmim][BF₄], [Mmim][PF₆], and [Emim][PF₆] on the Ag (111) surface from molecular dynamics simulations (T = 5K). (F) Scanning tunneling microscopy (STM) image of 2D [Emim][PF₆] (T = 5K, U = −1 V, I = 0.1 nA) [55]. Note: The scale bars in (B)−(F) represent the length of 1 nm.

The melting behaviors of the 2D ILs were then explored using MD simulations, revealing an abnormal multistage melting process. Three distinct movements were identified in the melting process: localized rotation of ions, out-of-plane flipping of a cation, and interfacial free diffusion. In contrast to 3D ILs, which have a single melting point, the 2D ILs exhibited multiple transition points due to the weakened HB network within the thin films. The researchers determined critical temperatures (TCRs) for each of these transitions through the evolution of the self-rotational diffusive coefficient, the flipped ratio of ions, and the 2D structure factors.

The multistage melting behaviors were found to be dependent on the interaction between the ILs and the solid substrate. By conducting MD simulations on various FCC metal surfaces with different lattice parameters and surface interaction energies, the researchers confirmed that the metal surface properties had little influence on the detailed structures of the 2D ILs, indicating that the arrangement of ILs primarily depended on the intrinsic interactions within the IL layers.

To further understand the intrinsic mechanism of the multistage melting behaviors, the researchers evaluated the hydrogen bonding within the 2D ILs at different temperatures. For 3D IL crystals, there was a single transition point, and the number of hydrogen bonds decreased significantly during melting. However, for 2D ILs, there were two obvious turning points in the number of hydrogen bonds, corresponding to the out-of-plane flipping and fully disordered transition, respectively. This confirmed the abnormal multistage melting behaviors of 2D ILs.

The researchers constructed TCR$-\gamma$ phase diagrams for 2D ILs based on their melting behaviors at different interaction energies with the substrate. They found that the critical transition points increased almost linearly with the interaction energy, suggesting that interface engineering could be used to precisely regulate the structural transitions of 2D ILs.

Based on the energy barriers for the movements observed during melting, the researchers developed a predictive model to describe the multistage structural transitions of 2D ILs on different solid surfaces. This model, derived from the MD simulations, showed good agreement with the experimental results and allowed for the estimation of critical transition temperatures for specific 2D ILs on various solid surfaces.

Finally, the study explored the ultrahigh CO_2 adsorption capacity of 2D ILs (Fig. 2.14). Due to their exposed and sparse hydrogen bond network, 2D ILs exhibited significantly enhanced CO_2 capture capabilities compared to thick IL films or 3D bulk cases. The adsorption$-$desorption process of CO_2 had little impact on the structure of 2D ILs, indicating their high robustness for CO_2 capture and fixation. This enhanced CO_2 adsorption capability of 2D ILs has significant implications for their potential applications in CO_2-capture-fixation chemistry and other related fields.

(A)

gas region

CO$_2$ flow

Cat.

2D IL Layer

dispersed CO$_2$

(B)

$\Delta G = -1.29$ eV

$\Delta G = -0.44$ eV

2D MmimPF$_6$

n_{CO_2} (nm^{-3})

d (nm)

(C)

CO$_2$@MmimPF$_6$ total ad.   ↓heating

n_{ad} (#/nm^2)

surface ad.

i:T_1 ~ 300 K

void ad.

ii:T_2~ 370 K

2D ↓heating

3D

iii:T_{IL}~655 K

t (ns)

(D)

i:T_1 ~ 300 K ii:T_2~ 370 K iii:T_{IL}~655 K

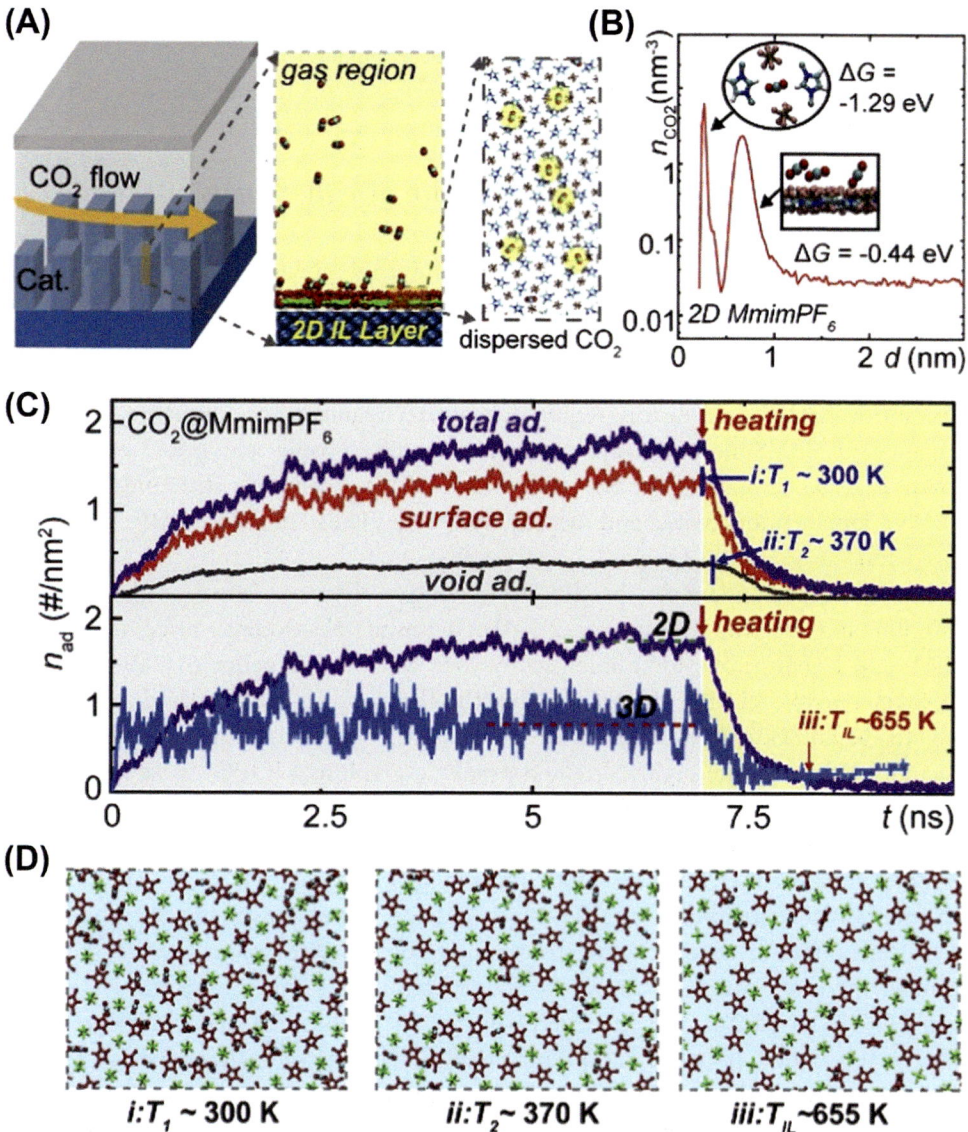

Figure 2.14 The high CO$_2$ adsorption capacity of 2D ILs. (A) Utilization of 2D ILs in the integrated CO$_2$-capture-fixation system and the dispersive absorbed CO$_2$. (B) The adsorption number density of CO$_2$ via the 2D [Mmim][PF$_6$]@M $\gamma = 1.71$ when the partial pressure of CO$_2$ (P_{CO2}) was ~5.87 bar. (C) Adsorption–desorption process of CO$_2$ as the temperature increased for the 3D and 2D ILs. (D) Snapshots of the atomic structure of 2D [Mmim][PF$_6$]@M $\gamma = 1.71$ in different desorption states [55].

Overall, the results section of the study provides a comprehensive and detailed analysis of the dynamic melting behaviors and functionalities of 2D ILs. It elucidates the intrinsic mechanisms behind their multistage melting processes, offers insights into the role of solid surface properties in regulating these behaviors, and demonstrates the potential of 2D ILs for ultrahigh CO_2 adsorption capacity. These findings not only advance our understanding of the unique properties of 2D ILs but also pave the way for their applications in a wide range of scientific and industrial contexts.

2.4.3 Electron transfer and friction feature of two-dimensional ionic liquids

It is well known that electron transfer is a universal phenomenon that can occur between any contact materials [56,57]. Therefore, for 2D ILs confined on the solid surface, there must be interfacial electron transfer, which may significantly affect the IL properties. In such a case, understanding the structural feature and electron transfer behavior of 2D ILs is necessarily required for better regulating and improving the performance of ILs. Simultaneously, most ILs are hygroscopic and prior research has shown that water can affect the properties of ILs, including structure, viscosity, polarity, ionic conductivity, and density [58−60]. Thus, the water effect must be considered.

Sun et al. developed a new method to investigate the water effect on the structural transition and electron transfer of 2D ILs by using electrostatic force microscopy (EFM) and Kelvin probe force microscopy (KPFM) measurements [61]. By introducing water into the 2D ILs, they found that the ILs changed from ordered stripe structure to agglomerated structure with increasing water content. Specifically, for $[C_{18}mim][NTf_2]$ on the highly oriented pyrolytic graphite (HOPG) surface, ILs firstly formed monolayer structures with a thickness of ~ 0.4 nm in the absence of water and then the microscale layer shrank to a smaller size with clear boundaries and finally the agglomerated structure appeared with increasing water content. A similar transition behavior was also observed in the ILs-MoS_2 system. The major difference is the critical water concentration at which the structure aggregated; the water content increased from 88.9% (mole fraction) to 93.8% (mole fraction) as the substrate was changed from HOPG to MoS_2. The increased capacity of water molecules comes from the stronger interaction between cation and MoS_2, which possesses negative charges on the surface compared to HOPG. XPS experiments showed that water molecules could interact with the anion through the O-H$\cdots$O hydrogen bond, which reduced the cation−anion interaction and caused the observed structural transition.

As for the electron transfer, the reversed EFM phase contrast at different polarities of the AFM tip proves that there exists electron transfer between 2D ILs and solid surface, as shown in Fig. 2.15A and B. The phase difference between ILs and HOPG decreased with increasing water content and finally stabilized at a smaller value (Fig. 2.15C), indicating that the addition of water weakened the electron transfer. That is because the water molecule mainly interacts with the anions and hence the water molecule could shield the electron transfer channel by wrapping around the

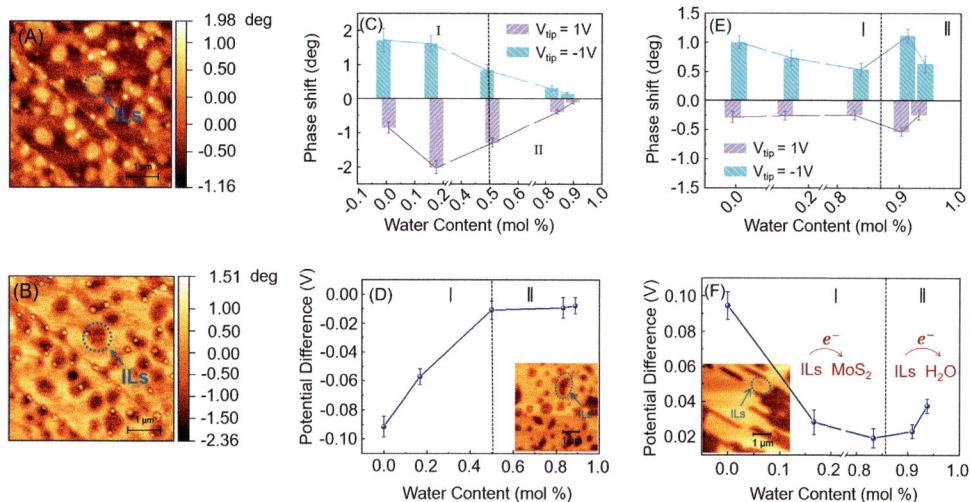

Figure 2.15 Typical EFM phase images of [C$_{18}$mim][NTf$_2$] ILs with the water content of 16.7%(mole fraction) at the HOPG surface with (A) $V_{tip} = -1$ V and (B) $V_{tip} = 1$ V. (C) The phase difference and (D) potential difference between ILs and HOPG at different water contents and different tip biases. The insert shows the surface potential image of pure ILs with the darker region for ILs and the brighter region for HOPG. (E) The phase difference and (F) potential difference between ILs and MoS$_2$ at different water contents and tip biases. The insert shows the surface potential image of pure ILs with the brighter region for ILs and the darker region for MoS$_2$ [61].

anions. At high water content, the ultralow phase difference implied the formation of hydrated anions. Meanwhile, the surface potential of ILs was lower than that of HOPG at different water contents (Fig. 2.15D), suggesting that the direction of electron transfer was from HOPG to ILs. For the hydrophilic MoS$_2$ substrate, the EFM phase reversal phenomenon was also observed and featured a different electron transfer of ILs to MoS$_2$, opposite to that of HOPG (Fig. 2.15F). Finally, they revealed the electric field dependence of electron transfer in the presence of water. It was found that it was the magnitude of voltage rather than the polarity that determined the amount of electron transfer. The larger voltage lead to stronger electrostatic attraction between ILs and substrate, which reduced the separation distance between ILs and substrate and made electron transfer more likely to occur. In contrary, the addition of water weakened the enhancement due to the shielding effect of water molecules.

Because of the ordered structure, 2D ILs are also good boundary lubricants. This blooming field has been extensively studied in structuring, modeling, and computations, among which the main observation is the strong dependence of the lubrication performance on the structure of surface ILs [62]. Lu et al. performed a system friction study of ILs with different alkyl chain lengths on the HOPG surface and revealed that the ordered structure had an amazingly ultralow friction coefficient of the order of 0.001, as shown in Fig. 2.16 [63]. They first found that the ordered stripe structure

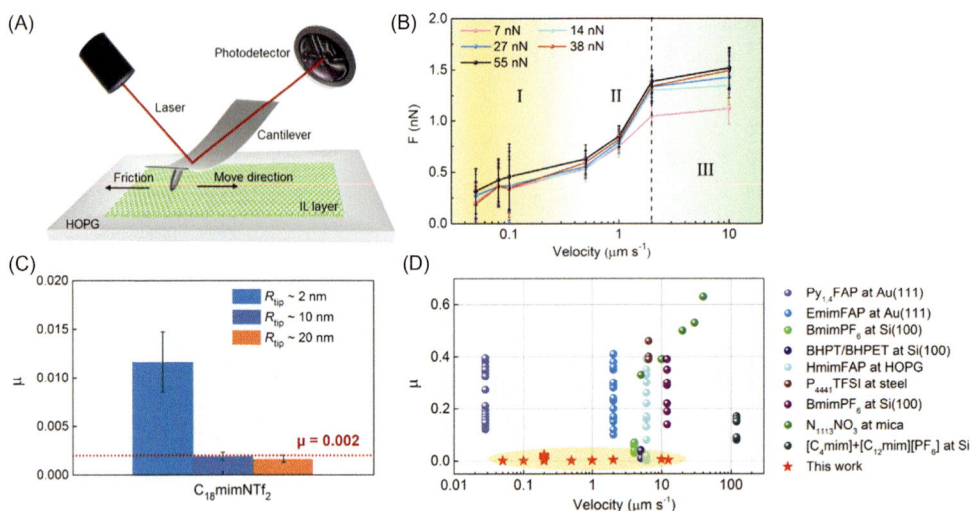

Figure 2.16 Experimental setup and friction feature of monolayer ILs on the HOPG surface. (A) Schematic diagram of friction experiments. (B) Velocity-dependent friction force of [C$_{18}$mim][NTf$_2$] at different normal forces with a tip radius of 20nm. (C) Friction coefficient of [C$_{18}$mim][NTf$_2$] at the velocity of 0.2 μm/s with different tip radii. (D) Friction coefficient of various ILs in references obtained by AFM experiments [63].

could only be formed as the carbon number of the side alkyl chain was larger than 12. Based on the simulation, a head–to–head structure model with the alkyl tails arranged in an all–trans mode lying flat on the HOPG surface was obtained, which coincided well with the experiments. The main reason of the critical behavior is the strong van der Waals interaction between longer alkyl chains and solid surface, which could stabilize the ordered structure and is absent for short alkyl chains, as observed in references [64,65]. Moreover, solvent also affected the structure by changing the periodicity of the structure with polar solvents, such as dimethyl sulfoxide and methanol, favored larger periods and contributed to more stable structures.

The stripe structure leads to fluctuations with a series of saw–tooth cycles in friction force. In one cycle, the ordered alkyl chain contributed to low friction, while the disordered warped imidazole ring contributed to high friction due to the firm pinned effect of the AFM tip on the rough structure. Moreover, the friction coefficient decreased with the increasing proportion of the ordered region in one period. Thus, the ultralow friction values were closely related to the ordered regions formed by alkyl chains. With the same ordered structure, the friction coefficient decreased as the tip radius increased. Specifically, the friction coefficient of different ILs dropped sharply from ~ 0.01 to ~ 0.002 as the tip radius increased from 2 nm to 10 nm, while a merely negligible decline was observed from 10 nm to 20 nm. Compared with the periodicity of the ordered structure of ~ 6 nm, it was concluded that the friction coefficient decreased remarkably only when the tip radius was larger than the structure periodicity. That is because a smaller tip radius responds to structure ups and downs sensitively and, hence, is more likely to get stuck in rough regions, which would require a larger force to drive the tip scanning. Once the tip radius is beyond the threshold, the stuck behavior will be greatly suppressed. The energy dissipation calculated from the friction loop showed that more energy was dissipated at a larger tip radius, hence contributing to a larger friction coefficient.

For the friction force, it could be divided into three stages with increasing velocity, as plotted in Fig. 2.16B. The friction force rose appreciably in the first stage, which was roughly a linear relationship with the logarithm of velocity. Then, it remained approximately constant in the third stage, and the maximum force exhibited a gradual increase with the rise in the normal force [66]. Importantly, in a velocity range of $0.1-10 \ \mu m/s$, the monolayer structure showed much lower friction coefficient values compared to those reported in the literature (Fig. 2.16D) [67−76]. Meanwhile, the monolayer exhibited robust stability and was kept stable during continuous scanning at pressure as high as ~ 78 MPa. These results demonstrated the superior lubrication performance of the ordered 2D ILs on the solid surface.

2.5 Prediction and control of ionic liquid structures

Many thermodynamic properties can be expressed as statistical averages of microscopic quantities, that is, time averages in MD. The most typical ones are the volume and energy of a simulated system. The latter includes kinetic and potential energies, which can be broken down into different contributions such as electrostatic, dispersion, and so forth, which can be used to explore their origination further. On the other hand, the fluctuation of these averages can be used to derive thermodynamic first-order response coefficients such as heat capacity, isothermal compressibility, and isobaric expansion.

Another important class of properties are transport coefficients, such as diffusion coefficients, viscosities, conductivities, and thermal conductivities, which originate from perturbations of the equilibrium state, for example, gradients in concentration correspond to diffusion, gradients in flow rate correspond to viscosity, and so on. There are two options for obtaining these properties in simulations: one is based on nonequilibrium simulations similar to real experiments, where the corresponding gradients are artificially introduced, and the coefficients are obtained according to the definitions, but specific algorithms need to be developed for simulations since the gradients generated in the micro-world tend to differ by several orders of magnitude from those of the macroscopic world. The other is based on the so-called linear response theory computed through equilibrium simulation. The trajectory obtained by MD simulation is mathematically a time series, and the time correlation function (TCF) is defined for a certain microscopic quantity (such as the velocity), which is the statistical correlation between the time after a moment t and the time at this moment. The above transport coefficient can be obtained by the Green—Kubo (GK) integral of the TCF, or by using the Einstein's equation to obtain the slope of the integral quantity with respect to the time.

For example, the reported self-diffusion coefficients usually calculated by Einstein's equation,

$$D_{\alpha,i} = \frac{1}{2}\lim_{t\to\infty}\frac{d}{dt}\left\langle\frac{1}{N}\sum_{i=1}^{N}|r_{\alpha,i}(t)-r_{\alpha,i}(0)|^2\right\rangle_{t_0} \tag{2.3}$$

The averaged value in angle bracket is called mean-squared displacement (MSD), which is calculated over all possible different originate times t_0 (usually half of the simulation time) with a same time span of t, and $r_{\alpha,i}$ is the position of particle i in the x, y or z direction. As shown in Eq. (2.3), all of the particles can be used in the average to achieve good sampling. In contrast, the shear viscosity is calculated by the GK integration of the nondiagonal components of the pressure tensor:

$$\eta = \frac{V}{k_B T}\int_0^\infty dt\langle\tau_{\alpha\beta}(t)\tau_{\alpha\beta}(0)\rangle_{t_0} \tag{2.4}$$

Unlike MSD, there are only 3 pressure tensors; τ_{xy} , τ_{yx}, and τ_{xz} can be used in the calculation at each moment. Thus it is not possible to enhance the sampling by increasing the size of simulation, but only by extending the simulation time. Typically several hundred ns of simulation time is required to calculate the viscosity, an order of magnitude longer than the time required to calculate the diffusion coefficient.

Currently, for pure IL, the above properties can be reasonably predicted by simulations using well-established classical FFs. The typical results are shown in Fig. 2.17, including the density, heat of vaporization, diffusion coefficient, and viscosity. The simulations are consistent with experimental results for a series of $[C_n\text{mim}][\text{NTf}_2]$ ionic liquids.

The advantage of molecular simulation lies not only in the ability to predict the macroscopic thermodynamic properties in a bottom-up way, which provides guidance for designing new materials and solvents, but also in the ability to obtain the microscopic structure and its dynamic evolution, so as to understand the mechanism behind the macroscopic phenomena. The most commonly used is the RDF, which characterizes the local density of an atom around another atom as a function of their distance. The RDF is relatively easy to calculate and therefore is widely used in microstructural analyses, and in combination with properly defined angular analyses, it is usually used to analyze the strength and orientation of hydrogen bonds. Integrating the RDF yields a coordination number, which is often used to study solvation

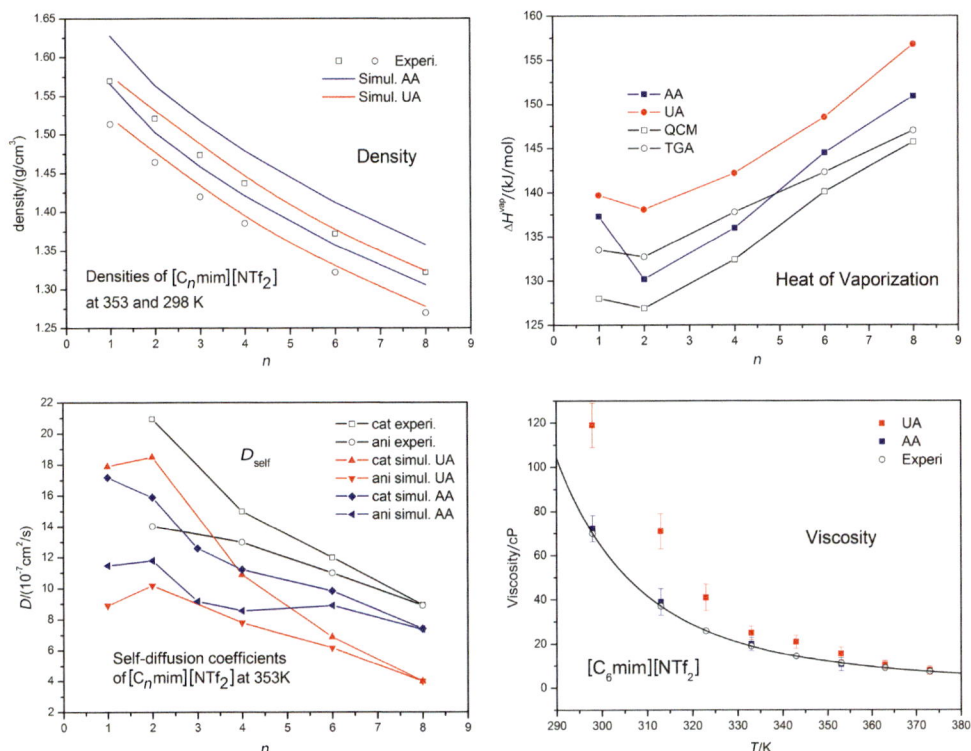

Figure 2.17 Experimental properties of a series of ionic liquids [C$_n$mim][NTf$_2$], compared with the predicted values in simulation(at 298.15K).

phenomena around molecules. The static structure factor can be obtained by the Fourier transform of the RDF. If the RDF of all atoms is calculated, the spectra of X-ray scattering or neutron scattering can be calculated by weighting these RDFs, which establishes a direct connection with the experimental phenomena and is used to examine the FF and to analyze the origin of nanostructures in ionic liquids. In addition to the RDF, the distribution of other atoms around a central molecule in three dimensions can be studied more intuitively by means of the spatial distribution function (SDF).

Here, we give an interesting result concerning the methylation of C$_2$-hydrogen in imidazolium cation. It is well known that there is hydrogen bonding between the C$_2$-H and the anion, but the viscosity increases upon methylation, which seems a counterintuitive. As shown in Fig. 2.18, the viscosities are not only well predicted by the simulations, but also explained by the significant change of micro-structure near the cations, as shown by the SDF of anions.

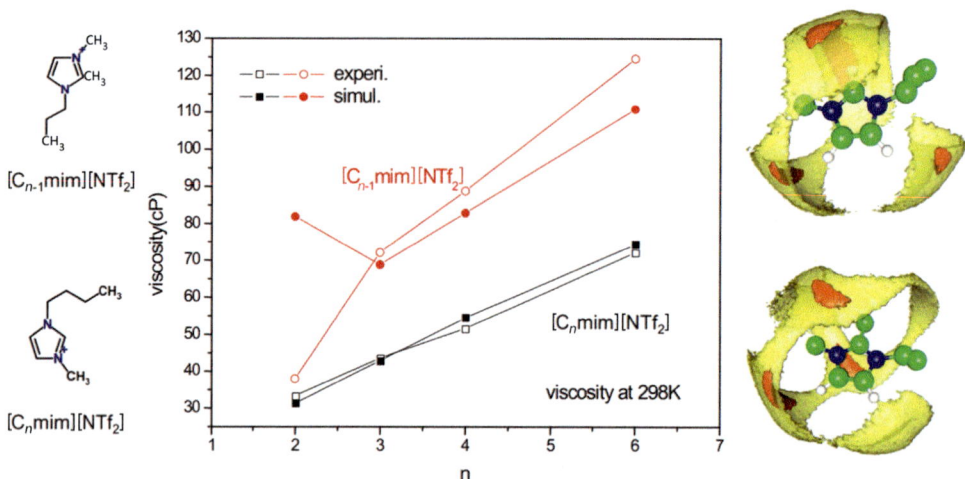

Figure 2.18 The change of viscosities after methylation of C_2-hydrogen in imidazolium cation for ionic liquids of $[C_n mim][NTf_2]$ and $[C_{n-1} mim][NTf_2]$.

References

[1]   Canongia LJ, Deschamps J, Pádua A. Modeling ionic liquids using a systematic all-atom force field. The Journal of Physical Chemistry B 2004;108:2038−47.

[2]   Gong Z, Sun H. Extension of TEAM force-field database to ionic liquids. Journal of Chemical and Engineering Data 2019;64:3718−30.

[3]   Doherty B, Zhong X, Gathiaka S, et al. Revisiting OPLS force field parameters for ionic liquid simulations. Journal of Chemical Theory and Computation 2017;13:6131−45.

[4]   Liu Z, Huang S, Wang W. A refined force field for molecular simulation of imidazolium-based ionic liquids. The Journal of Physical Chemistry B 2004;108:12978−89.

[5]   Wang L, Martinez T, Pande V. Building force fields: an automatic, systematic, and reproducible approach. The Journal of Physical Chemistry Letter 2014;5:1885−91.

[6]   Bedrov D, Piquemal J, Borodin O, et al. Molecular dynamics simulations of ionic liquids and electrolytes using polarizable force fields. Chemical Reviews 2019;119:7940−95.

[7]   Goloviznina K, Lopes J, Gomes M, et al. Transferable, polarizable force field for ionic liquids. Journal of Chemical Theory and Computation 2019;15:5858−71.

[8]   McDaniel J, Choi E, Son C, et al. Ab initio force fields for imidazolium-based ionic liquids. The Journal of Physical Chemistry B 2016;120:7024−36.

[9]   Yan T, Burnham C, Del Pópolo M, et al. Molecular dynamics simulation of ionic liquids: the effect of electronic polarizability. The Journal of Physical Chemistry B 2004;108:11877−81.

[10] Borodin O. Polarizable force field development and molecular dynamics simulations of ionic liquids. The Journal of Physical Chemistry B 2009;113:11463−78.

[11] Wang Y, Izvekov S, Yan T, et al. Multiscale coarse-graining of ionic liquids. The Journal of Physical Chemistry B 2006;110:3564−75.

[12] Wang Y, Lyubartsev A, Lu Z, et al. Multiscale coarse-grained simulations of ionic liquids: comparison of three approaches to derive effective potentials. Physical Chemistry Chemical Physics 2013;15:7701−12.

[13] Moradzadeh A, Motevaselian M, Mashayak S, et al. Coarse-grained force field for imidazolium-based ionic liquids. Journal of Chemical Theory and Computation 2018;14:3252−61.

[14] Uhlig F, Zeman J, Smiatek J, et al. First-principles parametrization of polarizable coarse-grained force fields for ionic liquids. Journal of Chemical Theory and Computation 2018;14:1471−86.

[15] Souza P, Alessandri R, Barnoud J, et al. Martini 3: a general purpose force field for coarse-grained molecular dynamics. Nature Methods 2021;18:382−8.

[16] Vazquez-Salazar L, Selle M, de Vries A, et al. Martini coarse-grained models of imidazolium-based ionic liquids: from nanostructural organization to liquid-liquid extraction. Green Chemistry 2020;22:7376−86.

[17] Perlt E, Ray P, Hansen A, et al. Finding the best density functional approximation to describe interaction energies and structures of ionic liquids in molecular dynamics studies. Journal of Chemical Physics 2018;148:193835.

[18] Wang C, Qian C, Li Z, et al. Molecular insights into the abnormal wetting behavior of ionic liquids induced by the solidified ionic layer. Industrial & Engineering Chemistry Research 2020;59:8028−36.

[19] Wang C, Liu G, Cao R, et al. Revealing the wetting mechanism of Li^+-doped ionic liquids on the TiO_2 surface. Chemical Engineering Science 2023;265:118211.

[20] Zhang S, Wang Y, He H, et al. A new era of precise liquid regulation: quasi-liquid. Green Energy & Environment 2017;2:329−30.

[21] Wang Y, Huo F, He H, et al. The confined [Bmim][BF_4] ionic liquid flow through graphene oxide nanochannels: a molecular dynamics study. Physical Chemistry Chemical Physics 2018;20:17773−80.

[22] Wang C, Wang Y, Lu Y, et al. Height-driven structure and thermodynamic properties of confined ionic liquids inside carbon nanochannels from molecular dynamics study. Physical Chemistry Chemical Physics 2019;21:12767−76.

[23] Lin R, Huang P, Ségalini J, et al. Solvent effect on the ion adsorption from ionic liquid electrolyte into sub-nanometer carbon pores. Electrochimica Acta 2009;54:7025−32.

[24] Tsai W, Taberna P, Simon P. Electrochemical quartz crystal microbalance (EQCM) study of ion dynamics in nanoporous carbons. Journal of the American Chemical Society 2014;136:8722−8.

[25] Forse A, Griffin J, Merlet C, et al. NMR study of ion dynamics and charge storage in ionic liquid supercapacitors. Journal of the American Chemical Society 2015;137:7231−42.

[26] Dong K, Liu X, Dong H, et al. Multiscale studies on ionic liquids. Chemical Reviews 2017;117:6636−95.

[27] Rogers R, Seddon K. Ionic liquids—solvents of the future? Science 2003;302:792−3.

[28] Xin B, Hao J. Imidazolium-based ionic liquids grafted on solid surfaces. Chemical Society Reviews 2014;43:7171−87.

[29] Polesso B, Bernard F, Ferrari H, et al. Supported ionic liquids as highly efficient and low-cost material for CO_2/CH_4 separation process. Heliyon 2019;5:e02183.

[30] Offner-Marko L, Bordet A, Moos G, et al. Bimetallic nanoparticles in supported ionic liquid phases as multifunctional catalysts for the selective hydrodeoxygenation of aromatic substrates. Angewandte Chemie International Edition 2018;57:12721−6.

[31] El Sayed S, Bordet A, Weidenthaler C, et al. Selective hydrogenation of benzofurans using ruthenium nanoparticles in lewis acid-modified ruthenium-supported ionic liquid phases. ACS Catalysis 2020;10:2124−30.

[32] Armstrong J, Hurst C, Jones R, et al. Vapourisation of ionic liquids. Physical Chemistry Chemical Physics 2007;9:982−90.

[33] Cremer T, Killian M, Gottfried J, et al. Physical vapor deposition of [Emim][Tf_2N]: a new approach to the modification of surface properties with ultrathin ionic liquid films. Chemphyschem 2008;9:2185−90.

[34] Uhl B, Buchner F, Alwast D, et al. Adsorption of the ionic liquid [BMP][TFSA] on Au(111) and Ag(111): substrate effects on the structure formation investigated by STM. Beilstein Journal of Nanotechnology 2013;4:903−18.

[35] Su Q, Qi Y, Yao X, et al. Ionic liquids tailored and confined by one-step assembly with mesoporous silica for boosting the catalytic conversion of CO_2 into cyclic carbonates. Green Chemistry 2018;20:3232−41.

[36] Xie W, Ji X, Feng X, et al. Mass-transfer rate enhancement for CO_2 separation by ionic liquids: theoretical study on the mechanism. AIChE Journal 2015;61:4437−44.

[37] Xie W, Ji X, Feng X, et al. Mass transfer rate enhancement for CO_2 separation by ionic liquids: effect of film thickness. Industrial & Engineering Chemistry Research 2016;55:366−72.

[38] Tang Z, Lu L, Dai Z, et al. CO_2 absorption in the ionic liquids immobilized on solid surface by molecular dynamics simulation. Langmuir 2017;33:11658—69.
[39] Li B, Wang C, Zhang Y, et al. High CO_2 absorption capacity of metal-based ionic liquids: a molecular dynamics study. Green Energy Environment 2021;6:253—60.
[40] Wang C, Wang Y, Gan Z, et al. Topological engineering of two-dimensional ionic liquid islands for high structural stability and CO_2 adsorption selectivity. Chemical Science 2021;12:15503—10.
[41] Dong K, Zhang S, Wang J, et al. Understanding the hydrogen bonds in ionic liquids and their roles in properties and reactions. Chemical Communications 2016;52:6744—64.
[42] Zhou X, Shu H, Li Q, et al. Electron-injection driven phase transition in two-dimensional transition metal dichalcogenides. Journal of Materials Chemistry C 2020;8:4432—40.
[43] Ding S, Guo Y, Hülsey M, et al. Electrostatic stabilization of single-atom catalysts by ionic liquids. Chem 2019;5:3207—19.
[44] Bi S, Banda H, Chen M, et al. Molecular understanding of charge storage and charging dynamics in supercapacitors with MOF electrodes and ionic liquid electrolytes. Nature Materials 2020;19:552—8.
[45] Dai C, Zhang J, Huang C, et al. Ionic liquids in selective oxidation: catalysts and solvents. Chemical Reviews 2017;117:6929—83.
[46] Watanabe M, Thomas M, Zhang S, et al. Application of ionic liquids to energy storage and conversion materials and devices. Chemical Reviews 2017;117:7190—239.
[47] Mezger M, Schröder H, Reichert H, et al. Molecular layering of fluorinated ionic liquids at a charged sapphire (0001) surface. Science 2008;322:424—8.
[48] Vishwakarma N, Singh A, Hwang Y, et al. Integrated CO_2 capture-fixation chemistry via interfacial ionic liquid catalyst in laminar gas/liquid flow. Nature Communications 2017;8:14676.
[49] Hayes R, Imberti S, Warr G, et al. The nature of hydrogen bonding in protic ionic liquids. Angewandte Chemie International Edition 2013;52:4623—7.
[50] Dong K, Huo F, Zhang S. Thermodynamics at microscales: 3D→2D, 1D and 0D. Green Enregy Environment 2020;5:251—8.
[51] Biedron AB, Garfunkel EL, Castner EW, et al. Ionic liquid ultrathin films at the surface of Cu(100) and Au(111). The Journal of Chemical Physics 2017;146:054704.
[52] Meusel M, Lexow M, Gexmis A, et al. Atomic force and scanning tunneling microscopy of ordered ionic liquid wetting layers from 110 K up to room temperature. ACS Nano 2020;14:9000—10.
[53] Elbourne A, McDonald S, Voïchovsky K, et al. Nanostructure of the ionic liquid-graphite stern layer. ACS Nano 2015;9:7608—20.
[54] Buchner F, Forster-Tonigold K, Uhl B, et al. Toward the microscopic identification of anions and cations at the ionic liquid|Ag(111) interface: a combined experimental and theoretical investigation. ACS Nano 2013;7:7773—84.
[55] Wang Y, Lu Y, Wang C, et al. Two-dimensional ionic liquids with an anomalous stepwise melting process and ultrahigh CO_2 adsorption capacity. Cell Reports Physical Science 2022;3:100979.
[56] Lin S, Chen X, Wang Z, et al. Contact electrification at the liquid-solid interface. Chemical Reviews 2022;122:5209—32.
[57] Lin S, Xu L, Wang A, et al. Quantifying electron-transfer in liquid-solid contact electrification and the formation of electric double-layer. Nature Communications 2020;11:399.
[58] Nickerson SD, Nofen EM, Chen H, et al. A combined experimental and molecular dynamics study of iodide-based ionic liquid and water mixtures. The Journal of Physical Chemistry B 2015;119:8764—72.
[59] Jeon J, Kim H, Goddard WA, et al. The role of confined water in ionic liquid electrolytes for dye-sensitized solar cells. The Journal of Physical Chemistry Letter 2012;3:556—9.
[60] Gliege ME, Lin W, Xu Y, et al. Molecular dynamics insight into the role of water molecules in ionic liquid mixtures of 1-butyl-3-methylimidazolium iodide and ethylammonium nitrate. The Journal of Physical Chemistry B 2022;126:1115—24.
[61] Sun T, Lu Y, Lu J, et al. Water-controlled structural transition and charge transfer of interfacial ionic liquids. The Journal of Physical Chemistry Letter 2022;13:7113—20.

[62] Gong X, Li L, et al. Nanometer-thick ionic liquids as boundary lubricants. Advanced Engineering Materials 2018;20:1700617.
[63] Lu Y, Wang Y, Huo F, et al. Ultralow friction and high robustness of monolayer ionic liquids. ACS Nano 2022;16:16471−80.
[64] Meusel M, Lexow M, Gezmis A, et al. Growth of multilayers of ionic liquids on Au(111) investigated by atomic force microscopy in ultrahigh vacuum. Langmuir 2020;36:13670−81.
[65] Galluzzi M, Bovio S, Milani P, et al. Surface confinement induces the formation of solid-like insulating ionic liquid nanostructures. The Journal of Physical Chemistry C 2018;122:7934−44.
[66] Riedo E, Gnecco E, Bennewitz R, et al. Interaction potential and hopping dynamics governing sliding friction. Physical Review Letters 2003;91:084502.
[67] Sweeney J, Hausen F, Hayes R, et al. Control of nanoscale friction on gold in an ionic liquid by a potential-dependent ionic lubricant layer. Physical Review Letters 2012;109:155502.
[68] Li H, Rutland MW, Atkin R, et al. Ionic liquid lubrication: influence of ion structure, surface potential and sliding velocity. Physical Chemistry Chemical Physics 2013;15:14616−23.
[69] Li H, Wood RJ, Rutland MW, et al. An ionic liquid lubricant enables superlubricity to be "switched on" in situ using an electrical potential. Chemical Communications 2014;50:4368−70.
[70] Elbourne A, Sweeney J, Webber GB, et al. Adsorbed and near-surface structure of ionic liquids determines nanoscale friction. Chemical Communications 2013;49:6797.
[71] Cooper PK, Wear CJ, Li H, et al. Ionic liquid lubrication of stainless steel: friction is inversely correlated with interfacial liquid nanostructure. ACS Sustainable Chemistry & Engineering 2017;5:11737−43.
[72] Cooper PK, Staddon J, Zhang SW. Nano- and macroscale study of the lubrication of titania using pure and diluted ionic liquids. Frontiers in Chemistry 2019;7:287.
[73] Palacio M, Bhushan B. Molecularly thick dicationic ionic liquid films for nanolubrication. Journal of Vacuum Science & Technology A 2009;27:986−95.
[74] Bhushan B, Palacio M, Kinzig B. AFM-based nanotribological and electrical characterization of ultrathin wear-resistant ionic liquid films. Journal of Colloid and Interface Science 2008;317:275−87.
[75] Mo Y, Huang F, Zhao F. Functionalized imidazolium wear-resistant ionic liquid ultrathin films for MEMS/NEMS applications. Surface and Interface Analysis 2011;43:1006−14.
[76] Pu J, Liu X, Wang L, et al. Formation and tribological properties of two-component ultrathin ionic liquid films on Si. Surface and Interface Analysis 2011;43:1332−40.

CHAPTER 3

Ionic liquids intensify reaction process

Contents

3.1 Overview	57
3.2 Ionic liquids regulate homogeneous catalysis reaction	58
3.2.1 Application in the synthesis of *n*-amyl alcohol acetate	59
3.2.2 Application in the synthesis of succinic acid diisopropyl ester	60
3.2.3 Application in the synthesis of dimethyl succinate	61
3.2.4 Application in the synthesis of diethyl succinate	61
3.2.5 Application in the synthesis of dimethyl carbonate	62
3.3 The multiphase reaction based on ionic liquids	63
3.3.1 Application in methyl methacrylate	64
3.3.2 Application in biodiesel production	65
3.3.3 Application in the synthesis of butyl acetate	66
3.3.4 Application in the synthesis of glycerol monolaurate	66
3.3.5 Application in the synthesis of dimethyl carbonate	67
3.3.6 Application in hydroxyl condensation reaction	70
3.3.7 Application in butyl citrate	71
3.4 Bioionic liquids: tunable microenvironment for biocatalysis	71
3.4.1 Introduction	71
3.4.2 Ionic liquids and free enzymes	72
3.4.3 Ionic liquids and whole cell catalysis	75
3.5 Ionic liquids intensified biomass conversion	79
3.5.1 Introduction	79
3.5.2 Ionic liquids intensify cellulose conversion	81
3.5.3 Ionic liquids intensify hemicellulose conversion	85
3.5.4 Ionic liquids intensify lignin conversion	86
References	90

3.1 Overview

Ionic liquids (ILs) have gained significant attention in recent times due to their unique and potential applications in various fields of science and technology. One of the most promising and important applications of ILs is their use as solvents and catalysts in chemical reactions. The unique properties of ILs, such as their high thermal stability, low volatility, and wide liquid range, make them suitable for a wide range of homogeneous catalytic reactions. This has led to the development of IL-intensified reactions, where ILs are used to enhance the efficiency and selectivity of chemical processes. Homogeneous catalytic reactions, such as hydrogenation, oxidation, carbonylation, biomass conversion, and multiphase reactions, have been successfully intensified using ILs.

These reactions typically involve the use of transition metal complexes as catalysts, which are dissolved in the IL to facilitate the desired chemical transformations. The use of ILs in these reactions has been shown to improve reaction rates, selectivity, and overall efficiency compared to traditional molecular solvents. One of the key advantages of using ILs in catalytic reactions is their ability to provide a unique environment for catalysts and reactants. The tunable nature of ILs allows for the modification of their properties, such as polarity, acidity, and basicity, which can be tailored to match the requirements of specific chemical reactions. This enables researchers to design task-specific ILs that are optimized for a particular catalytic process, leading to improved reaction performance. Furthermore, ILs have been found to stabilize and activate catalysts, leading to increased catalytic efficiency and lifetime. The unique solvation properties of ILs allow for the formation of well-defined catalyst complexes, which can lead to improved control over the reaction mechanism and product distribution. These attributes make ILs an attractive choice for intensifying catalytic reactions in various industrial and academic settings. In addition to their role as solvents and catalysts, ILs have also been employed as cosolvents, extraction agents, and electrolytes in various chemical processes. Their ability to dissolve a wide range of organic and inorganic compounds, along with their negligible vapor pressure and nonflammability, makes ILs desirable for applications where conventional solvents may pose environmental or safety concerns.

This chapter aims to provide a comprehensive understanding of the current state of research on IL-intensified reactions, with a focus on their applications in homogeneous catalytic processes. The key advantages and challenges associated with using ILs in catalytic reactions, along with recent developments and promising future directions in this field, are presented. Additionally, the potential impact of IL-intensified reactions in the context of sustainable and green chemistry, as well as their relevance in industrial applications, is expressed. As research in this area continues to expand, it is expected that IL-intensified reactions will play an increasingly important role in driving innovation and sustainability in the field of chemical synthesis and processing.

3.2 Ionic liquids regulate homogeneous catalysis reaction

Homogeneous phase reaction refers to the chemical reaction in which all the substances participating in the reaction are in the same phase, and there is no interphase mass transfer. The reactants, reaction products, solvents, and catalysts can be considered uniformly distributed in any of the differential volumes, although the material concentrations may vary considerably at different spatial locations of the reaction system. The research of homogeneous catalyst has been widely paid attention to by the scientific community and industry; the active center of homogeneous catalyst is relatively uniform, the selectivity is high, the side reaction is less, and the characterization is easy. Commonly used homogeneous catalytic reaction catalysts can be divided into acid-base catalysts, a few nonmetallic molecular catalysts such as I_2 and NO, catalysts for soluble transition metal compounds (salts and complexes).

When ILs participate in catalytic reactions, the role of ILs can be roughly divided into two categories: one is as a green reaction solvent. Using their special ability to dissolve the reaction substrate and organometallic catalyst, the reaction is carried out in the IL phase, and at the same time, the product enters the organic solvent phase by taking advantage of its insoluble characteristics with some organic solvents, so that the separation of the product can be well realized, and the recovery and reuse of the catalyst in the IL phase can be achieved simply by physical phase separation. The other type is the functionalized IL, that is, the IL is also used as the catalyst of the reaction in addition to the green reaction medium, such as the use of IL inherent Lewis acidity to catalyze esterification reaction, alkylation reaction, isomerization, etherification, and so forth. It can also be used to purposefully synthesize catalysts with special catalytic properties. In these reactions involving IL, not only is the IL used as a green reaction medium or catalyst, but because of the "designability" of its structure, the selection of appropriate IL can often play a synergistic catalytic role, so that the catalytic activity and selectivity are improved.

3.2.1 Application in the synthesis of *n*-amyl alcohol acetate

Mengshuai Liu et al. [1] synthesized three types of imidazole-based dicationic $[C_2(mim)_2]$, $[C_3(mim)_2]$, and $[C_4(mim)_2]$ ILs with traditional counter anion HSO_4. The key physicochemical properties of these ILs, such as thermal stability, solubility in common solvents, acidic property, and corrosion of Austenitic stainless steel 316, were determined. Esterification of alcohols by carboxylic acids was carried out at room temperature using a group of imidazole dicationic ILs. The catalytic performance was found to be much better than that of conventional non-cation-functionalized ILs, and the corrosion was five times weaker than sulfuric acid. The esters were easily recovered due to immiscibility with the IL. The reported imidazole dicationic ILs are promising catalysts for esterification reactions.

The synthesis approach described in the study is both concise and practical, thanks to the commercial availability of the starting materials and the convenient reaction conditions. The dicationic ILs were employed individually as catalysts for Fischer esterification, resulting in high yields of esters obtained at room temperature. This is attributed to the effective solvation of water, a byproduct of the reaction, by the IL. In many cases, the IL and water form a single phase, while the ester separates into a distinct top phase. This allows for easy separation of the product from the catalyst.

Table 3.1 Recycling of $[C_4(mim)_2][HSO_4]_2$ for the esterification of acetic acid with 1-pentanol.

Cycle number	Cumulative recovery of IL/%	Yield/%
1	100	86
2	99.2	87
3	96.7	85
4	95.8	85
5	94.2	83
6	90.8	81
7	87.5	80
8	84.2	80

Consequently, the accessibility of these ILs, along with the clear separation of the product phase and the absence of leaching of the ILs into the product, demonstrates their potential for being developed as industrial esterification catalysts and media.

The imidazole dicationic ILs exhibit not only superior catalytic activity compared to previously reported systems but also the ability to be reused for at least eight cycles without any pretreatment or significant loss of activity. Additionally, the use of these ILs eliminates the need for volatile organic solvents typically employed for removing water formed during esterification through azeotropic processes. As a result, ILs offer an environmentally friendly alternative to conventional solvents. Furthermore, it is worth noting that $[C_2(mim)_2][HSO_4]_2$ demonstrates superior thermal stability in comparison to $[C_3(mim)_2][HSO_4]_2$ and $[C_4(mim)_2][HSO_4]_2$. This characteristic makes it suitable for various other reactions or operations that involve the production of water requiring high-temperature removal. The potential of this approach will be further explored in future research endeavors (Table 3.1).

3.2.2 Application in the synthesis of succinic acid diisopropyl ester

Diisopropyl succinate, also known as diisopropyl succinate, is an aliphatic carboxylic acid derivative. It is mainly used in organic pigments, medicine, spice intermediates, food additives, agricultural products processing, gas chromatography fixed liquids, and other fields. In the early 1980s, Ciba Refining Company applied diisopropyl succinate (diisopropyl succinate) for the first time to a new high-performance organic pigment containing 1, 4-diketo pyrrole and pyrrole (DPP pigments). Diester succinate compounds have good light and heat resistance and solvent resistance, and their coloring power is high after pigment treatment. The color is bright, and their dispersion is very good.

Dishun Zhao et al. [2] synthesized nine different kinds of Lewis acid ILs through a two-step process. The structures of these liquids were characterized using proton nuclear magnetic resonance spectroscopy (^{1}H NMR) and Fourier-transform infrared spectroscopy (FT-IR) techniques. The prepared Lewis acid ILs were used as catalysts in the esterification reaction between succinic acid and isopropyl alcohol. As the dosage of halide increased, the acidity of the ILs became stronger. Among them, [Bmim] Br–Fe$_2$Cl$_6$ exhibited the best catalytic performance. The optimal conditions for synthesizing succinic acid diisopropyl ester were as follows: the catalyst amount was 10.0% (g/g) of succinic acid, the reaction temperature was 100°C, the reaction time was 4 hours, and the ratio of succinic acid to isopropyl alcohol was 1:5. Under these optimal conditions, the yield of succinic acid diisopropyl ester reached 88.9%, and the esterification rate was 92.7%. The catalyst could be recycled up to 6 times, with only a 1.7% decrease in the yield of succinic acid diisopropyl ester.

The reuse performance of IL was tested, and it was found that the yield of IL was only reduced by 1.7% after 6 times of reuse. The reasons are as follows: on the one hand, the mass of the adhesive bottle wall decreases during the IL recovery process; on the other hand, after water is used as an extractant to separate the catalyst and product, the residual water of the IL phase is not completely removed.

3.2.3 Application in the synthesis of dimethyl succinate

Dimethyl succinate, also known as dimethyl succinate, is an important synthetic flavor and food additive and can be used as an edible preservative; at the same time, it is also an important chemical intermediate and is widely used in the preparation of a variety of chemicals.

Zhao et al. [3] designed and synthesized seven kinds of ILs, N-methyl imidazolium bisulfate ([Hmim][HSO$_4$]), 1-butyl-3-methyl imidazole phosphate ([Bmim][H$_2$PO$_4$]), 1-buty-3-methyl imidazole bisulfate ([Bmim][HSO$_4$]), 1-butyl-3-methyl imidazole tetrafluoroborate ([Bmim][BF$_4$]), N-butylpyridine bisulfate([Bpy][HSO$_4$]), N-butylpyridine phosphate([Bpy][H$_2$PO$_4$]), and N-ethylpyridine bisulfate ([Epy][HSO$_4$]). The synthesis of dimethyl succinate catalyzed by ILs was studied for the first time. It was shown that [Epy][HSO$_4$] had the best catalytic performance, and optimal conditions for the synthesis of dimethyl succinate were obtained: succinic acid and methanol molar ratio 1:3.0, amount of catalyst 5%(g/g) of succinic acid, reaction temperature 70°C, and reaction time 2 h. Under the optimal conditions, the yield of dimethyl succinate was up to 91.83%, and the esterification rate was 96.32%. The catalyst was recycled seven times without substantial decrease in activity. Moreover, compared to conventional industrial catalysts, the IL catalyst had the superiority of smaller catalyst dosage, less byproduct, mild reaction conditions, high yield, and recycled catalyst usage. Finally, the mechanism of esterification catalyzed by ILs was discussed.

3.2.4 Application in the synthesis of diethyl succinate

Liu et al. [4] synthesized four functionalized dicationic ILs, including bis-(3-methyl-1-imidazolium) butyl ditoluenesulfonate (Im-PTSA), bis-(3-methyl-1-imidazolium) butyl hydrogen sulfate (Im-HSO$_4$), bis-(1-pyridinium) butyl ditoluenesulfonate

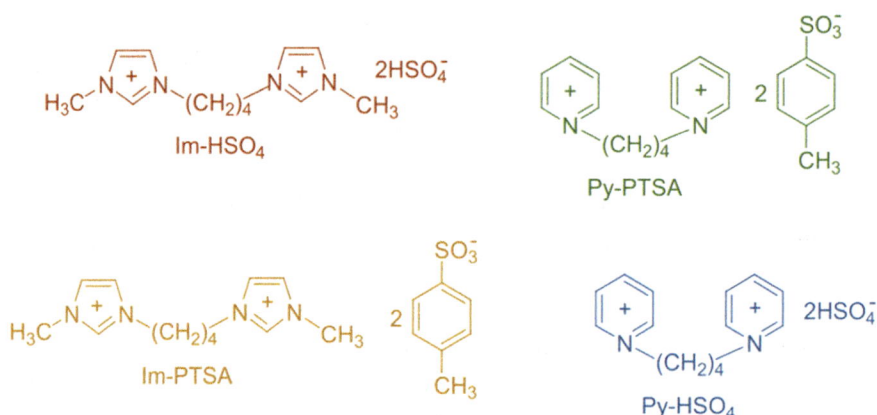

Figure 3.1 Four binuclear ionic liquids.

(Py-PTSA), and bis-(1-pyridinium) butyl hydrogen sulfate (Py-HSO$_4$). The synthesized ILs were structurally analyzed using FT-IR and ^{1}H NMR. The thermal stability of the ILs was tested by thermogravimetric analysis (TGA). Additionally, the acidity and solubility of the ILs were examined. The catalytic activity of the four ILs was studied in the esterification reaction between succinic acid and ethanol. The results showed that when the molar ratio of $n(C_4H_6O_4)$: $n(C_2H_5OH)$ was 1:3, the catalyst Im-PTSA accounted for 1.90% of the total mass, the reaction temperature was 70°C, and the reaction time was 2.5 hours; the ester yield reached 93.6%, with a selectivity of 100%. Furthermore, the IL could be reused for eight cycles without significant decrease in catalytic activity. The corrosion behavior of the dicationic functionalized ILs on austenitic stainless steel 316 L was investigated and compared with concentrated sulfuric acid. The corrosion rate of the steel specimen was less than 1/10 of that in concentrated sulfuric acid. Using bis-(3-methyl-1-imidazolium) butyl ditoluenesulfonate (Im-PTSA) as the catalyst, the esterification reactions of monobasic organic acid and dibasic organic acid with various alcohols were investigated, showing high ester yields and selectivity. After the reaction, the products automatically separated from the catalyst, simplifying the separation process. This study suggests that these dicationic functionalized ILs have the potential to be developed as promising esterification catalysts. The resultant four binuclear ILs are shown in Fig. 3.1.

3.2.5 Application in the synthesis of dimethyl carbonate

Wei et al. [5] reported that three nucleophilic and structurally unique quaternary ammonium ILs were synthesized using the 1,2-propanediol anion ($CH_3CH_2OHCH_2O^-$) obtained from ester exchange products as raw materials. The synthesis process resulted in high yields and high purity of the ILs. These synthesized ILs were then used in a one-step reaction for the production of dimethyl carbonate (DMC) using propylene oxide (PO),

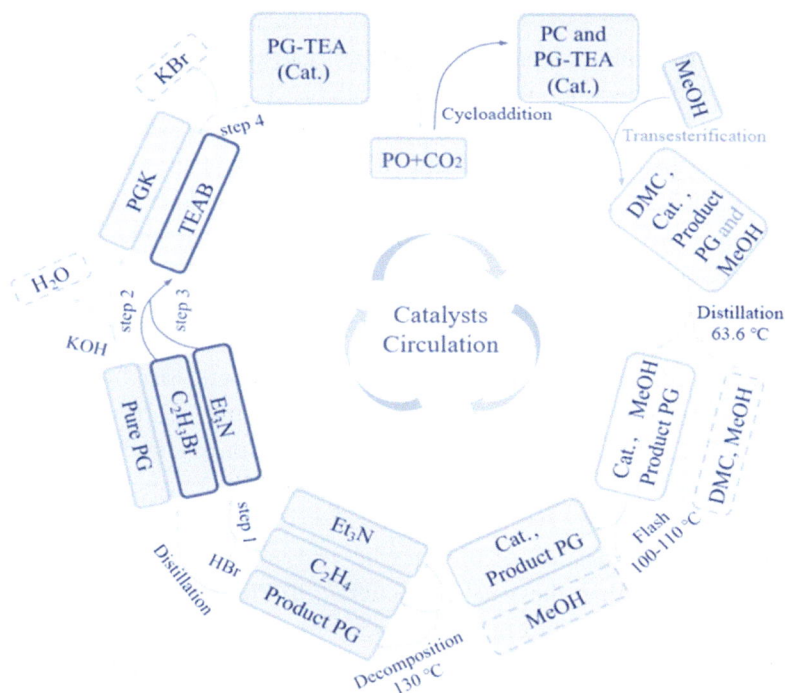

Figure 3.2 Catalyst recycling route.

carbon dioxide (CO_2), and methanol (MeOH) as raw materials. The catalytic reaction involved two stages: PO-CO_2 ring addition and PC-MeOH ester exchange, both performed without separating any catalyst. Under mild conditions, the conversion rate of PO reached 99.0%, and the yield of PC reached 99.0%, with a turnover number (TON) value of 119.5. After the ring addition, methanol was immediately introduced, resulting in a PC conversion rate of 71.2% and a selectivity for DMC of over 99%. The TON value for the ester exchange reaction was 57.4. The effects of temperature and time on the ring addition and ester exchange reactions were investigated. Different epoxy compounds were studied for their performance in CO_2 ring addition, all showing good reactivity. By utilizing the decomposability of the prepared ILs and the recyclability of the decomposition products, a novel catalyst recycling process was designed to achieve complete recycling without introducing any impurities into the product system. Fig. 3.2 shows the route of catalyst recycling in the whole reaction process.

3.3 The multiphase reaction based on ionic liquids

Polymerized ILs (PILs) are formed by polymerizing IL monomers that contain double bonds or other polymerizable chains. The development of polymeric ILs (PILs) began in the 21st century. In this chapter, our research group has successfully synthesized numerous polymeric IL catalysts and applied them to various reactions, such as esterification, aldol condensation, transesterification, and more. These catalysts have exhibited excellent catalytic performance. Furthermore, we have also synthesized a polymeric IL adsorbent for the desulfurization of fuel oil. This adsorbent has demonstrated remarkable effectiveness in removing dibenzothiophene and achieving desulfurization goals. Overall, our research efforts have resulted in the successful synthesis of diverse polymeric IL catalysts and an effective adsorbent for fuel oil desulfurization, showcasing their potential application in various reactions and environmental purification processes.

3.3.1 Application in methyl methacrylate

Zhang et al. [6] synthesized uniform large particles (millimeter-sized) of acid polymeric IL (PIL-A) as a catalyst. This catalyst was characterized using FT-IR, X-ray diffraction, TGA/dynamic scanning calorimetry, and scanning electron microscopy (SEM) techniques. The thermostability and mechanical strength of PIL-A were found to be much higher than those of commercial resin. The catalytic properties of PIL-A were investigated through the esterification of methacrylic acid and methanol as the target reaction. Results showed that the catalytic activity of PIL-A was superior to that of commercial resin. The optimum reaction conditions were determined to be as follows: 5%(mass fraction) catalyst, 95°C reaction temperature, 3–hour reaction time, and a methacrylic acid to methanol mole ratio of 1:1.2. Under these conditions, the yield of methyl methacrylate (MMA) reached 100%.

Furthermore, the catalytic activities of PIL-A in different esterifications were also explored, demonstrating its higher catalytic activity across various reactions. Importantly, the catalytic activity of PIL-A did not decrease even after it was reused five times. Additionally, continuous reaction to produce MMA using a fixed-bed reactor was investigated, with the percentage of MMA in the product exceeding 96%.

It can be seen from Fig. 3.3 that the polymerized ion features are spherical particles, and SEM images show that they are tightly layered structures, so they have high mechanical strength.

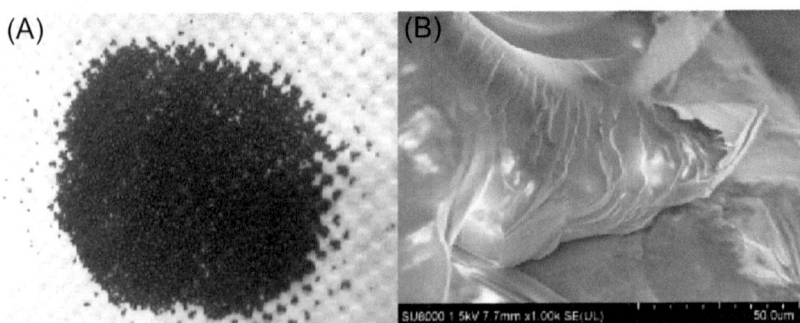

Figure 3.3 Photos of polymerized ionic liquid (A) and SEM images (B).

3.3.2 Application in biodiesel production

Bian et al. [7] successfully prepared a novel Brønsted acidic poly IL (PIL-M) through free-radical polymerization. The heterogeneous catalyst boasted a surface area of 301.1 m^2/g and demonstrated good thermal stability, decomposing at 410°C as confirmed by TGA. The optimal reaction conditions were determined to be: 6%(mass fraction) catalyst amount, a reaction temperature of 80°C, a methanol to oleic acid ratio of 9:1, and a reaction time of 3 hours. Under these optimal conditions, the yield of methyl oleate reached an impressive 95.9%. PIL-M's high surface area (301.1 m^2/g) and hierarchical nanopores contributed to its excellent catalytic activity in biodiesel production. Compared to commercial resin (Amberlyst 15) and other catalysts, PIL-M exhibited superior catalytic performance, maintaining a yield of over 90% even after four reuses. Furthermore, FT-IR analysis confirmed that the structure of PIL-M remained stable after these reuses. Fig. 3.4 shows the synthesis of poly IL (PIL-M).

Bian et al. [8] developed another novel polymerized IL catalyst for the synthesis of biodiesel. The IL monomer was synthesized by quaternary ammonization of N, N-dimethyl-3-aminophenol with 1,3-propanesultone and p-hydroxybenzenesulfonic acid. Formaldehyde was then used to synthesize the poly acidic ILs (FCPIL). The catalyst was characterized using infrared spectroscopy, SEM, TGA, and nuclear magnetic carbon spectroscopy. Results indicated that the catalyst displayed good thermal stability, with a decomposition temperature of

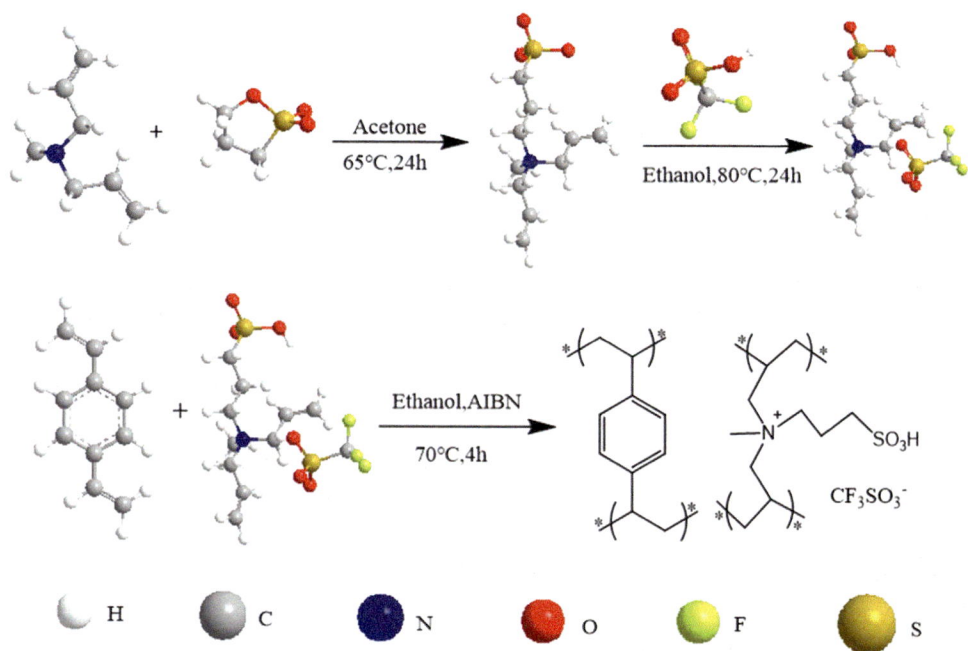

Figure 3.4 Synthesis of PIL-M.

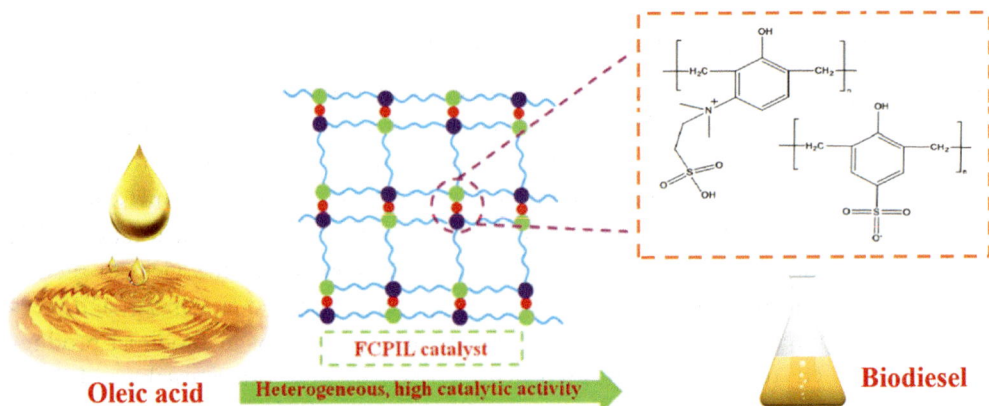

Figure 3.5 Biodiesel synthesis process.

260°C, and was highly suitable for oleic acid esterification. The catalytic activity of the FCPIL was evaluated by esterifying oleic acid with methanol, with four parameters (reaction time, methanol amount, temperature, and catalyst amount) optimized for the maximum yield. Under optimal conditions [5%(mass fraction) catalyst amount, 9:1 methanol/oleic acid ratio at 80°C for 1.5 hours], an impressive yield of 93.3% was achieved. Remarkably, even after four cycles of use, the catalytic activity of the FCPIL remained relatively unchanged. Fig. 3.5 shows the biodiesel synthesis process.

3.3.3 Application in the synthesis of butyl acetate

Bian et al. [9] synthesized a novel poly-IL for esterification through phenolic condensation. Various characterization techniques were applied, and the results showed that the catalyst had high acidity (4.5 mmol/g) and good thermal stability. The decomposition temperature of the polyionic liquid catalyst (PIL-S) was 240°C, which was suitable for the esterification product. Due to these excellent properties, PIL-S demonstrated outstanding catalytic activity in esterification. The optimal conditions were found to be 6%(mass fraction) catalyst amount, and the ratio of acetic acid to *n*-butanol was 0.8:1 at 95°C for 3 hours. Under these conditions, the yield of ester was 97.1%. PIL-S also showed promising catalytic effects in other esterification systems. Compared to commercial resins (Amberlyst 15) and other catalysts, PIL-S exhibited superior catalytic activity. Even after eight reuses, the yield of ester did not noticeably decrease (over 94%). Furthermore, after eight reuses, the structure of PIL-S remained stable. Fig. 3.6 was the route of synthesis of butyl acetate from acetic acid and butanol.

3.3.4 Application in the synthesis of glycerol monolaurate

Yajuan Wang et al. [10] developed a new solid acid catalyst called porous polyaminobenzenesulfonic acid using the one-pot method. Characterization of the catalysts was conducted using SEM, transmission electron microscopy, Brunauer–Emmett–Teller (BET), FT-IR, elemental analysis (EDS), and TGA, revealing that the catalyst is both porous and

Solid acid

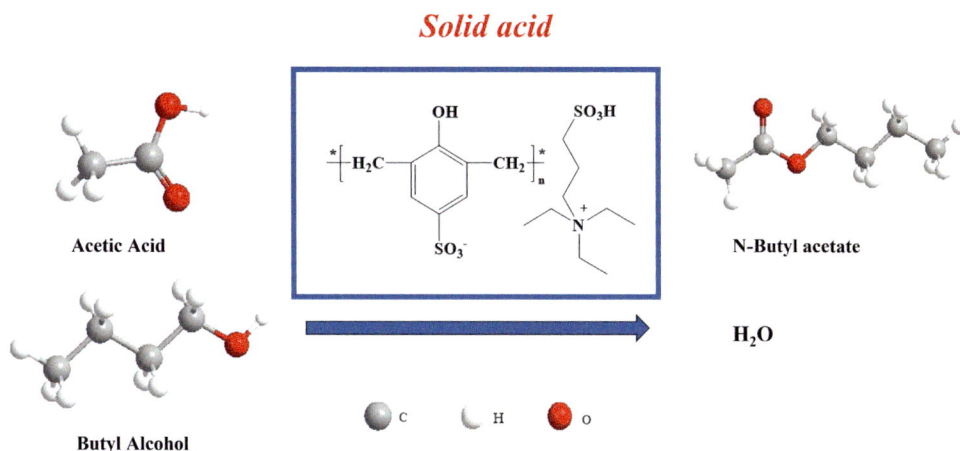

Figure 3.6 The route of synthesis of butyl acetate from acetic acid and butanol.

stable. Its catalytic performance was evaluated through the synthesis of glycerol monolaurate using lauric acid and glycerin. The reaction process was thoroughly investigated, and optimal reaction conditions were determined as follows: a molar ratio of lauric acid to glycerol of 1:2, 6% (mass fraction) catalyst amount, a reaction temperature of 443 K, and a reaction time of 3.5 h. Under these conditions, the conversion rate of lauric acid reached 94.89%, and the selectivity was 61.66%. Furthermore, the catalytic activity remained consistent over five cycles of reuse. Since the monomer used has an HSO_3 group, there is no need for a sulfonation step, making the polymer structure more stable and the preparation process more simple. Therefore, this solid acid catalyst holds great promise for use in esterification reactions. Fig. 3.7 shows the synthesis of monoglycerides of laurate.

3.3.5 Application in the synthesis of dimethyl carbonate

Liu et al. [11] successfully synthesized a large particle spherical polyionic liquid–solid base catalyst through suspension polymerization. The catalyst was characterized by infrared spectroscopy, SEM, EDS, and BET. The characterization results indicate that PILs-FL possesses a unique core structure, a specific surface area size ($153 \ m^2/g$), a pore size of 6.3 nm, and high alkalinity (2.4 mmol/g). The catalyst performance was evaluated by the reaction between vinyl carbonate and methanol to form DMC. The optimal reaction conditions were found to be a molar ratio of methanol to EC of 9:1, a catalyst dosage of 3%(mass fraction), a reaction time of 5 hours, and a temperature of 90°C. Under these conditions, the yield of DMC was up to 93.7%, and the selectivity of DMC was 100%. Additionally, the catalyst can be easily separated by simple filtration, and its activity did not significantly decrease after being reused five times. The catalyst shape remained intact, indicating that it possessed high mechanical strength. Therefore, PILs-FL is a promising solid base catalyst. Fig. 3.8 shows the synthesis process of DMC catalyzed by PILs-FL.

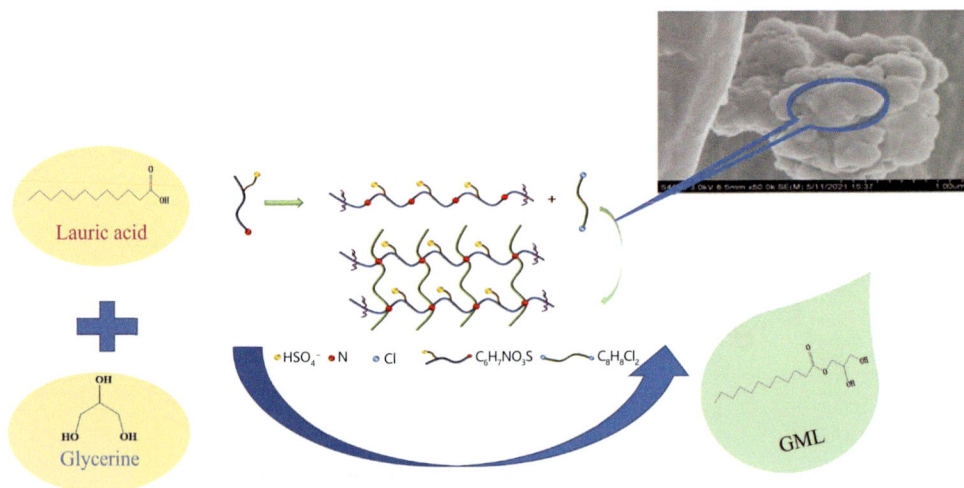

Figure 3.7 Synthesis of monoglycerides of laurate.

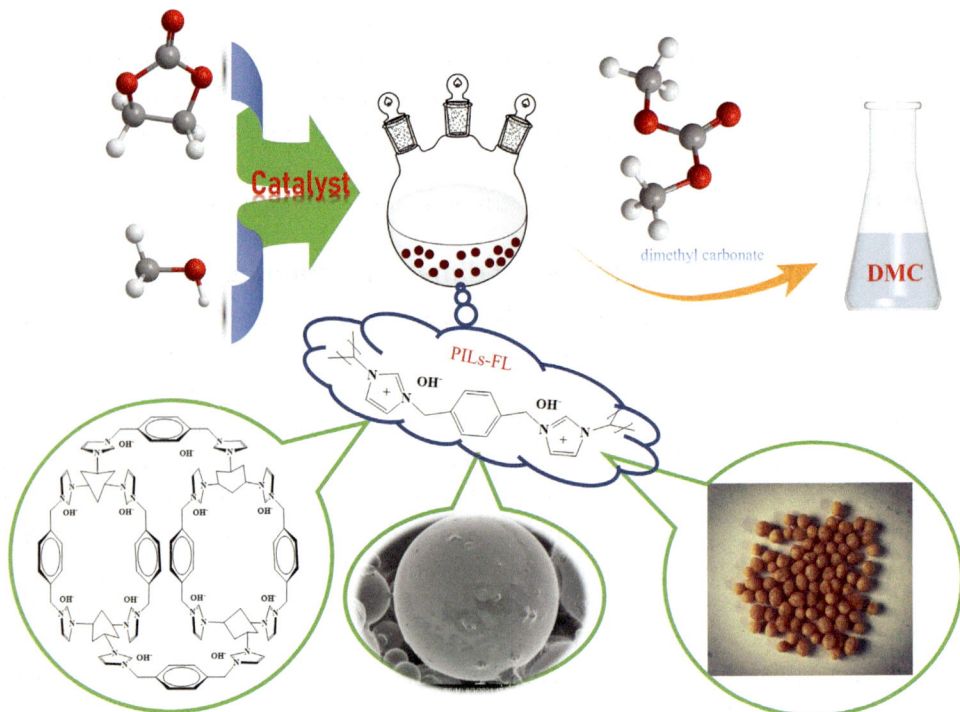

Figure 3.8 Synthesis of dimethyl carbonate by PILs-FL.

DMC is widely used due to its nontoxic nature, making it a safer alternative to more hazardous substances such as dimethyl sulfate or methyl chloride. In a study by Liu et al. [12], they synthesized spherical particle catalysts (PILs–XSS) through suspension polymerization. The catalyst was characterized using SEM-EDS, FT-IR, TGA, and BET-BJH analysis. The results of the characterization revealed that PILs–XSS possessed a specific core structure, a pore size of 14.3 nm, a surface area of 71.5 m^2/g, and demonstrated stability at 293.4°C. The performance of the catalyst was evaluated through a one-pot synthesis of DMC, with detailed optimization of the process conditions. The optimal reaction conditions were determined as follows: a dosage of 2.5%(mass fraction) PILs–XSS, a CO_2 pressure of 1.5 MPa, a reaction temperature of 100°C, a reaction time of 4 hours, a molar ratio of methanol to PO of 3:1, and a dosage of 3% (mass fraction) cocatalyst Na_2CO_3. Under these optimized conditions, the conversion of PO reached 98.8%, and the yield of DMC was 53.7%. Furthermore, the combined catalyst PILs–Na_2CO_3 could be easily recovered by filtration within the reaction system. After five repeated uses, the catalyst maintained its activity and shape, demonstrating remarkable mechanical strength. This study highlighted not only the efficient transesterification capability of Na_2CO_3, but also its promotion of the CO_2 reaction. Fig. 3.9 shows the synthesis process of DMC catalyzed by PILs–XSS.

Figure 3.9 Synthesis of dimethyl carbonate by PILs–XSS.

3.3.6 Application in hydroxyl condensation reaction

Wang et al. [13] developed a novel catalyst composed of IL porous organic polymers. The catalyst was thoroughly characterized using SEM, BET measurements, FT-IR, and TGA. The characterization results indicated that the synthesized catalyst exhibited a porous polymer structure and displayed excellent thermal stability. The catalytic performance of the catalyst was evaluated by synthesizing methacrolein through the reaction of formaldehyde and propionaldehyde. The optimal reaction conditions were determined as follows: the catalyst amount was 10% (mass fraction) of the mass of propionaldehyde, the reaction temperature was set at 85°C, the reaction time lasted for 1.5 hours, and the ratio of propionaldehyde to formaldehyde was maintained at 1:1.2. Under these optimal conditions, the conversion rate of propionaldehyde reached an impressive 99.57%, with a selectivity of 98.77% toward methacrolein. Furthermore, the catalytic activity of the catalyst remained relatively stable even after being reused for five cycles. Additionally, the researchers conducted a study on the reaction kinetics of the methacrolein synthesis. The results revealed that the reaction followed a first-order kinetics with a reaction order of 1.6. The activation energy was determined to be 50.28 kJ/mol, and the preexponential factor was calculated to be 4.17×10^7 $(mol/L)^{-1.456}/s$. Overall, the findings demonstrated the promising potential of the synthesized ionic porous organic polymers catalyst in the production of methacrolein, highlighting its remarkable catalytic activity, stability, and favorable reaction kinetics. Fig. 3.10 shows the synthesis of methacrolein.

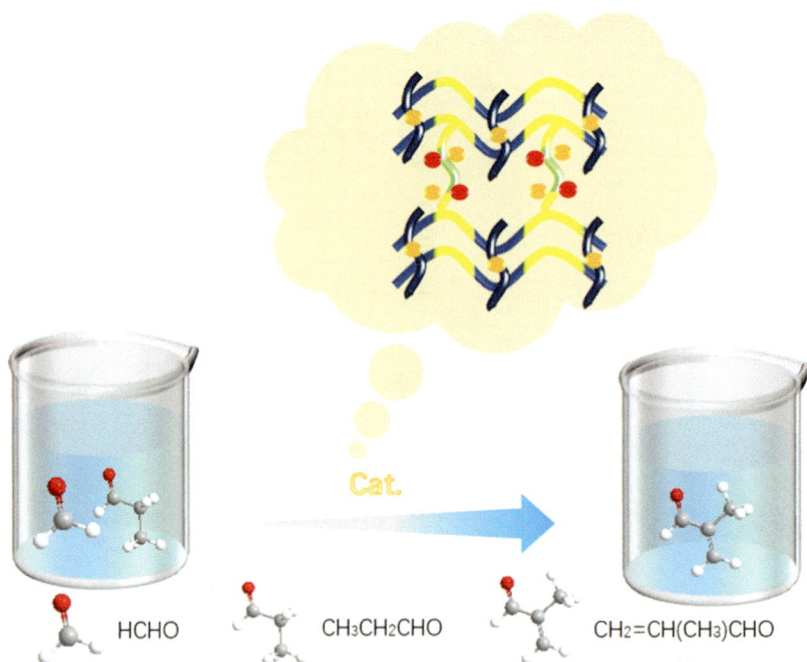

Figure 3.10 Synthesis of methacrolein.

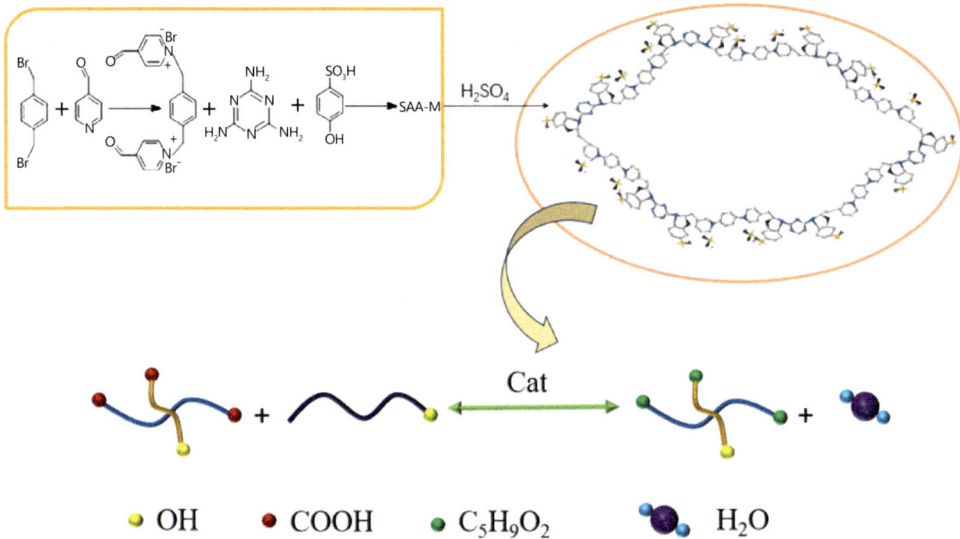

Figure 3.11 Synthesis of butyl citrate.

3.3.7 Application in butyl citrate

Wang et al. [14] developed a novel porous macromolecule oxazine poly-IL solid acid catalyst. This catalyst was characterized using SEM, BET, FT-IR, EDS, and TGA. The characterization results indicated that the synthesized catalyst is a porous polymer with good heat stability. The catalytic performance of the catalyst was assessed through the synthesis of tributyl citrate from citric acid and n-butanol. The optimal reaction conditions were determined as follows: 9% (mass fraction of citrate) catalyst amount, a reaction temperature of 150°C, a reaction time of 3.5 hours, and a citric acid to n-butanol ratio of 1:3. Under these optimal conditions, the conversion rate of citric acid reached 91.74%. Furthermore, the catalytic activity did not significantly diminish after being reused five times. The reaction conditions were optimized using a surface optimization method. The fitted optimum conditions were found to be: a catalyst amount of 8.881% (the mass of citric acid), a reaction temperature of 148.259°C, a reaction time of 3.331 hours, and a citric acid to n-butanol ratio of 1:2.958. Under these conditions, the conversion rate of citric acid reached 92.998%. Importantly, the simulation results were in good agreement with the experimental findings. Fig. 3.11 shows the synthesis of butyl citrate.

3.4 Bioionic liquids: tunable microenvironment for biocatalysis

3.4.1 Introduction

ILs have gained great attention over the past decade due to their outstanding properties [15,16]. Usually, ILs are completely composed of cations and anions presented as liquid at temperatures lower than 100°C [17], which the ions are designable that could meet the criteria to perform all kinds of expected reactions [18]. In comparison to traditional organic solvents [19], ILs offer more favorable properties such as nonvolatility, nonflammability, high ionic conductivity, and stable thermal stability [19−21]. With the interest increasing in ILs, researchers have investigated the physiochemical properties and wider tunable properties of ILs [22,23]. Because of their polarity, hydrophobicity, miscibility, and hydrogen bond basicity [24], ILs show great potential in synthesis and catalysis [25].

There is a family of ILs called biocompatible ILs (Bio-ILs), which are safe, bio-friendly and biorenewable [26]. The Bio-ILs are usually task-specifically designed and synthesized using environment-friendly, natural, or biocompatible compounds [27]. Owing to their noted physicochemical properties, Bio-ILs are tailored to comply with various applications [28]. Many studies work on choline and amino acid-based Bio-ILs, which are able to catalyze or synthesize the reactions with less cytotoxicity [29]. These Bio-ILs could be used in biomedical applications, bioavailability enhancers, improvement of drug solubility, and biocatalysis [30].

Biocatalysis is a rapidly expanding area in both academia and industry, which can be generally defined as the use of enzymes to perform chemical conversion of organic compounds [21]. Usually, enzymes work better in aqueous solvents, while the catalytic reactions performing in a biphase system, in which the organic layer contains the substrate and the aqueous layer contains enzymes, offer more advantages [31]. In this case, the tunable properties of ILs could provide the enzymes with a more favorable micro-environment to work. ILs could provide enzymes with good thermal stability, structural stability, and operational stability. At the same time, ILs could provide high conversion rate, enantioselectivity, recoverability, and recyclability [32]. The first successful series of examples of applying ILs in enzyme-catalyzed reactions was in 2000. The studies reported that 1-butyl-3-methylimidazolium [Bmim] could be used in bioconversion [33]. From 2000 onward, a number of enzymes and microorganisms have been tested in ILs and many advantages have been investigated.

The applications of Bio-ILs for biocatalysis could be divided into two reaction systems. The first one is the free enzyme conversion. Our group investigated the particular ionozyme, in which the ILs were involved as solvent and stabilizer for efficient bioactivation of CO_2 [34]. The second one is whole cell biocatalysis, in which microbial cells are involved. During the reaction, the enzymes work as part of the microorganisms [35]. In this section, we discuss the tunable properties of the Bio-ILs, their effect in performing in enzyme and whole cell reactions, and their applications in biocatalysis.

3.4.2 Ionic liquids and free enzymes

Enzyme plays an important role in all kinds of biochemical reactions and drug delivery system. Increasing the stability, activity, and efficiency of the enzyme is imperative [36].

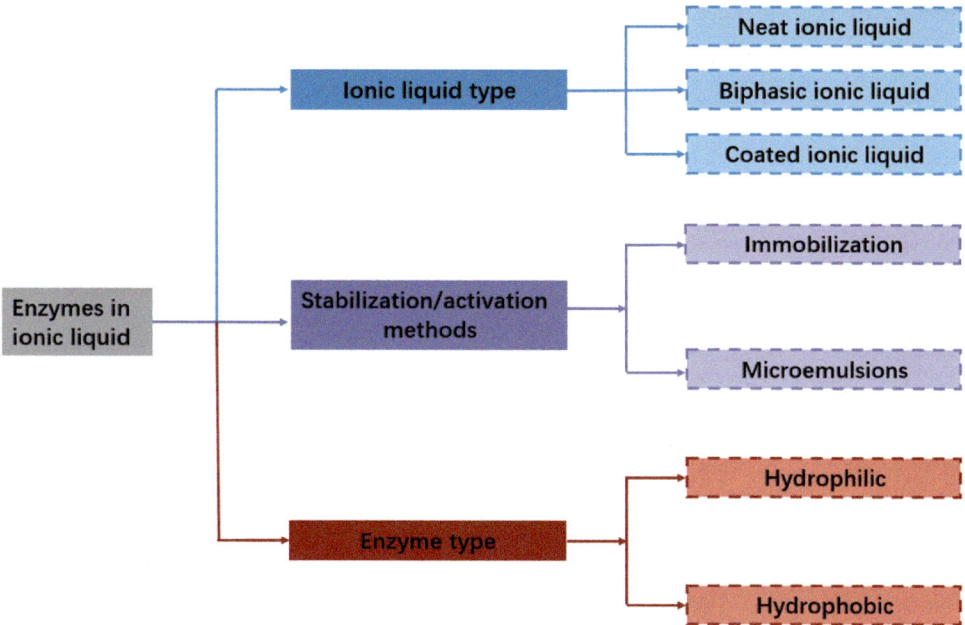

Figure 3.12 Different classifications approaching in IL biocatalysis of enzyme.

The ILs are designed to possess unique physicochemical properties, which are developed based on the need for physically and chemically specific reactions [26,37]. Also, ILs are investigated in improving the thermal stability of the proteins [23,38–40], protecting the 3D structure of proteins [41,42], and maintaining the biological activities of the proteins and enzymes [43,44]. Kumar et al. [45] showed that the ILs could prevent the self-aggregation of insulin. The ammonium-based ILs, triethylammonium dihydrogen phosphate [TEAP], could take the function in this case. Todinova et al. [46] reported that imidazole-based ILs, 1-butyl-3-methylimidazolium acetate ([Bmim][Ac]), could increase the thermal stability of insulin. In addition, there were many other enzymes that ILs could be incorporated with, such as lipases, proteases, oxidoreductases, and peroxidases [30]. To introduce the enzyme/ILs systems, several classification methods of different approaches are proposed, as shown in Fig. 3.12. In this section, the ILs and enzyme systems are introduced by the hydropathy properties of the enzymes.

3.4.2.1 Applications of ionic liquids in the hydrophilic enzyme-based catalysis system

The purity, polarity, stability, viscosity, solvent properties, and miscibility make ILs a favorable solvent for most enzymes. The enzymes are usually hydrophilic and work better in aqueous solutions, in which the enzyme structure–activity information is determined in aqueous media and recorded in the protein data bank [47]. However,

Table 3.2 Hydrophilic enzymes and IL microenvironment.

Enzyme	IL buffer	Reaction system	Reference
Alcohol dehydrogenase	[Bmim][NTf₂]/buffer	Biphasic system, manipulation	[20]
Catalase	[Bmim][TfO], [Ch][Cl]	ILs	[49,50]
Chloroperoxidase	[Bmim][MeOSO₃]	Water-miscible ILs	[51]
Dehalogenase,	[Omim][BF₄]	ILs	[52]
Formate dehydrogenase	[Mmim][MeSO₄]	Water-miscible ILs	[53]
β-Glucuronidase	[Bmim][PF₆]/buffer	Immobilized on zinc oxide nanoparticles (NPs)	[54]
Histidase	[Bmim][NTf₂]	ILs	[55]
Hydroxynitrile lyase	[Pmim][BF₄]	Biphasic system	[56]
Laccase	[BMPyr][BF₄]	Water-miscible ILs	[57]
Lumbrokinase	[Emim][Br]	Water-miscible ILs	[58]
Lysozyme	[DDmim][DBS], [Mor1,4][Ala]	Water-miscible ILs	[59,60]
Thermolysin	[Bmim][PF₆]	Neat ILs	[61]
Tyrosinase	[Bmim][BF₄], [Emim][NTf₂]	Water-miscible ILs	[62]
Xylanase	[Emim][Ac], [DBNH][EtCO₂]	Hydrophilic ILs	[63,64]

general aqueous solutions might cause enzyme hydrolysis and loosing activity. The studies of the performance of the enzymes in neat ILs, water/ILs biphasic media, or coated ILs could help solve the current limitations and improve the activity of the enzymes [48]. Usually, the enzyme is composed of a hydrophobic pocket and a hydrophilic surface. The hydropathy properties of the enzyme introduced are not absolute. Table 3.2 gives the recent famous hydrophilic enzymes modified with ILs. The hydrophilic enzymes could be involved in hydrolase-catalyzed reactions, and some of these reactions could be performed in water. Chloroperoxidases and laccases both achieve higher activity in the ILs/water mixture [51,57].

3.4.2.2 Applications of ionic liquids in the hydrophobic enzyme-based catalysis system

The hydrophobic enzymes work better in organic solvents, in which they could achieve higher activity. By applying the ILs with a buffer solution, the solvent system become a two-phase system. Under the biphasic condition, most hydrophobic enzymes could work at the surface between the ILs and aqueous layer. In this system, ILs work as a cosolvent to increase the solubility of the enzymes. As shown in Table 3.3, lipases, proteases, and esterases are successful examples to be presented. In these reactions, either the enzymes or substrates are hard to be dissolved in water, in

Table 3.3 Hydrophobic enzymes and IL microenvironment.

Enzyme	IL buffer	Reaction system	Reference
Alkaline phosphatase	[H$_3$NEt][NO$_3$]	Immobilization	[65]
L-Asparaginase	[Ch][Cl]	ILs	[66]
Candida antarctica lipase A (CALA)	[Bmim][Tf], [Omim][BF$_4$]	Immobilization	[67]
Candida antarctica lipase B (CALB)	[BMPY][NTf$_2$], [Omim][BF$_4$]/MTBE	Biphasic system, immobilization	[68,69]
Cellulase	[IM]/[Ac]	ILs	[70]
α-Chymotrypsin	[Omim][PF$_6$], [Emim][(CF$_3$SO$_2$)$_2$N]	Biphasic system,	[71,72]
Esterase	[Bmim][PF$_6$]	Biphasic system,	[73]
Lactic dehydrogenase	[Omim][Br], [Bmim][CF$_3$SO$_3$]	ILs	[74]
Lipase	[Bmim][PF$_6$]	Biphasic system	[75,76]
Naringinase	Various ILs/buffer	Multiple phase system	[77]
Penicillin G acylase	[Bmim][NTf$_2$], [Emim][NTf$_2$]	Biphasic system,	[78]
Proteinase K	[Bmim][BF$_4$], [Bmim][OTf]	Immobilization, biphasic system,	[79]
Pseudomonas cepacia lipase (PCL)	[Omim][NTf$_2$]	Microemulsion	[80]
Trametes versicolor laccase	[Bmim][PF$_6$]	Microemulsion	[81]

which the cosolvents are needed. The famous lipase CALB could be dissolved and obtain a higher stability in ILs [68,69]. It works similarly in proteases, in which the ILs could help increase the solubility of the substrate or the product. The protease K, trypsin, and α-chymotrypsin can catalyze the reaction with a 78% yield compared to a traditional buffer solution [29,61].

Another common way of improving the enzyme stabilization and activation is immobilization. The coating of an isolated enzyme in an IL has proven to be a very efficient method to support the enzyme activity and increase the stability and recyclability. Many enzymes like lipase, proteinase, and some others could also work in this catalytic way with ILs [82].

3.4.3 Ionic liquids and whole cell catalysis

Whole cell catalysis is another attractive field in biocatalysis, which refers to the process of chemical transformation using complete biological organisms (whole cells, tissues, or even individuals) as catalysts. Compared with immobilized and free enzyme catalyze systems, whole cell catalysis is relatively cost-effective and easier to perform due to its high level of sustainability, excellent chemical selectivity, low cost, and easy self-replication. However, due to the limitation of cell membrane permeability, the efficiency of whole cell catalysis is usually lower than that of enzyme catalysis [83, 84]. Taking advantage of the unique properties of ILs, the application of ILs in whole cell catalysis to improve their catalytic performance is a promising new direction of biocatalysis. There are relatively few reports on whole cell biocatalysts with ILs in recent years. In 2000, Cull et al. [33] firstly reported the application of ILs to whole cell catalytic reactions. *Rhodococcus* R312 was used to catalyze the whole cell reduction of 1,3-phenyldicyanide to 3-cyanobenzamide in biphasic systems of [Bmim][PF$_6$]/water and toluene/water, respectively.

ILs commonly used for whole cell catalysis are shown in Fig. 3.13 [85]. Cations of ILs are mainly composed of imidazolyl, pyridinyl, and pyrrolidine groups with large volume, low symmetry, weak molecular interaction, and low charge density, especially imidazole derivatives. While anions mainly include multinuclear anions such as [Al$_2$Cl$_7$]$^-$, [Al$_3$Cl$_{10}$]$^-$, and [Au$_2$Cl$_7$]$^-$ and mononuclear anions such as [BF$_4$]$^-$, [PF$_6$]$^-$, [NTf$_2$]$^-$, and [E$_3$FAP]$^-$, the application of mononuclear anions is more common.

Generally, the whole cell catalysis containing ILs is carried out in the IL—water bidirectional system. The hydrophobic IL acts as the second liquid phase, and the

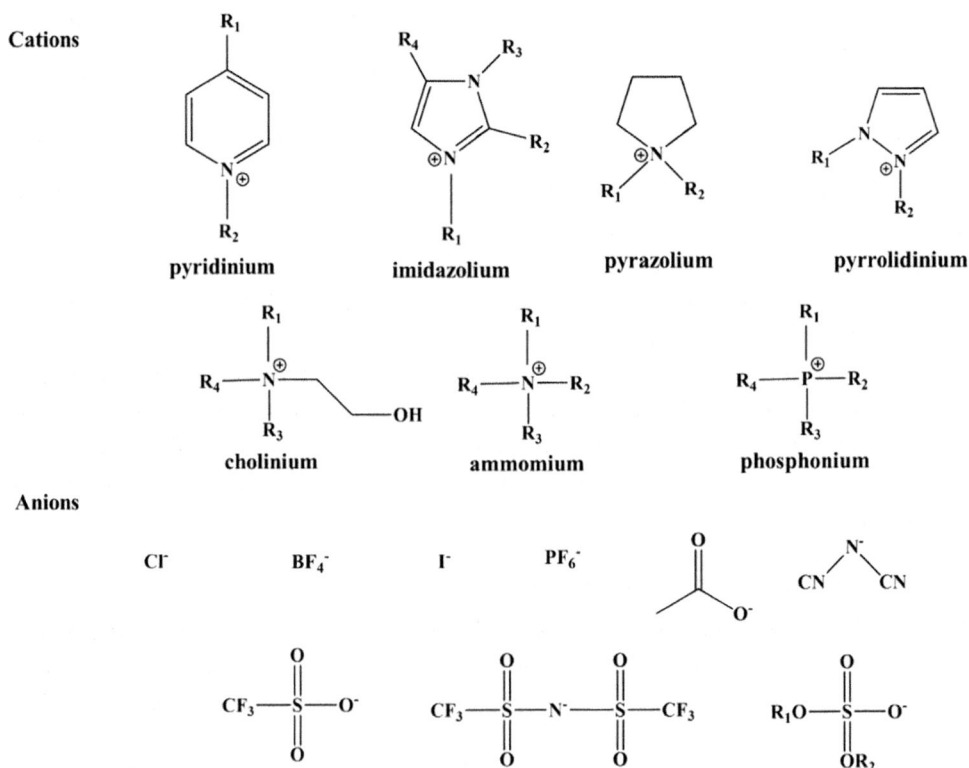

Figure 3.13 Commonly used ILs in whole cell catalysis.

microbial cells are usually dispersed in the aqueous phase. During the reaction, the substrates and products are mostly kept in the IL phase, which can be transported to the aqueous phase, making the reaction more efficient. In addition, the interaction between the IL itself and the microbial cells is one of the decisive factors affecting the performance of whole cell catalysis. Some ILs can change the permeability of the cell membrane, which makes the catalysis easier. However, many ILs have certain toxicity to cells, which greatly limits the application of ILs in whole cell catalysis. A deeper understanding of the mechanisms underlying the effects of ILs on microbial cells are favorable to enhance the biological activities of whole cell catalysis.

3.4.3.1 *Effects of ionic liquids on microbial cells*

Although many ILs are biocompatible and green solvent, it is inevitable that the microbial cells will be affected in the IL system. Understanding the effects of ILs on microbial cells is the key to improve their utilization in whole cell catalysis, where both positive and negative effects are involved.

ILs are considered as a new green solvent and their toxicity to cells as a disadvantage in whole cell catalysis cannot be ignored. Studies have shown that the toxicity of ILs may be manifested by their interaction with the cell membrane, while another possibility for toxicity is that ILs have a certain ability to penetrate into the nucleus. Nicola et al. [86] studied the toxicity of more than 90 ILs to *Escherichia coli* and identified a large number of relatively nontoxic ILs. In general, imidazolium salts with short alkyl chains are relatively nontoxic, especially when paired with alkyl sulfate anions. Methyl pyrrolidonium salts are also desirable, whereas quaternary ammonium salts miscible with water are usually toxic. In conclusion, reducing alkyl chain length would reduce their toxicity. This conclusion was also demonstrated by the study of Ranke et al. [87], through investigating the toxicity of imidazole ILs using the leukemic promyelocytic cell line IPC-81 and the glioma cell line C6. They found that the length of the alkyl chain in imidazole ILs had a great effect on the toxicity of IL. With the increase of alkyl chain length, the inhibitory effect of IL on cells increased sharply, that is, the toxicity of IL itself increased with the increase of alkyl chain length.

Lee et al. [88] found that hydrophilic ILs were less toxic to cells compared with hydrophobic ILs. $[Emim]^+$ was less toxic than other cations. $[BF_4]^-$ also showed relatively less toxicity than other anions, and the combination of these 2 ions [Emim][BF$_4$] showed lower toxicity than other ILs. When using ILs in whole cell catalysis, special attention should be paid to the comprehensive consideration of both catalytic performance and cytotoxicity. It is important to develop ILs with low toxicity and high catalytic efficiency.

Apart from the inhibitory effect of toxicity, many ILs can promote catalysis by acting on cells, mainly by changing cell permeability, which could promote exchange of matter. The alteration of cell permeability may be attributed to the hydrophilic IL, which interacts with the charged groups on the substrate (hydrogen bond or covalent bond interaction) and effectively transfers the charge of the cell membrane. This eventually leads to changes in the ionic state of the cell membrane and affects membrane permeability. Li et al. [89] synthesized [HOEtN1,1,1][BF$_4$],which showed excellent properties of low substrate toxicity to microbial cells and moderate improvement of cell membrane permeability, yielding 96.7% of ethyl (S)-4-chloro-3-hydroxybutanoate under optimal conditions. Zhang et al. [90] treated *E. coli* BL21-pET21a-rhaB1 cells with choline chloride-urea [ChCl]/[U] and deep eutectic solvent to produce isoquercitrin by biotransformation of rutin. The whole cell catalysis showed the highest catalytic activity after treatment at different concentrations. Under the optimal conditions of rutin concentration of 0.05 g/L, pH 6.5, and 40°C, the yield of isoquercitrin was up to 93.05% ± 1.3%.

With the increasing research on Bio-ILs, more studies are focused on achieving a balance between toxicity and efficiency. Researchers have discovered that some ILs have positive effects on microorganisms. While altering the permeability, certain ILs have been found to affect enzymes within cells. Martins et al. [91] first discovered the impact of ILs on the proteome of cells and observed that [Amim][BF$_4$] influenced the expression of enzyme genes present in *Vibrio* sp. Q67 cells. Specifically, the expression of genes encoding luciferase, superoxide dismutase, and catalase was upregulated. In addition, some ILs have the effect of enhancing cell activity. Dipeolu et al. [92] showed that [EtOHNMe$_3$][Me$_2$PO$_4$] and DMEAA increased the growth rate of *Clostridium sporogenes* by 28%, in which case these ILs were often metabolized, or increased nutrient availability.

To date, more detailed mechanisms of IL action on cells and their influence on catalytic activity need to be further investigated and explored.

3.4.3.2 Application of whole cell catalysis with ionic liquids

Whole cell catalysis has a wide range of applications in the synthesis of chiral compounds due to its excellent stereoscopic, regional, and chemical selectivity. Chiral alcohols are important intermediates in the synthesis of pesticides, liquid crystals, fragrances, and drugs. The enantioselective reduction of ketones catalyzed by whole cell catalysis is an effective and reliable way to obtain optically active alcohols. Therefore, the most important field of application of ILs in whole cell catalysis focuses on the reduction reaction to produce chiral alcohols. In addition, ILs have shown promising applications in the production of biofuels such as biodiesel and bioethanol from biomass. Due to their unique solubility properties, ILs have been used for the pretreatment of biomass.

Ou et al. [93] used immobilized *Acetobacter* to catalyze efficient asymmetric reduction of 2-octenone to (R) -2-octenol. The introduction of IL [Bmim][SAc] has been used to accelerate the biotransformation process. In the buffer/*n*-tetradecane biphasic system containing [Bmim][SAc], the optimal substrate concentration was increased by 83 times (500 mmol/L) compared to the single-phase buffer, resulting in 53.4% yield (267 mmol/L) and 99% product. Li et al. [94] developed a system for the efficient production of (R)-3, 5-bis trifluoromethyl phenyl ethanol [(R)-BTPE] from 3, 5-bis trifluoromethyl acetophenone [3, 5-BTAP] using the *Trichoderma asperellum* ZJPH0810 as a catalyst in a [N1,1,1,1][PF$_6$]-distilled water reaction system. After selection and optimization, the highest substrate and product concentrations observed for reduction were 75 mmol/L and 65.4 mmol/L, respectively, which were 1.5-and 1.4-fold increases, respectively, compared to monophasic aqueous media.

Other types of reactions can be catalyzed by whole cell catalysis, such as oxidation, decomposition, and transesterification, while these reactions are rarely reported by now. In the study of Fu et al. [95], *Armillaria luteo-virens Sacc* ZJUQH100−6 cells were used to establish an IL system for efficient production of betulinic acid by oxidation of terpene betulinic acid to betulinic acid. In the established system containing ILs, the highest product yield of 11.14% was observed at 18 h, which was higher than that observed in the single-phase aqueous system. Ester exchange reaction can be used for biodiesel synthesis. Aral et al. [96] used four different microorganisms in the [Bmim][BF$_4$]/[Emim][BF$_4$] system for whole cell catalysis to produce fatty acid methyl esters from methanol, among which the wild-type *Rhizobium oryzae* had the highest yield of fatty acid methyl esters. In this system, the IL also has the function of methanol storage, which can reduce the toxic effect of methanol on cells.

3.5 Ionic liquids intensified biomass conversion

3.5.1 Introduction

At present, the development of human society is closely related to the exploitation of fossil resources such as coal, petroleum, and natural gas, by which the fundamental chemicals and materials could be effectively offered. However, the continuously increasing demand for energy nowadays has led to excessive consumption of the nonrenewable resources and global warming. Hence, it is urgent to develop renewable resources to be served as a promising substitute for fossil resources, which could liberate our heavy dependence on the limited energy and reduce the serious damage of traditional energy to the environment, promote the harmonious and sustainable development of human society.

As the abundant organic carbon resource in nature, biomass has been regarded as a profusely available renewable resource to provide value-added chemicals as the replacement for the limited fossil carbon reserves. Biomass has various types and is mainly represented by lignocellulose, chitin, starch, and so on, among which lignocellulose represents the most abundant form of biomass [97]. Currently, approximately 2×10^{11} t/a of lignocellulosic biomass could be yielded from sources such as agricultural residues, bioenergy crops, and woody materials [98]. In spite of its abundance, less than 5% of the lignocellulose resource has been exploited for further upgradation [99]. Therefore, the valorization of lignocellulose holds promise to meet the inevitable and foreseeable progress of our society. Typically, lignocellulose is mainly composed of cellulose (35%−50%), hemicellulose (25%−35%), and lignin (10%−30%), the contents of which depend on the source of feedstock [100]. High value-added chemicals have been produced from lignocellulose, such as 5-hydroxymethylfurfural (HMF), 2,5-dimethylfuran (DMF), and sorbitol from cellulose, furfural, xylose, and arabinose from hemicellulose, and vanilline, guaiacol, and benzene from lignin, as shown in Fig. 3.14. During the conversion procedures mentioned above, various catalytic technologies like oxidation, hydrolysis, and hydrogenation have been delved in this field to develop the industrial potential of the underutilized but valuable renewable resource [97]. In this context, traditional volatile organic solvents like acetonitrile and acetone have been commonly involved, which are environmentally unfriendly and may lead to the hurdle of harsh operating conditions. Hence, investigating efficient green and sustainable reaction system to solve these issues represents an important issue to be taken into account for both industrial and academic research.

In this regard, ILs have been for a long time considered as an ideal candidate to substitute traditional volatile organic solvents, which are prone to provide an effective

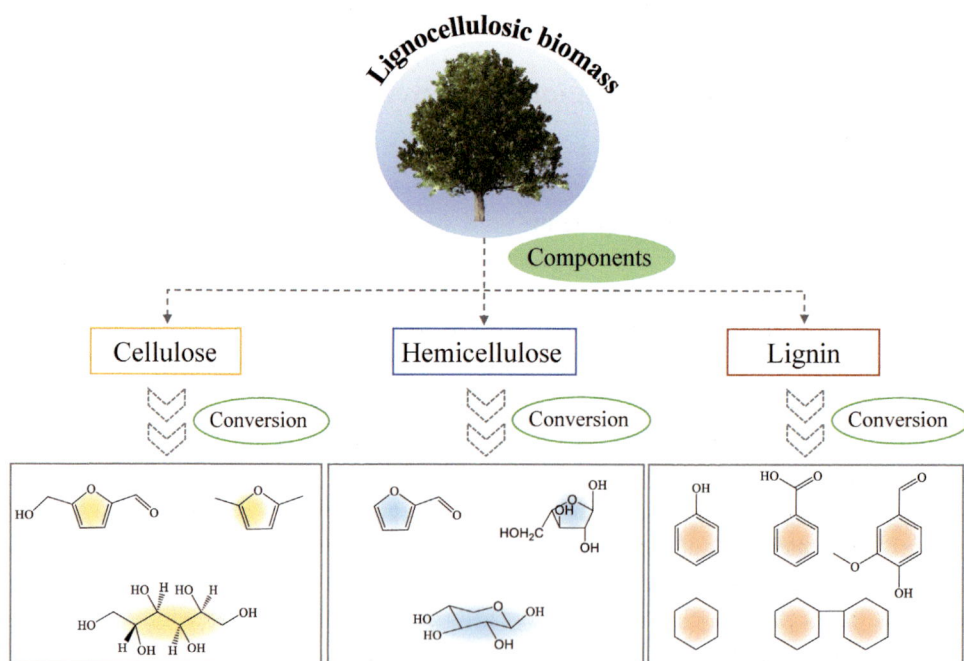

Figure 3.14 Components of lignocellulosic biomass and the products from lignocellulosic biomass conversion.

and green reaction system for lignocellulose conversion in an environmentally benign route. In addition, high solubility of cellulose, hemicellulose, and lignin in ILs has been discovered upon the hydrogen bonds between ILs and lignocellulose, which could further contribute to promoting the conversion procedure [101]. In general, ILs have been defined as liquid salts composed of a cation and an anion. Their attractive features such as near-zero vapor pressure, nonflammability, designability, and high solubility of substrates make them an intense focus in various fields of green catalysis, separation, electrochemistry, and so on [102]. In recent years, much effort has been put into using ILs as both efficient solvents and catalysts in lignocellulose conversion [103,104]. On the one hand, the operating conditions in ILs could be much milder than those conducted in traditional volatile organic solvents as the low vapor pressure of ILs, conducive to the exploitation of energy-conserving conversion. On the other hand, the cation and anion of ILs could be designed for the targeted properties to adjust the interaction between ILs and lignocellulose/ reaction intermediate, upon which the reactivity of the reactants would be further boosted and the reaction intermediate could be stabilized. Consequently, efficiency of lignocellulose conversion that is conducted in ILs has been anticipated to be significantly improved. In addition, the high solubility of ILs for lignocellulose contributes to providing a homogeneous reaction environment, by which the effective collision could be enhanced and thus improves the reaction efficiency [105]. In brief, ILs, as the green and functional reaction media, have exhibited promising performance in lignocellulose upgradation, which will intensify biomass conversion in an effective and environmentally benign procedure.

3.5.2 Ionic liquids intensify cellulose conversion

Cellulose (35%−50%) is the most abundant component in lignocellulose, which is a linear polymer comprised of D-glucose units linked by β-1,4-glycosidic bonds, as shown in Fig. 3.15. Various important value-added chemicals can be yielded via cellulose conversion by oxidation, hydrolysis, hydrogenation, and so on [106−108]. Additionally, some of these cellulose-derived chemicals are regarded as platform chemicals to be further valorized. In this respect, the conversion procedure can be effectively intensified by ILs as both solvent and catalyst, which not only dissolve cellulose but also promote its conversion effectively.

(1) *Cellulose monomer and cellulose macromolecule conversion intensified by ionic liquids*

HMF has been considered as significant platform compound because of its potential application for replacing voluminously consumed petroleum-based building blocks [109]. Producing HMF from cellulose is of great significance owing to the abundance and renewability of the raw material. Generally, hydrolysis of cellulose firstly proceeds to yield glucose, which could be isomerized to form fructose, followed by the

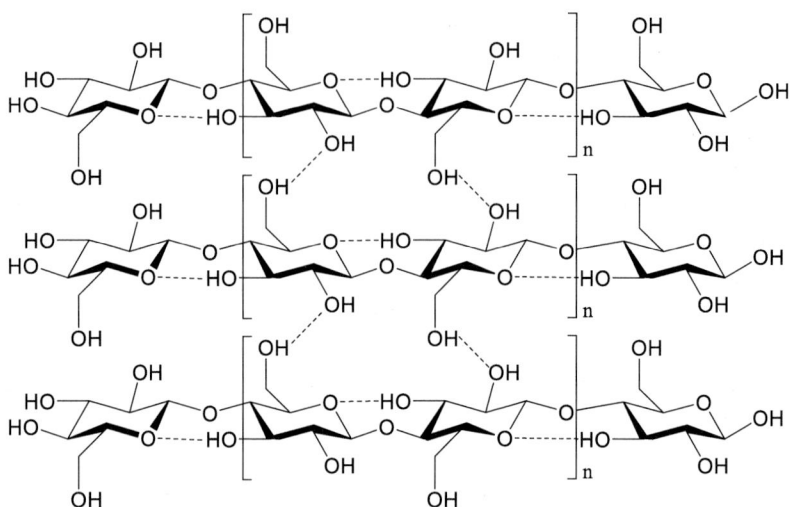

Figure 3.15 Schematic illustration of cellulose.

dehydration to produce HMF. Zhang and coworkers suggested that chromium (II) chloride in 1-alkyl-3-methylimidazolium chloride ([Emim]Cl) could serve as an effective catalytic system for the conversion of glucose into HMF with a yield near 70% [110]. The formation of $[Emim]^+CrCl_3^-$ was supposed to be the key step for the catalytic reaction, in which the $CrCl_3^-$ anion played a critical role in the effect of a formal hydride transfer and therefore contributed to isomerization of glucose to fructose. Subsequently, fructose underwent rapid dehydration to form HMF in the presence of the catalyst in IL. In addition to forming the key catalytic species with chromium chloride $[Emim]^+CrCl_3^-$, the IL [Emim]Cl also facilitates the reaction by dissolving cellulose through forming hydrogen bonds with cellulose. Han and coworkers developed an effective catalytic system composed of the ILs 1-ethyl-3-methylimidazolium tetrafluoroborate ([Emim][BF$_4$]) and the Lewis acid SnCl$_4$, in which the yield of HMF reached 62% [111]. In the [Emim][BF$_4$] system, Sn atoms interact with the hydroxyl group of glucose to form a five-membered–ring chelate structure and significantly catalyze the conversion of glucose into HMF. The Cl atoms in SnCl$_4$ were supposed to interact with hydrogen atoms and transfer the hydrogen atoms. The Sn atom was proposed to interact with oxygen atoms and facilitate the generation of the enediol intermediate, which was converted into fructose and subsequently underwent dehydration to yield HMF.

To elucidate the role of structure of ILs in the production of HMF in depth, Zhang and coworkers studied the preparation of HMF from fructose in the ILs containing various cations and anions, the structure of which is shown in Fig. 3.16 [112]. It was found that both the aggregations of cations and the hydrogen bonds between anions and fructose exerted an influence on the production of HMF. Specially, the cations of ILs with

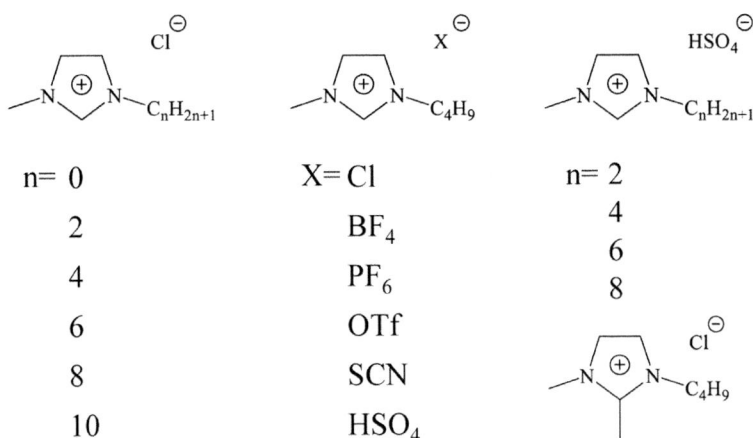

Figure 3.16 The structure of the ILs used.

alkyl chains shorter than 4 showed excellent catalytic effects to promote fructose dehydration. The aggregation caused by long alkyl groups of the imidazolium cations hindered the transfer of protons and electrons and prevented the interaction of hydrogen at the C_2 position with hydroxyl of fructose to dehydrate to yield HMF. Meanwhile, the anions of ILs like Cl^- which could form stronger hydrogen bonds with fructose resulted in better reactivity in HMF preparation than those anions like OTf^- forming weak hydrogen bonds with fructose. In addition, high acidity of IL could also contribute to high catalytic activity of fructose dehydration to yield HMF.

In addition to converting the cellulose monomer, the direct upgradation of the cellulose macromolecule into value-added chemicals is also attractive but challenging due to the inexpensive raw material and simple procedure without separation of intermediates. Sorbitol is an important industrial chemical with a wide application in medicine, daily chemicals, food, and so on [113]. The key step for cellulose conversion into sorbitol is the hydrolysis that is hampered by the robust glycosidic bonds and hydrogen bonds existing in cellulose. Zhang and coworkers investigated the direct conversion of the ball-milled microcrystalline cellulose into sorbitol in the ILs 1-allyl-3-methylimidazolium chloride ([Amim]Cl) [114]. The strong inter- and intramolecular hydrogen bonds network between hydroxyl groups in cellulose could be effectively destroyed by [Amim]Cl dissolution. In addition, the formation of water-soluble ionic cellulose enhanced the effective collision of cellulose and catalyst remarkably, consequently intensifying the acquisition of sorbitol from cellulose conversion.

(2) Cellulose-derived chemicals conversion intensified by ionic liquids

Cellulose-derived chemicals like HMF and DMF have been for a long time regarded as significant platform chemicals for various useful value-added chemicals and fuels, which make the conversion of cellulose-derived chemicals an intense focus for developing biorefineries [115,116]. In this procedure, ILs with the promising properties such as high solubility for the reactants/products and forming hydrogen bonds with the reactants play a key role in intensifying the conversion.

The conversion of HMF to 2,5-furandicarboxylic acid (FDCA), an important bio-derived substitute for fossil-based polyethylene terephthalate, is of great importance for cellulose valorization [117]. Zhang and coworkers investigated the base-free catalytic conversion of HMF into FDCA intensified by ILs in depth [118]. Almost complete HMF conversion was obtained with 44.2% yield of FDCA using catalyst $Ce_{0.5}Fe_{0.15}Zr_{0.35}O_2$ in ILs. The IL 1-butyl-3-methylimidazolium chloride ([Bmim]Cl) showed the best performance for its good ability to generate hydrogen bonds and thermal stability. Meanwhile, the short alkyl chain of the IL cation (≤ 4) made the ILs not form definite aggregates, which was beneficial for the conversion. Furthermore, another work conducted by Zhang and coworkers involved the conversion of HMF into FDCA applying Fe-Zr-O as a catalyst in [Bmim]Cl [119]. The promotion of [Bmim]Cl to the reaction was due to the high solubility of FDCA in ILs upon hydrogen bond formation, leading to the desorption of FDCA from the catalyst and recovery of the catalyst.

An IL/heteropoly acid catalytic system was delved and found to effectively convert HMF to FDCA by Zhang and coworkers [120]. The ILs [Bmim]Cl exerted the best performance as both solvent and cocatalyst, in which the yield of FDCA achieved as high as 89% with 99% conversion of HMF. On the one hand, the affinity between HMF and catalyst was strengthened by the high solubility of FDCA in ILs, which in addition avoided over oxidation of the furan ring. On the other hand, multiple hydrogen bonds were generated between IL anion (Cl^-) /cation ($[Bmim]^+$) and hydroxy groups of HMF, catalyzing the formation of aldehyde groups and thereby facilitating the conversion of HMF into FDCA. Furthermore, Lu and coworkers studied the conversion of HMF into FDCA applying SBA-15-supported heteropoly acids in [Bmim] Cl [121]. 76% yield of FDCA was obtained, which was maintained at 60% after five runs, in which [Bmim]Cl showed good stability during the reaction.

DMF is a significant furanic compound derived from HMF via hydrogenation, which could be converted into useful industrial chemicals like the commonly petro-derived p-xylene (PX). The production of PX from DMF, including three steps of Diels–Alder, dehydration and decarboxylation reactions, could be achieved in one step at mild conditions upon the promotion of ILs [122]. The acidic ILs were used both as solvent and catalyst, the characteristics of which significantly affected the reactivity and selectivity of the conversion. The neutral ILs did not play a catalytic role, while the ILs with strong acidity would facilitate further hydrolytic decomposition of DMF into 2,5-hexanedione. The Brønsted acid IL (BAIL) 1-butyl-3-methylimidazolium hydrogen sulfate ([Bmim][HSO$_4$]) was verified to show the best performance compared to the other ILs used, affording high yield of PX and 2,5-methylbenzoic acid with up to 89% aromatic selectivity in one step at room temperature [122].

Furthermore, many metal triflates are proved to be an effective catalyst for the Diels−Alder and dehydration of HMF, while the moisture sensitivity of this kind of catalyst greatly limits their application. Zhang and coworkers applied Sc(OTf)$_3$ as the catalyst for the synthesis of PX from DMF in ILs [123]. The hydrophobic ILs 1,3-diethylimidazolium bis(trifluoromethylsulfonyl)imide ([Emim]NTf$_2$) were used as solvent to remove the water formed during the reaction to prevent Sc(OTf)$_3$ from deactivation, and the catalytic system provided 68% aromatic selectivity in one step.

3.5.3 Ionic liquids intensify hemicellulose conversion

Hemicellulose, another important component of biomass, is a branched heteropolysaccharide polymer mainly consisting of pentoses, hexoses, and glucuronic acids, as shown in Fig. 3.17 [124]. Conversion of hemicellulose is anticipated to offer essential value-added chemicals like furfural and sugars. The application of ILs in the field of hemicellulose conversion has also attracted much attention as the reaction media.

Hydrolysis of hemicellulose is the mainstream strategy for the production of furfural, while the commonly used acids like inorganic acids and solid acids unfortunately suffer from poor selectivity and deactivation issues. The BAILs have been regarded as green and efficient catalysts to catalyze the hydrolyzation of hemicellulose into furfural. Wu and coworkers achieved highly efficient conversion of xylose and hemicellulose into furfural catalyzed by various [HSO$_4$]-based BAILs [125]. The results indicated that the acidity of BAILs played a key role in catalyzing the reaction, the increase of which was beneficial for the hydrolyzation. Changing the number of methyl groups on the pyridine ring could tune the acidity of BAILs, and [Hpy][HSO$_4$] with the strongest acidity showed the best catalytic performance, offering furfural with the yield of 75.4% and 80.4% from hemicellulose and xylose, respectively.

Three superacid SO$_4$H-functionalized ILs were synthesized by Tao and coworkers [126]. The [Ch-SO$_4$H][CF$_3$SO$_3$] with super acid strength exerted the best catalytic performance to afford more than 80% yield of furfural. Further study showed that the catalytic activity of ILs [Ch-SO$_4$H][CF$_3$SO$_3$] is higher than those of CH$_3$SO$_3$H and TsOH due to the higher acid strength of the ILs. Meanwhile, Amberlyst-15 was found to display lower reaction activity than the ILs in the work for the large mass transfer resistance.

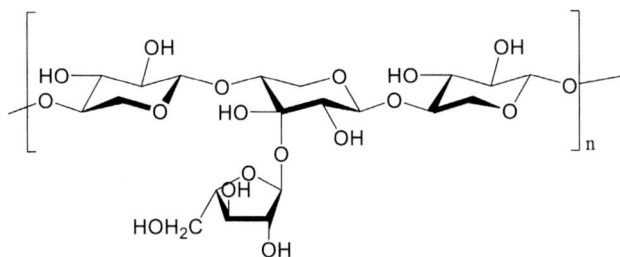

Figure 3.17 Schematic illustration of hemicellulose.

In addition to furfural, the BAILs were also used to catalyze the hydrolysis of hemicellulose to afford sugars. Dhepe and coworkers achieved 87% yield of C_5 sugars (xylose and arabinose) from conversion of hardwood hemicellulose catalyzed by BAILs [127]. The catalytic activity increased with the acidity of the BAILs, that is, 1-methyl-3-IJ-(3-sulfopropyl)-imidazolium hydrogensulfate ([C_3SO_3Hmim][HSO_4]) > 1-methyl-3-(3-sulfopropyl)-imidazolium p-toluenesulfonate ([C_3SO_3Hmim][PTS]) > 1-methyl-3-(3-sulfopropyl)-imidazolium chloride ([C_3SO_3Hmim][Cl]) > [Bmim][Cl]. In comparison, mineral acid (H_2SO_4, HCl) displayed lower performance than BAILs, which is relevant to the ion-dipole-type interaction of the BAILs with the reactants.

3.5.4 Ionic liquids intensify lignin conversion

Lignin is regarded as the most abundant aromatic resource which accounts for 10%–30% of lignocellulose. Structurally, lignin derives from three monolignol (guaiacyl propanol, p-hydroxyphenyl propanol, and syringyl alcohol), which connects to each other with various C-C/C-O bonds, mainly involving β-O-4, α-O-4, 4-O-5, β-1, β-5, β-β, and 5−5 interunit linkages, as shown in Fig. 3.18.

Due to its functionalized aromatic rich feature, the conversion of lignin resource into high value-added aromatics is currently of tremendous interest as an ideal substitute for limited fossil. In recent years, ILs have been widely applied in various lignin transformation technologies such as oxidation, hydrodeoxygenation (HDO), photocatalysis, and so on, promoting the conversion and selectivity of lignin upgradation.

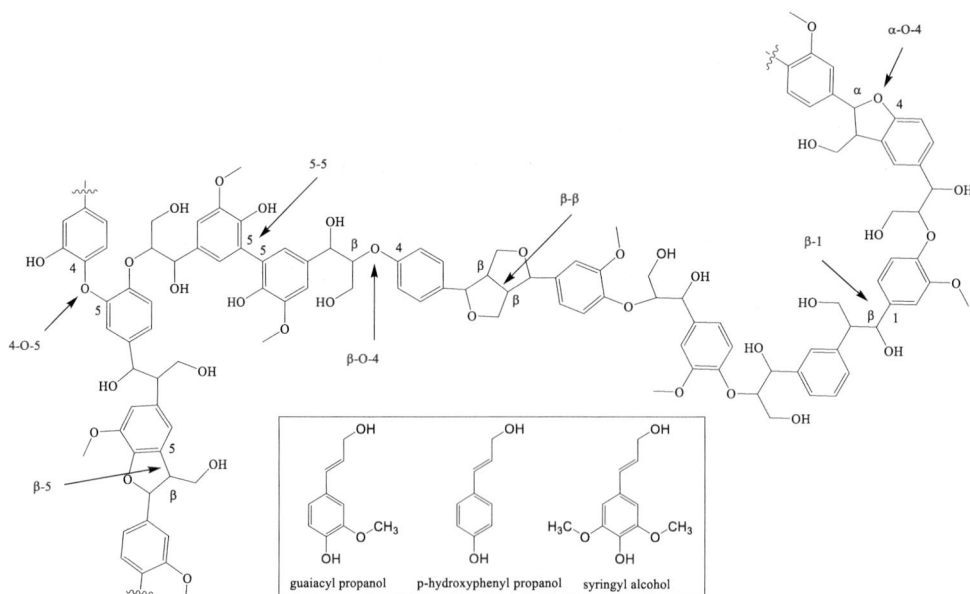

Figure 3.18 Schematic illustration of lignin.

(1) *Oxidation of lignin and lignin model compounds intensified by ionic liquids*

Oxidation remains one of the most important technologies as lignin aromatic structures could be preserved to a great extent to offer functional aromatic products. Due to the complex structure of lignin, lignin model compounds are usually studied to be oxidized in depth by researchers to investigate the reaction mechanism [128]. Recently, much attention has been paid to lignin oxidation in ILs, which are regarded as ideal reaction media due to the ideal dissolution of oxygen and lignin in ILs, as well as the good catalytic performance that enables the oxidation conducted under mild conditions.

Metal-based catalysts such as iron, manganese, copper, and vanadium are commonly used to promote the efficiency of lignin conversion. Kumar and coworkers conducted the oxidation of a lignin model compound, veratryl alcohol, in IL 1-butyl-3-ethylimidazolium hexafluorophosphate ([Bmim][PF$_6$]) catalyzed by iron(III) porphyrins to afford aromatics like veratraldehyde [129]. [Bmim][PF$_6$]was supposed to stabilize the intermediates generated in the reaction upon the noncoordinating nature and weak nucleophilicities of [Bmim][PF$_6$], which led to much higher reactivity in ILs than water. Ragauskas and coworkers reported an effective vanadium-catalyzed aerobic oxidation of the lignin model compound aromatic alcohol in [Bmim][PF$_6$], in which the product yield of 90% could be achieved with the selectivity of aromatic aldehyde as high as 99% [130]. The reaction efficiency was significantly facilitated by [Bmim][PF$_6$] for the enhanced solubility of catalysts and reactants in ILs. An effective conversion of alkali lignin was achieved to selectively afford the products like phenols and aromatic ketone by An and coworkers [131]. The acidic ILs and polyoxometalate-IL (POM-IL) synergistically catalyzed the conversion of alkali lignin and affected the aromatics distribution. In addition, the intermolecular hydrogen bonds and π−π stacking of lignin with the IL imidazole ring and POM anion made lignin more flexible and facilitated its conversion.

In spite of the high efficiency that the metal-based catalysts exhibited, the harmful metal waste produced and the difficulties in posttreatment hindered their wide application in lignin oxidation. Hence, much attention has been paid to the development of a metal-free catalytic system. Han and coworkers reported an effective oxidation of a β-O-4 lignin model compound in IL 1-benzyl-3-methylimidazolium bis(trifluoromethylsulfonyl)imide ([Bnmim][NTf$_2$]), which served as both solvent and catalyst [132]. Especially, the anion [NTf$_2$]$^-$ played a key role in facilitating OOH free radicals generation due to the delocalization of electron pairs promoted by the strong electronegativity of the heteroatoms bearing in the anion. The lignin C-C/C-O bonds cleavage was subsequently initiated and yielded aromatics including benzoic acid and phenol. In addition to IL anions, the cations of ILs were also proved to effectively catalyze lignin oxidation by forming multiple hydrogen bonds. The conversion of a β-O-4 lignin model compound 2-phenoxyacetophenone into benzoic acid and phenol was studied by Lu and coworkers [133]. The C-C bond of β-O-4 linkage was effectively cleaved by oxidation in the ILs 1-methyl-3-(3-sulfopropyl)-imidazolium chloride ([CPmim][NTf$_2$]). Further

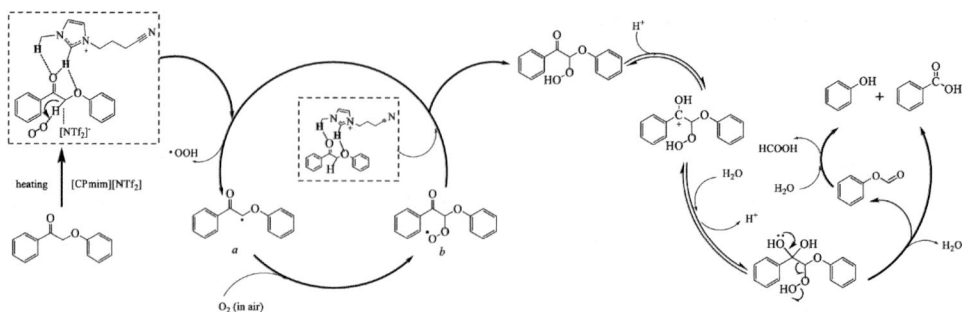

Figure 3.19 Reaction pathway of 2-phenoxyacetophenone in [CPmim][NTf$_2$].

study showed that the [CPmim]$^+$ of the ILs formed three ipsilateral hydrogen bonds with lignin β-O-4 linkages at both sides of the C-C bond. Upon this kind of weak bond joint effects, free radicals are generated by the cleavage of adjacent C-H bond and subsequently initiated the strong C-C bond fragmentation to harvest benzoic acid and phenol with a yield of more than 90%, as shown in Fig. 3.19.

(2) *Hydrodeoxygenation of lignin and lignin model compounds intensified by ionic liquids*

HDO is an important strategy for lignin conversion to biofuels like arene and cycloparaffin under reduction conditions, which is anticipated to offer less mixed products. Generally, two steps of HDO, that is hydroprocessing and deoxygenation processes, are involved to cleave lignin linkages and remove oxygen element from the products. To date, ILs have been considered as ideal reaction media for HDO of lignin to achieve higher reactivity and efficiency, which are closely related to the unique characteristics of ILs like designability and lignin solubility.

Noble metal NPs are known as highly active catalysts for the HDO procedure. In this context, ILs could be used as promising media to stabilize and protect NPs due to the structural directionality (IL effect), self-organization, and electrostatic stabilization effect, leading to higher reactivity of metal NPs [134]. Currently, the combination of ILs and metal NPs that is regarded as a pseudohomogeneous catalytic system has been proved to enable the HDO process with high efficiency. Zhang and coworkers developed a pseudohomogeneous catalytic system in ILs for selective C-O bond fragmentation and HDO of lignin model compounds [134]. With the synergism of phosphoric acid, almost 100% conversion of monomer and dimer lignin model compound was achieved, and the catalytic system composed of Pt NPs and [Bmim][PF$_6$] displayed the best performance for lignin model compounds with maximum 97% selectivity. The catalysts involving Pd, Pt, Rh, and Ru NPs were proved to be stabilized and well-distributed in ILs without aggregation. It is supposed that the hydrophobic ILs exerted better performance in the reaction due to the moisture sensitivity of metal NPs to aggregate. Furthermore, the higher solubility of hydrogen in ILs than other solvents like water was beneficial for the effective collision of hydrogen and catalyst and thus promoted the catalytic efficiency.

Obviously, the Brønsted acids (BAs) are usually used to synergistically catalyze the HDO procedure. However, given the possible corrosion of reactors, it is necessary to develop a BA-free catalytic system with high HDO catalytic efficiency. The HDO conversion of monomer and dimer lignin model compounds to alkanes was investigated in ILs without the addition of BA by Zhang and coworkers [135]. The catalytic system of [Bmim][PF$_6$] with well-dispersed Ru NPs supported on SBA-15 exerted the best HDO activity with over 99% conversion and 98% selectivity. It is interesting to find that the anions of ILs significantly affect the deoxidization process. The IL anion [PF$_6$]$^-$ showed obvious advantages in harvesting deoxidization products compared to other anions like [BF$_4$]$^-$ and [NTf$_2$]$^-$ and organic solvents like isopropanol and *n*-heptane. Meanwhile, the undesirable side reactions were suppressed due to the absence of BA, resulting in high selectivity of products. The protic ILs were found to be effective in catalyzing deoxygenation of the HDO process, which also have the advantage of low cost. Lu and coworkers synthesized the low-cost and efficient protic ILs 2-hydroxy-*N*-(2-hydroxyethyl)-*N*-methylethanaminium trifluoromethanesulfonate ([BHEM]CF$_3$SO$_3$) with two hydroxy on the cation, which coordinated with Rh/C to catalyze HDO of lignin model compounds (phenol) to offer alkane [136]. In the catalytic system, 100% conversion of phenol could be achieved with 93.3% yield of cyclohexane. It was proposed that [BHEM]CF$_3$SO$_3$ shows good performance in catalyzing the deoxidation process, while Rh/C catalyzes the hydrogenation of HDO.

(3) *Photocatalysis of lignin and lignin model compounds intensified by ionic liquids*

Photocatalysis of lignin is regarded as a clean, effective, and energy-saving technology for harvesting value-added aromatics, which usually occurs in milder conditions (room temperature and atmospheric pressure), and there is no need for strong oxidants or reducing agents and renewable source of energy [137]. Generally, free radicals are generated upon the irradiation of light to the photocatalysts like semiconductors and subsequently lead to the conversion of lignin by reacting with substrates. However, the fast electron-hole pair recombination of photocatalysts and leaching issue of metals in doped photocatalysts hinder the application of photocatalysis in lignin conversion. Actually, ILs have been proved to promote free radical generation of lignin and hence are anticipated to intensify lignin valorization in metal-free conditions with high efficiency.

Zhang and coworkers demonstrated a self-initiated radical photocatalytic conversion of lignin model compounds (2-phenoxyacetophenone) via IL induction, which led to 98.4% conversion of the reactant with 93.3% yield of benzoic acid and 68.8% yield of phenol at room temperature after 150 min UV light irradiation [138]. The IL 1-propyl-3-methylimidazolium bis(trifluoromethylsulfonyl)imide ([Pmim][NTf$_2$]) was applied as both solvent and catalyst to facilitate the breakage of the C$_\beta$-H bond to form free radicals, followed by the C-O bond fragmentation catalyzed by the BAIL [PrSO$_3$Hmim]

[OTf]. In addition to promoting the formation of free radicals, it was found that ILs could change the charge distribution of lignin and stabilize the reaction intermediates in the photocatalytic reaction, upon which lignin conversion was further intensified by ILs. Zhang and coworkers studied the C-C bond cleavage of β-O-4 and β-1 lignin model compounds promoted by the IL 1-butyl-3-methylimidazolium bis(trifluoromethylsulfonyl)imide ([Bmim][NTf$_2$]) with the synergism of UV light and heating to afford aromatic aldehydes. The conversion for the β-O-4 lignin model compound (2-phenoxy-1-phenylethanol) was proved to undergo two steps, namely, protonation and the Norrish type I reaction, which could be able to provide phenylacetaldehyde and benzaldehyde, respectively. Interestingly, the electron density around the oxygen of the -OH group of 2-phenoxy-1-phenylethanol increased upon its interactions with anion, cation, and ion pair of [Bmim][NTf$_2$], facilitating the protonation of the hydroxy group. In addition, the formed carbocation intermediate could be stabilized by the IL anion [NTf$_2$]$^-$, in turn facilitating the protonation procedure. Hence, the production of phenylacetaldehyde was intensified by ILs, followed by the C-C bond cleavage to yield benzaldehyde by photothermal synergism. Meanwhile, the C-C bond of β-1 lignin model compound (bibenzyl) could also be cleaved in [Bmim][NTf$_2$] under photothermal conditions to offer benzaldehyde and phenol.

References

[1] Zhao D, Liu M, Zhang J, et al. Synthesis, characterization, and properties of imidazole dicationic ionic liquids and their application in esterification. Chemical Engineering Journal 2013;221:99—104.

[2] Zhao D, Ge J, Zhai J, et al. Synthesis of succinic acid diisopropyl ester catalyzed by Lewis acid ionic liquids. CIESC Journal 2014;65(2):561—9.

[3] Zhao D, Liu M, Xu Z, et al. Synthesis of dimethyl succinate catalyzed by ionic liquids. CIESC Journal 2012;63(4):1089—94.

[4] Zhao D, Liu M, Ge J, et al. Synthesis of binuclear ionic liquids and their catalytic activity for esterification. Chinese Journal of Organic Chemistry 2012;32:2382—9.

[5] Wei W, Wang Y, Yan Z, et al. One-step DMC synthesis from CO$_2$ under catalysis of ionic liquids prepared with 1,2-propylene glycol. Catalysis Today 2023;418:114052.

[6] Zhang J, Zhang SJ, Han JX, et al. Uniform acid poly ionic liquid-based large particle and its catalytic application in esterification reaction. Chemical Engineering Journal 2015;271:269—75.

[7] Bian YH, Zhang J, Liu CZ, et al. Synthesis of cross-linked poly acidic ionic liquids and its application in biodiesel production. Catalysis Letters 2020;150:969—78.

[8] Bian YH, Shan QW, Zhang J, et al. Biodiesel production over esterification catalyzed by a novel poly(acidic ionic liquid)s. Catalysis Letters 2021;151:3523—31.

[9] Bian YH, Zhang J, Zhang SJ, et al. Synthesis of polyionic liquid by phenolic condensation and its application in esterification. ACS Sustainable Chemistry & Engineering 2019;7:17220—6.

[10] Wang YJ, Zhang J, Shan QW, et al. Preparation of porous polyaminobenzenesulfonic acid and synthesis of glycerol monolaurate. Chemical Papers 2022;76(4):2431—45.

[11] Liu WQ, Wang YJ, Zhang J, et al. Large particle spherical poly-ionic liquid-solid base catalyst for high efficiency transesterification of ethylene carbonate to prepare dimethyl carbonate. Fuel 2022;324:124580.

[12] Liu W, Li J, Xie Y, et al. Synthesis of dimethyl carbonate from CO$_2$ catalyzed by spherical polymeric ionic liquid catalyst. Chemical Papers 2023;78:1553-65.

[13] Wang YJ, Liu WQ, Zhang J, et al. Synthesis of novel ionic porous organic polymers and its application in hydroxyl condensation reaction. Catalysis Letters 2023;153:1797−806.

[14] Wang YJ, Liu WQ, Li JX, et al. Preparation of porous macromolecule oxazine poly-ionic liquid solid acid catalyst and its application in an esterification reaction. Journal of Chemical Technology & Biotechnology 2023;98:129−39.

[15] Kragl U, Eckstein M, Kaftzik N. Enzyme catalysis in ionic liquids. Current Opinion in Biotechnology 2002;13(6):565−71.

[16] Wasserscheid P, Keim W. Ionic liquids—new "solutions" for transition metal catalysis. Angewandte Chemie-International Edition 2000;39(21):3772−89.

[17] Yang Z, Pan WB. Ionic liquids: green solvents for nonaqueous biocatalysis. Enzyme and Microbial Technology 2005;37(1):19−28.

[18] Ha SH, Koo YM. Enzyme performance in ionic liquids. Korean Journal of Chemical Engineering 2011;28(11):2095−101.

[19] Weltn T. Room-temperature ionic liquids: solvents for synthesis and catalysis. Chemical Reviews 1999;99(8):2071−83.

[20] Dreyer S, Kragl U. Ionic liquids for aqueous two-phase extraction and stabilization of enzymes. Biotechnology & Bioengineering 2008;99(6):1416−24.

[21] Zhao H, Malhotra SV. Applications of ionic liquids in organic synthesis. Aldrichimica Acta 2002;35 (3):75−83.

[22] Jain N, Kumar A, Chauhan S, et al. Chemical and biochemical transformations in ionic liquids. Tetrahedron 2005;61(5):1015−60.

[23] Zhao H. Methods for stabilizing and activating enzymes in ionic liquids—a review. Journal of Chemical Technology and Biotechnology 2010;85(7):891−907.

[24] Moniruzzaman M, Kamiya N, Goto M. Activation and stabilization of enzymes in ionic liquids. Organic & Biomolecular Chemistry 2010;8(13):2887−99.

[25] Naushad M, ALOthman ZA, Khan AB, et al. Effect of ionic liquid on activity, stability, and structure of enzymes: a review. International Journal of Biological Macromolecules 2012;51(4):555−60.

[26] Gomes JM, Silva SS, Reis RL. Biocompatible ionic liquids: fundamental behaviours and applications. Chemical Society Reviews 2019;48(15):4317−35.

[27] Xu P, Liang S, Zong MH, et al. Ionic liquids for regulating biocatalytic process: achievements and perspectives. Biotechnology Advances 2021;51:107702.

[28] Sheldon R. Catalytic reactions in ionic liquids. Chemical Communications 2001;23:2399−407.

[29] da Silva VG, de Castro RJS. Biocatalytic action of proteases in ionic liquids: improvements on their enzymatic activity, thermal stability and kinetic parameters. International Journal of Biological Macromolecules 2018;114:124−9.

[30] Li Z, Han Q, Wang K, et al. Ionic liquids as a tunable solvent and modifier for biocatalysis. Catalysis Reviews-Science and Engineering 2022; 66(2):484-530.

[31] Meyer LE, von Langermann J, Kragl U. Recent developments in biocatalysis in multiphasic ionic liquid reaction systems. Biophysical Reviews 2018;10(3):901−10.

[32] Schindl A, Hagen ML, Muzammal S, et al. Proteins in ionic liquids: reactions, applications, and futures. Frontiers in Chemistry 2019;7:347.

[33] Cull SG, Holbrey JD, Vargas-Mora V, et al. Room-temperature ionic liquids as replacements for organic solvents in multiphase bioprocess operations. Biotechnology & Bioengineering 2000;69 (2):227−33.

[34] Ji XL, Xue YJ, Li Z, et al. Ionozyme: ionic liquids as solvent and stabilizer for efficient bioactivation of CO_2. Green Chemistry 2021;23(18):6990−7000.

[35] Imam HT, Krasnan V, Rebros M, et al. Applications of ionic liquids in whole-cell and isolated enzyme biocatalysis. Molecules 2021;26(16):4791.

[36] Sivapragasam M, Moniruzzaman M, Goto M. Recent advances in exploiting ionic liquids for biomolecules: solubility, stability and applications. Biotechnology Journal 2016;11(8):1000−13.

[37] Wilkes JS, Zaworotko MJ. Air and water stable 1-ethyl-3-methylimidazolium based ionic liquids. Journal of the Chemical Society-Chemical Communications 1992;13:965−7.

[38] Kumar A, Venkatesu P. Does the stability of proteins in ionic liquids obey the Hofmeister series. International Journal of Biological Macromolecules 2014;63:244–53.
[39] Weingartner H, Cabrele C, Herrmann C. How ionic liquids can help to stabilize native proteins. Physical Chemistry Chemical Physics 2012;14(2):415–26.
[40] Patel R, Kumari M, Khan AB. Recent advances in the applications of ionic liquids in protein stability and activity: a review. Applied Biochemistry and Biotechnology 2014;172(8):3701–20.
[41] Byrne N, Rodoni B, Constable F, et al. Enhanced stabilization of the tobacco mosaic virus using protic ionic liquids. Physical Chemistry Chemical Physics 2012;14(29):10119–21.
[42] Geng F, Zheng LQ, Liu J, et al. Interactions between a surface active imidazolium ionic liquid and BSA. Colloid and Polymer Science 2009;287(11):1253–9.
[43] Vrikkis RM, Fraser KJ, Fujita K, et al. Biocompatible ionic liquids: a new approach for stabilizing proteins in liquid formulation. Journal of Biomechanical Engineering-Transactions of the Asme 2009;131(7):074514.
[44] Zhao H, Baker GA, Song Z, et al. Designing enzyme-compatible ionic liquids that can dissolve carbohydrates. Green Chemistry 2008;10(6):696.
[45] Kumar A, Venkatesu P. Prevention of insulin self-aggregation by a protic ionic liquid. RSC Advances 2013;3(2):362–7.
[46] Todinova S, Guncheva M, Yancheva D. Thermal and conformational stability of insulin in the presence of imidazolium-based ionic liquids. Journal of Thermal Analysis and Calorimetry 2016;123(3):2591–8.
[47] Berman HM, Battistuz T, Bhat TN, et al. The protein data bank. Acta Crystallographica Section D Biological Crystallography 2002;58(6):899–907.
[48] Bihari M, Russell TP, Hoagland DA. Dissolution and dissolved state of cytochrome C in a neat, hydrophilic ionic liquid. Biomacromolecules 2010;11(11):2944–8.
[49] Dong X, Fan YC, Yang P, et al. Ultraviolet-visible (UV-Vis) and fluorescence spectroscopic investigation of the interactions of ionic liquids and catalase. Applied Spectroscopy 2016;70(11):1851–60.
[50] Ghobadi R, Divsalar A. Enzymatic behavior of bovine liver catalase in aqueous medium of sugar based deep eutectic solvents. Journal of Molecular Liquids 2020;310:113207.
[51] Chiappe C, Neri L, Pieraccini D. Application of hydrophilic ionic liquids as co-solvents in chloroperoxidase catalyzed oxidations. Tetrahedron Letters 2006;47:5089–93.
[52] Wu M, Hu J, Wu YX, et al. Enhanced dechlorination of an enzyme-catalyzed electrolysis system by ionic liquids: Electron transfer, enzyme activity and dichloromethane diffusion. Chemosphere 2021;281:130913.
[53] Kaftzik N, Wasserscheid P, Kragl U. Use of ionic liquids to increase the yield and enzyme stability in the beta-galactosidase catalysed synthesis of N-acetyllactosamine. Organic Process Research & Development 2002;6(4):553–7.
[54] Kaleem I, Rasool A, Lv B, et al. Immobilization of purified beta-glucuronidase on ZnO nanoparticles for efficient biotransformation of glycyrrhizin in ionic liquid/buffer biphasic system. Chemical Engineering Science 2017;162:332–40.
[55] Reilly JT, Coats MA, Reardon MM, et al. Study of biocatalytic activity of histidine ammonia lyase in protic ionic liquids. Journal of Molecular Liquids 2017;248:830–2.
[56] Gaisberger RP, Fechter MH, Griengl H. The first hydroxynitrile lyase catalysed cyanohydrin formation in ionic liquids. Tetrahedron-Asymmetry 2004;15(18):2959–63.
[57] Hinckley G, Mozhaev VV, Budde C, et al. Oxidative enzymes possess catalytic activity in systems with ionic liquids. Biotechnology Letters 2002;24:2083–7.
[58] Fan YC, Dong X, Zhong YY, et al. Effects of ionic liquids on the hydrolysis of casein by lumbrokinase. Biochemical Engineering Journal 2016;109:35–42.
[59] Rather MA, Dar TA, Singh LR, et al. Structural-functional integrity of lysozyme in imidazolium based surface active ionic liquids. International Journal of Biological Macromolecules 2020;156:271–9.
[60] Rakowska PW, Kloskowski A. Impact of the alkyl side chains of cations and anions on the activity and renaturation of lysozyme: A systematic study performed using six amino-acid-based ionic liquids. Chemistryselect 2021;6(13):3089–95.
[61] Erbeldinger M, Mesiano AJ, Russell AJ. Enzymatic catalysis of formation of Z-aspartame in ionic liquid—an alternative to enzymatic catalysis in organic solvents. Biotechnology Progress 2000;16(6):1129–31.

[62]   Goldfeder M, Egozy M, Ben-Yosef VS, et al. Changes in tyrosinase specificity by ionic liquids and sodium dodecyl sulfate. Applied Microbiology and Biotechnology 2013;97(5):1953−61.

[63]   Hebal H, Parviainen A, Anbarasan S, et al. Inhibition of hyperthermostable xylanases by superbase ionic liquids. Process Biochemistry 2020;95:148−56.

[64]   Manna B, Ghosh A. Understanding the conformational change and inhibition of hyperthermophilic GH10 xylanase in ionic liquid. Journal of Molecular Liquids 2021;332(1):115875.

[65]   Saiyed ZM, Sharma S, Godawat R, et al. Activity and stability of alkaline phosphatase (ALP) immobilized onto magnetic nanoparticles (Fe_3O_4). Journal of Biotechnology 2007;131(3):240−4.

[66]   Magri AP, Pereira T, Cilli MM, et al. Enhancing the biocatalytic activity of l-asparaginase using aqueous solutions of cholinium-based ionic liquids. ACS Sustainable Chemistry & Engineering 2019;7:19720−31.

[67]   Schofer SH, Kaftzik N, Wasserscheid P, et al. Enzyme catalysis in ionic liquids: lipase catalysed kinetic resolution of 1-phenylethanol with improved enantioselectivity. Chemical Communications 2001;5:425−6.

[68]   Lau RM, van Rantwijk F, Seddon KR, et al. Lipase-catalyzed reactions in ionic liquids. Organic Letters 2000;2(26):4189−91.

[69]   Sandig B, Michalek L, Vlahovic S, et al. A monolithic hybrid cellulose-2,5-acetate/polymer bioreactor for biocatalysis under continuous liquid-liquid conditions using a supported ionic liquid phase. Chemistry-A European Journal 2015;21(44):15835−42.

[70]   Grewal J, Ahmad R, Khare SK. Development of cellulase-nanoconjugates with enhanced ionic liquid and thermal stability for in situ lignocellulose saccharification. Bioresource Technology 2017;242:236−43.

[71]   Laszlo JA, Compton DL. Alpha-chymotrypsin catalysis in imidazolium-based ionic liquids. Biotechnology and Bioengineering 2001;75(2):181−6.

[72]   Lozano P, de Diego T, Guegan JP, et al. Stabilization of alpha-chymotrypsin by ionic liquids in transesterification reactions. Biotechnology and Bioengineering 2001;75(5):563−9.

[73]   Persson M, Bornscheuer UT. Increased stability of an esterase from *Bacillus stearothermophilus* in ionic liquids as compared to organic solvents. Journal of Molecular Catalysis B-Enzymatic 2003;22(1−2):21−7.

[74]   Dong X, Fan Y, Zhang H, et al. Inhibitory effects of ionic liquids on the lactic dehydrogenase activity. International Journal of Biological Macromolecules 2016;86:155−61.

[75]   Itoh T, Akasaki E, Kudo K, et al. Lipase-catalyzed enantioselective acylation in the ionic liquid solvent system: reaction of enzyme anchored to the solvent. Chemistry Letters 2001;3:262−3.

[76]   Nara SJ, Harjani JR, Salunkhe MM, et al. Lipase-catalysed polyester synthesis in 1-butyl-3-methylimidazolium hexafluorophosphate ionic liquid. Tetrahedron Letters 2003;44(7):1371−3.

[77]   Temme H, Dethloff O, Pitner WR, et al. Identification of suitable ionic liquids for application in the enzymatic hydrolysis of rutin by an automated screening. Applied Microbiology and Biotechnology 2012;93(6):2301−8.

[78]   Jiang Y, Xia H, Guo C, et al. Enzymatic hydrolysis of penicillin in mixed ionic liquids/water two-phase system. Biotechnology Progress 2007;23(4):829−35.

[79]   Eker B, Asuri P, Murugesan S, et al. Enzyme-carbon nanotube conjugates in room-temperature ionic liquids. Applied Biochemistry and Biotechnology 2007;143(2):153−63.

[80]   Moniruzzaman M, Kamiya N, Nakashima K, et al. Water-in-ionic liquid microemulsions as a new medium for enzymatic reactions. Green Chemistry 2008;10(5):497−500.

[81]   Khlupova ME, Lisitskaya KV, Amandusova AH, et al. Dihydroquercetin polymerization using laccase immobilized into an ionic liquid. Applied Biochemistry and Microbiology 2016;52(4):452−6.

[82]   Wang SF, Chen T, Zhang ZL, et al. Direct electrochemistry and electrocatalysis of heme proteins entrapped in agarose hydrogel films in room-temperature ionic liquids. Langmuir 2005;21(20):9260−6.

[83]   Scopel R, da Silva CF, Lucas AM, et al. Fluid phase equilibria and mass transfer studies applied to supercritical fluid extraction of Illicium verum volatile oil. Fluid Phase Equilibria 2016;417:203−11.

[84]   Kang Y, Yao X, Yang Y, et al. Metal-free and mild photo-thermal synergism in ionic liquids for lignin C_α−C_β bond cleavage to provide aldehydes. Green Chemistry 2021;23(15):5524−34.

[85]   Fan LL, Li HJ, Chen QH. Applications and mechanisms of ionic liquids in whole-cell biotransformation. International Journal of Molecular Sciences 2014;15(7):12196−216.

[86] Wood N, Ferguson JL, Gunaratne HQN, et al. Screening ionic liquids for use in biotransformations with whole microbial cells. Green Chemistry 2011;13(7):1843−51.

[87] Ranke J, Molter K, Stock F, et al. Biological effects of imidazolium ionic liquids with varying chain lengths in acute Vibrio fischeri and WST-1 cell. Ecotoxicology and Environmental Safety 2004;58(3):396−404.

[88] Lee SM, Chang WJ, Choi AR, et al. Influence of ionic liquids on the growth of *Escherichia*. Korean Journal of Chemical Engineering 2005;22(5):687−90.

[89] Li J, Fan M, Zhang R, et al. New ionic liquid increase the catalytic efficiency of recombinant *Escherichia coli* cells-mediated asymmetric. Journal of Chemical Technology & Biotechnology 2019;94(1):159−66.

[90] Zhang F, Zhu CT, Peng QM, et al. Enhanced permeability of recombinant *E. coli* cells with deep eutectic solvent for transformation of rutin. Journal of Chemical Technology & Biotechnology 2019;95(2):384−93.

[91] Martins I, Hartmann DO, Alves PC, et al. Proteomic alterations induced by ionic liquids in *Aspergillus nidulans* and *Neurospora crassa*. Journal of Proteomics 2013;94:262−78.

[92] Dipeolu O, Green E, Stephens G. Effects of water-miscible ionic liquids on cell growth and nitro reduction using *Clostridium sporogenes*. Green Chemistry 2009;11(3):397−401.

[93] Ou XY, Wu XL, Peng F, et al. Highly efficient asymmetric reduction of 2-octanone in biphasic system by immobilized *Acetobacter* sp. CCTCC M209061 cells. Journal of Biotechnology 2019;299:37−43.

[94] Jun L, Feng Q, Wang P. Exploiting benign ionic liquids to effectively synthesize chiral intermediate of NK-1 receptor antagonists catalysed by *Trichoderma asperellum* cells. Biocatalysis and Biotransformation 2021;39(1):124−9.

[95] Fu ML, Jing L, Dong YC, et al. Effect of ionic liquid-containing system on betulinic acid production from betulin biotransformation by cultured *Armillaria luteo-virens* Sacc cells. European Food Research and Technology 2011;233(3):507−15.

[96] Arai S, Nakashima K, Tanino T, et al. Production of biodiesel fuel from soybean oil catalyzed by fungus whole-cell biocatalysts in ionic liquids. Enzyme and Microbial Technology 2010;46(1):51−5.

[97] Zhang Z, Song J, Han B. Catalytic transformation of lignocellulose into chemicals and fuel products in ionic liquids. Chemical Reviews 2017;117(10):6834−80.

[98] Amini E, Valls C, Roncero MB. Ionic liquid-assisted bioconversion of lignocellulosic biomass for the development of value-added products. Journal of Cleaner Production 2021;326:129275.

[99] Avelino C, Sara I, Velty A. Chemical routes for the transformation of biomass into chemicals. Chemical Reviews 2007;107:2411−502.

[100] Bohre A, Modak A, Chourasia V, et al. Recent advances in supported ionic liquid catalysts for sustainable biomass valorisation to high-value chemicals and fuels. Chemical Engineering Journal 2022;450:138032.

[101] Haykir NI, Nizan SZ, Harirchi S, et al. Applications of ionic liquids for the biochemical transformation of lignocellulosic biomass into biofuels and biochemicals: a critical review. Biochemical Engineering Journal 2023;193:108850.

[102] Liu M, Zhang Z, Liu H, et al. Transformation of alcohols to enisters promoted by hydrogen bonds using oxygen as the oxidant under metal-free conditions. Science Advances 2018;4(10):1−8.

[103] Yang Y, Fan H, Meng Q, et al. Ionic liquid [Omim][OAc] directly inducing oxidation cleavage of the beta-O-4 bond of lignin model compounds. Chemical Communications 2017;53(63):8850−3.

[104] Gao H, Wang J, Liu M, et al. Enhanced oxidative depolymerization of lignin in cooperative imidazolium-based ionic liquid binary mixtures. Bioresource Technology 2022;357:127333.

[105] Suzuki S, Takahashi K. Ionic liquids as organocatalysts and solvents for lignocellulose reactions. The Chemical Record 2023;23(8):e202200264.

[106] Wei GH, Lu T, Liu HY, et al. Exploring the continuous cleavage-oxidation mechanism of the catalytic oxidation of cellulose to formic acid: a combined experimental and theoretical study. Fuel 2023;341:127667.

[107] Frecha E, Torres D, Remón J, et al. Catalytic hydrolysis of cellulose to glucose: on the influence of graphene oxide morphology under microwave radiation. Journal of Environmental Chemical Engineering 2023;11(2):109290.

[108] Liu Y, Chen L, Zhang W, et al. Recyclable Cu salt-derived Brønsted acids for hydrolytic hydrogenation of cellulose on Ru catalyst. CCS Chemistry 2022;4(9):3162−9.
[109] Fan L, Bai X, Wang Y, et al. From corn stover to 5-hydroxymethylfurfural by ball milling-microwave hydrothermal (BM-MHT). Biomass Conversion and Biorefinery 2022;14:15069-78.
[110] Zhao H, Holladay JE, Brown H, et al. Metal chlorides in ionic liquid solvents convert sugars to 5-hydroxymethylfurfural. Science 2007;316(5831):1597−600.
[111] Hu S, Zhang Z, Song J, et al. Efficient conversion of glucose into 5-hydroxymethylfurfural catalyzed by a common Lewis acid SnCl₄ in an ionic liquid. Green Chemistry 2009;11(11):1746.
[112] Shi C, Zhao Y, Xin J, et al. Effects of cations and anions of ionic liquids on the production of 5-hydroxymethylfurfural from fructose. Chemical Communications 2012;48(34):4103−5.
[113] Zhou Y, Liang Y, Liu X, et al. Efficient glucose hydrogenation to sorbitol by graphene-like carbon-encapsulated Ru catalyst synthesized by evaporation-induced self-assembly and chemical activation. ACS Sustainable Chemistry & Engineering 2023;11(32):12052−64.
[114] Gao K, Xin J, Yan D, et al. Direct conversion of cellulose to sorbitol via an enhanced pretreatment with ionic liquids. Journal of Chemical Technology and Biotechnology 2018;93:2617−24.
[115] Endot NA, Junid R, Jamil MSS. Insight into biomass upgrade: a review on hydrogenation of 5-hydroxymethylfurfural (HMF) to 2,5-dimethylfuran (DMF). Molecules 2021;26(22):6848.
[116] Lim HY, Rashidi NA. Lignocellulosic biomass conversion into 5-hydroxymethylfurfural and 2,5-dimethylfuran, and role of the "Green" solvent. Current Opinion in Green and Sustainable Chemistry 2023;41:100803.
[117] Jia W, Chen J, Yu X, et al. Toward an integrated conversion of fructose for two-step production of 2,5-furandicarboxylic acid or furan-2,5-dimethylcarboxylate with air as oxidant. Chemical Engineering Journal 2022;450:138172.
[118] Yan D, Xin J, Shi C, et al. Base-free conversion of 5-hydroxymethylfurfural to 2,5-furandicarboxylic acid in ionic liquids. Chemical Engineering Journal 2017;323:473−82.
[119] Yan D, Xin J, Zhao Q, et al. Fe-Zr-O catalyzed base-free aerobic oxidation of 5-HMF to 2,5-FDCA as a bio-based polyester monomer. Catalysis Science & Technology 2018;8(1):164−75.
[120] Chen R, Xin J, Yan D, et al. Highly efficient oxidation of 5-hydroxymethylfurfural to 2,5-furandicarboxylic acid with heteropoly acids and ionic liquids. ChemSusChem 2019;12(12):2715−24.
[121] Chen R, Zhao Q, Yan D, et al. Base-free synthesis of bio-derived 2,5-furandicarboxylic acid using SBA-15 supported heteropoly acids in ionic liquids. Chemistryselect 2022;7(25):1−7.
[122] Ni L, Xin J, Jiang K, et al. One-step conversion of biomass-derived furanics into aromatics by Bronsted acid ionic liquids at room temperature. ACS Sustainable Chemistry & Engineering 2018;6:2541−51.
[123] Ni L, Xin J, Dong H, et al. A simple and mild approach for the synthesis of p-xylene from bio-based 2,5-dimethyfuran by using metal triflates. ChemSusChem 2017;10(11):2394−401.
[124] Zhu R, Liu X, Li L, et al. Valorization of industrial xylan-rich hemicelluloses into water-soluble derivatives by in-situ acetylation in [Emim]Ac ionic liquid. International Journal of Biological Macromolecules 2020;163:457−63.
[125] Xu G, Tu Z, Hu X, et al. Protic Brønsted acidic ionic liquids with variable acidity for efficient conversion of xylose and hemicellulose to furfural. Fuel 2023;339:127334.
[126] Hui W, Zhou Y, Dong Y, et al. Efficient hydrolysis of hemicellulose to furfural by novel superacid SO₄H-functionalized ionic liquids. Green Energy & Environment 2019;4(1):49−55.
[127] Matsagar BM, Dhepe PL. Brönsted acidic ionic liquid-catalyzed conversion of hemicellulose into sugars. Catalysis Science & Technology 2015;5(1):531−9.
[128] Zhang J, Lei P, Yu D, et al. Oxidative cleavage of beta-O-4 linkage in lignin via Co nanoparticles embedded in 3DNG as catalyst. Chemistry 2023;29(12):e202203144.
[129] Chauhan S, Kumar A, Jain N. Biomimetic oxidation of veratryl alcohol with H₂O₂ catalyzed by iron(III) porphyrins and horseradish peroxidase in ionic liquid. Synlett 2007;3:411−14.
[130] Jiang N, Ragauskas AJ. Selective aerobic oxidation of activated alcohols into acids or aldehydes in ionic liquids. The Journal of Organic Chemistry 2007;72:7030−3.

[131] Zhang J, Zhu X, Xu X, et al. Cooperative catalytic effects between aqueous acidic ionic liquid solutions and polyoxometalate-ionic liquid in the oxidative depolymerization of alkali lignin. Journal of Environmental Chemical Engineering 2022;10(5):1—11.

[132] Yang Y, Fan H, Song J, et al. Free radical reaction promoted by ionic liquid: a route for metal-free oxidation depolymerization of lignin model compound and lignin. Chemical Communications 2015;51(19):4028—31.

[133] Kang Y, Yang Y, Yao X, et al. Weak bonds joint effects catalyze the cleavage of strong C-C bond of lignin-inspired compounds and lignin in air by ionic liquids. ChemSusChem 2020;13 (22):5945—53.

[134] Chen L, Xin J, Ni L, et al. Conversion of lignin model compounds under mild conditions in pseudo-homogeneous systems. Green Chemistry 2016;18(8):2341—52.

[135] Yang S, Lu X, Yao H, et al. Efficient hydrodeoxygenation of lignin-derived phenols and dimeric ethers with synergistic [Bmim][PF_6]-Ru/SBA-15 catalysis under acid free conditions. Green Chemistry 2019;21(3):597—605.

[136] Yang S, Cai G, Lu X, et al. Selective deoxygenation of lignin-derived phenols and dimeric ethers with protic ionic liquids. Industrial & Engineering Chemistry Research 2020;59(11):4864—71.

[137] Colmenares JC, Varma RS, Nair V. Selective photocatalysis of lignin-inspired chemicals by integrating hybrid nanocatalysis in microfluidic reactors. Chemical Society Reviews 2017;46 (22):6675—86.

[138] Kang Y, Lu X, Zhang G, et al. Metal-free photochemical degradation of lignin-derived aryl ethers and lignin by autologous radicals through ionic liquid induction. ChemSusChem 2019;12:4005—13.

CHAPTER 4

Ionic liquids intensify separation process

Contents

4.1 Overview   97
4.2 Ionic liquids intensify gas separation   98
    4.2.1 NH$_3$ separation with ionic liquids   98
    4.2.2 CO$_2$ capture with ionic liquids   105
4.3 Application of ionic liquids in liquid—liquid extraction   108
    4.3.1 Liquid—liquid extraction with hydrophobic ionic liquids   109
    4.3.2 Ionic liquid—molecular solvent complex liquid—liquid extraction   110
    4.3.3 Ionic liquid-based aqueous biphasic system extraction   111
    4.3.4 Ionic liquid extraction separation with a similar structure compound   112
4.4 Ionic liquids for protein and protein complex extraction   117
    4.4.1 Ionic liquids for protein extraction   117
    4.4.2 Ionic liquids for protein complex extraction   126
4.5 Membrane separation process with ionic liquids   128
    4.5.1 Introduction   128
    4.5.2 Ionic liquid membranes and preparation strategy   129
    4.5.3 Application in liquid separation   131
    4.5.4 Transport mechanism   135
References   135

4.1 Overview

Due to the benign stability and versatility of ionic liquids (ILs), the publication concerning the separation process intensified by ILs followed exponential growth. ILs exhibit the potential of industrial application in many separation processes, such as the gas separation process with ILs as absorbents, the liquid—liquid extraction separation process with ILs as extraction agents or solvents, and the IL-based membranes (ILMs) separation process with ILs as one of membrane materials.

This chapter aims to provide a comprehensive understanding of the current state of research. With the gas separation process intensified by ILs as absorbents, the effect of different kinds of functional ILs on the absorption performances of NH$_3$ or CO$_2$ was introduced and summarized systematically. Especially, a strategy of selective separation and efficient recovery of NH$_3$ from melamine tail gas containing about 70% NH_3 and 29% CO_2 using the nonaqueous IL system was elaborated. With the liquid—liquid extraction separation process with ILs as mediums, the conventional organic compounds or similar structures compounds could be separated by the different extraction systems, that is, the hydrophobic ILs, the hybrid solvents of ILs complexed with molecular solvents, or the IL-based biphasic aqueous. With the ILM separation process for gas or liquid, the topic of separation of gases, such as carbon capture, utilization and storage and volatile organic compounds (VOCs), which make up the majority, was also discussed. In addition, the applications of ILMs on liquid separation including the recovery or removal of metal ions and organic substances from aqueous solutions and seawater desalination are of great significance in view of resource recycling and environmental remediation. This section intends to provide a preliminary database for ILMs and their performance in liquid separation. It is expected to guide the exploration and design of more ILMs with excellent separation performance in the respective fields by providing insights into the existing structures and functionality.

4.2 Ionic liquids intensify gas separation

4.2.1 NH_3 separation with ionic liquids

4.2.1.1 Hydroxyl ionic liquids

Palomar et al. [1] carried out a systematic study to select the optimized ILs for NH_3 absorption. The gas—liquid equilibrium data and Henry's law coefficient of NH_3 over 272 ILs were done by the COSMO-RS method, and thermal stability, liquid-phase window, and NH_3 solubility were simultaneously considered to evaluate the suitability of ILs for NH_3 absorption by experimental studies. Finally, two hydroxyl ILs including 1−2(-hydroxyethyl)-3-methylimadazolium tetrafluoroborate ([EtOHmim][BF$_4$]) and choline bis(trifluoromethylsulfonyl) imide ([Choline][NTf$_2$]) were chosen as candidates for NH_3 absorption with 0.135 and 0.082 g NH_3/g IL at 100 kPa and 293 K, respectively. On the contrast, NH_3 solubility in 1-butyl-3-methylimidazolium tetrafluoroborate ([Bmim][BF$_4$]) under the same conditions was only 0.024 g NH_3/g IL. These results indicated that the introduction of hydroxyl groups can obviously improve NH_3 absorption performances owing to the interaction between hydroxyl groups and NH_3 molecules. Ruiz et al. [2] found that two hydroxyl ILs [EtOHmim][BF$_4$] and [Choline][NTf$_2$] give a high coefficient of performance and have low solution circulation ratios and simultaneously exhibit better NH_3 absorption capacities compared with the conventional ILs.

Bedia et al. [3] further selected four hydroxyl ILs including [EtOHmim][BF$_4$], [Choline][NTf$_2$], tris(2-hydroxyethyl) methyl-ammonium methylsulfate ([MTEOA][MeOSO$_3$]), and 1-(2-hydroxyethyl)-3-methylimidazolium dicyanamide ([EtOHmim]

[DCA]) as potential solvents with high NH_3 capacity and established an efficiency criterion for screening absolvents by considering comprehensively the thermodynamics and kinetics. It was found that the four hydroxyl ILs show lower Henry's constant values of NH_3 than the conventional IL [Bmim][BF$_4$]. Therefore, the hydroxyl-functionalized cations significantly increase NH_3 solubility of ILs due to the strong hydrogen-bond donor ability. Among them, [MTEOA][MeOSO$_3$] showed the highest NH_3 solubility of 0.219 g NH_3/g IL at 293 K and 100 kPa due to three hydroxyl groups tethering with the cation, but the increased numbers of hydroxyl groups also result in high viscosity and apparent diffusion coefficients. Meanwhile, the anion [MeOSO$_3$]$^-$ is easily degraded into sulfate or bisulfate under heating conditions and low thermal stability [4]. Hence, [Choline][NTf$_2$] was considered as the most suitable solvent for NH_3 absorption.

Li et al. [5] synthesized a series of hydroxyl-functionalized imidazolium ILs with different anions ([EtOHmim]X, X = [NTf$_2$], [PF$_6$], [BF$_4$], [DCA], [SCN] and [NO$_3$]) and studied the effects of hydroxyl cations, anionic structures, pressures, and temperatures on NH_3 solubility. Compared with the conventional ILs [Emim]X, [EtOHmim]X with the same anions showed higher NH_3 solubility due to the presence of the hydroxyl group and the highest NH_3 solubility; the mole fraction of 0.56 at 298 K and 159 kPa was achieved in 1−2(-hydroxyethyl)-3-methylimadazolium bis (trifluoromethylsulfonyl)imide([EtOHmim][NTf$_2$]), along with great recyclability. Meanwhile, the NH_3 absorption mechanism with the hydroxyl ILs was investigated by combining spectral analysis and quantum chemistry calculations. Comparing ^{1}H NMR spectra of fresh [EtOHmim][NTf$_2$], the peak at 5.16 ppm ascribed to the H atom on the hydroxyl groups disappeared for [EtOHmim][NTf$_2$] after absorbing NH_3 and reappeared after desorption, which may be the strong interaction between the electronegative N atom of NH_3 and the H atom of the hydroxyl group. Quantum chemistry calculations results further found that the hydrogen bonding of -56.32 kJ/mol is formed between the N atom of NH_3 and the H atom of hydroxyl group of [EtOHmim]$^+$, while the interaction energy between [Emim]$^+$ and NH_3 is only -20.98 kJ/mol, confirming the dominant role of hydroxyl groups in NH_3 absorption.

4.2.1.2 Protic ionic liquids

The introduction of protic hydrogen into ILs to form protic ILs (PILs) can enhance the interaction between NH_3 and ILs for improved NH_3 absorption performances. A new strategy of incorporating acidic protic H as a strong hydrogen-bond donor into the cation was developed. Shang et al. [6] designed and synthesized the PIL 1-butyl imidazolium bis (trifluoromethylsulfonyl)imide ([Bim][NTf$_2$]), which was compared with two other ILs, 1-butyl-3-methyl imidazolium bis(trifluoromethylsulfonyl) imide ([Bmim][NTf$_2$]) and 1-n-butyrate-3-methyl imidazolium bis(trifluoromethylsulfonyl) imide ([HOOC(CH$_2$)$_3$mim] [NTf$_2$]). The results showed that the hydrogen bond-donating abilities are related to their thermodynamic dissociation constants (pK_a). [Bim][NTf$_2$] with a moderate pK_a value had the highest NH$_3$ absorption capacity up to 0.113 g NH$_3$/g IL at 313 K and 100 kPa, which was attributed to the interactions between H-3 on the imidazole ring and the NH$_3$ molecule and NH$_3$ molecules sintering themselves in [Bim][NTf$_2$] through experimental characterization and theoretical calculations. Considering the effect of anions with the same cation, Shang et al. [7] synthesized 10 kinds of protic and conventional ILs and measured their densities, viscosities, thermal decomposition temperatures, and NH$_3$ solubility. The order of NH$_3$ solubility in protic ILs was [Bim][NTf$_2$] > [Bim][SCN] > [Bim][NO$_3$], which is consistent with the acidity order of the acid corresponding to the anions. The viscosity of PILs decreased during the NH$_3$ absorption, which was caused by the interaction between NH$_3$ and PILs. Zhao et al. [8] elucidated the roles of the ion structure. The IL−gas interaction and interface behavior in absorption of gases (NH$_3$, N$_2$, and H$_2$) by [Bim][NTf$_2$] and [Bmim][NTf$_2$] were studied through molecular dynamics (MD) simulations. The results showed that NH$_3$ interacts with the N3-H site of the [Bim]$^+$ cation forming a strong N3-H$\cdots$N(NH$_3$) hydrogen bond with the energy of − 79.0 kJ/mol, which is twice as much as the hydrogen bonding between C2-H of [Bmim]$^+$ and NH$_3$ (− 33.2 kJ/mol). The enrichment of cations at the PILs−gas interface penetrated NH$_3$ deeply into the bulk of PILs due to the strong interaction between NH$_3$ and the PILs, thus achieving the selective absorption of NH$_3$ from gases containing N$_2$ and H$_2$.

In order to further improve the NH$_3$ absorption performance, the functionalized sites were introduced into the PILs. Yuan et al. [9,10] designed and synthesized a novel type of dual-functionalized PILs with both acidity proton and hydroxyl group, including pyridinium-based PILs and imidazolium-based PILs. The NH$_3$ absorption isotherms of these dual-functionalized PILs were computed using the gas−liquid equilibrium method at 303.15−343.15 K and 600 kPa. The results indicated that the PILs exhibit high NH$_3$ solubility by simultaneously introducing hydroxyl sites into PILs, and the NH$_3$ solubility of 1−2(-hydroxyethyl) bis (trifluoromethylsulfonyl)imide [EtOHim][NTf$_2$] was up to 0.221 g NH$_3$/g IL, which was over 30-fold greater than that of [Emim][NTf$_2$] and fourfold greater than that of the functionalized IL [EtOHmim][NTf$_2$]. The mechanism of high NH$_3$ absorption performance in dual-functionalized PILs was ascribed to the synergistical interaction between weak acidity protons, hydroxyl groups, and NH$_3$ molecules. Considering the advantage of NH$_3$ absorption by hydrogen bonding and the flexible designability of ILs, Luo et al. [11] designed a number of cation-functionalized PILs that mainly consisted of cooperative hydrogen bonding to form reversible construction of ionic flexible networks and achieved sigmoidal NH$_3$ absorption isotherm. The heat released benefited the endothermic breakage of hydrogen bonding during NH$_3$ absorption. The PILs 2-aminopyridine bis(trifluoromethylsulfonyl)imide ([2PyH][NTf$_2$]) showed NH$_3$

absorption with a threshold pressure of 4 kPa and exhibited high NH_3 capacity of 0.173 g NH_3/g IL at 303.15 K and 100 kPa. [BzAm][NTf_2] with absorbed NH_3 could be striped rapidly at 323 K and 1 kPa within 30 min during the regeneration. Similarly, Deng et al. [12] synthesized six protic ethanolamine-based ILs for efficient and reversible NH_3 uptake by constructing multiple binding sites. Among them, the ethanolamine thiocyanate ([EtA][SCN]) illustrated the highest capacity of 0.359 g NH_3/g IL at atmospheric pressure and 293.15 K, which was originated from multiple hydrogen bonding between acidic proton, hydroxyl group, and thiocyanate with NH_3. The absorption enthalpy of NH_3 in [EtA][SCN] was -35.5 kJ/mol, meaning low energy consumption for NH_3 desorption.

The number of acidic protic in PILs plays an important role in NH_3 absorption because of strong hydrogen bonding. Shang et al. [13] designed a series of PILs, such as 2-methylimidazolium bis(trifluoromethylsulfonyl)imide ([2-mim][NTf_2] and imidazolium bis(trifluoromethylsulfonyl) imide([Im][NTf_2]), with multiple active sites by introducing two acidic protons on cations. The NH_3 capacity of [Im][NTf_2] was 3.46 mol NH_3/mol IL, which was ascribed to the hydrogen bonding between three NH_3 molecules and two acidic protons of PILs. Li et al. [14] chose the ionic salt 1,3,5-tri(imidazolium-1-yl) ([Ph_3ImH]) paired with bis (trifluoromethylsulfonyl)imide to synthesize [Ph_3ImH][NTf_2]$_2$ by constructing an ionic framework for NH_3 uptake. The results indicated that [Ph_3ImH] [NTf_2]$_2$ can reversibly capture 15.65 mmol/g IL at 298.15K and 100 kPa and desorbed completely at 353.15 K under vacuum, suggesting excellent recycle performance. The NH_3 molecules would be captured by the frustration of [Ph_3ImH][NTf_2]$_2$ through hydrogen bonding and physical interactions.

To maintain high NH_3 mass capacity, NH_3 selectivity, and great reversibility, Zeng et al. [15] designed and synthesized the triazole cation-functionalized ILs by introducing multiple protic H sites into N-heterocyclic cations for enhancing NH_3 separation. Considering triazoles with low molecular weight and one protic hydrogen themselves, triazoles reacted with acids to form novel PILs with three protic hydrogen sites on the triazole ring, which could absorb NH_3 by multiple hydrogen bonding. The results showed that 1, 2, 3-triazolium nitrate [1, 2, 3-TrizH_2][NO_3]$_2$ exhibits the high NH_3 mass capacity of 0.365 g NH_3/g IL at 303.15 K and 101 kPa, which is comparable to water, along with the outstanding NH_3/CO_2 selectivity of up to 182. The cations in PILs with protic hydrogens play an important role in high NH_3 mass capacity due to the coupled hydrogen bonding [16−20].

4.2.1.3 Metal ionic liquids

Besides NH_3 absorption by hydrogen bonding between cations and NH_3, another strategy to greatly improve NH_3 absorption performance is the introduction of metal centers into ILs through chemical complexation with NH_3. In general, metal chlorides ($CaCl_2$, $SrCl_2$, LiCl,and $ZnCl_2$) have excellent NH_3 absorption capacity, even higher than water, because NH_3 contains an unshared electron pair that can be complexed with metal ions as ligands. However, the mass transfer of metal chloride-NH_3 systems is poor [21]. Therefore, the introduction of metal centers to form metal ILs (MILs) in the liquid state may be an effective method to improve NH_3 capacity and avoid poor mass transfer.

Chen et al. [22] synthesized the MIL-containing zinc complex anions 1-ethyl-3-methylimidazolium zinc dichloride ([Bmim][Zn_2Cl_5]) and found that NH_3 solubility in [Bmim] [Zn_2Cl_5] at 323 K and 100 kPa is up to 0.305 g NH_3/g IL, which is much higher than those of hydroxyl ILs and PILs ever reported. Kohler et al. [23] loaded the MILs containing copper complex anions such as 1-methyl-3-octylimidazolium chloride ([C_8C_1Im]Cl)/ copper chloride($CuCl_2$) on activated carbon to form supported IL phase materials for adsorption of low concentration NH_3, which showed NH_3 adsorption capacity of 0.027 g NH_3/g IL in a dry gas stream at 303 K and 0.12 kPa. However, only 56% NH_3 could be desorbed at 353 K and N_2 purge, indicating that copper ions can interact NH_3 to form a stable $[Cu(NH_3)_4]^{2+}$ complex and result in poor recyclability.

Is it possible to select appropriate metal ions and ligands to adjust the complexation between metal ions and NH_3 to achieve efficient and reversible absorption? Zeng et al. [24,25] designed and synthesized a series of novel MILs with different metal centers including Zn, Ni, Mn, Cu, Sn, and Co for NH_3 absorption. The results indicated that the cobalt ILs [C_nmim]$_2$[$Co(NCS)_4$] ($n = 2$, 4, and 6) not only have better thermal stability (higher than 553.15 K), but also very high NH_3 absorption capacity of 0.198 g NH_3/g IL and NH_3/CO_2 selectivity, which are more than 10 times and 8 times than those of conventional ILs [C_nmim][SCN], respectively. After five times of absorption and desorption, the cobalt ILs showed stable performance of NH_3 absorption and could be completely recycled and reused. Experimental characterizations and quantum chemical calculations demonstrated that the superior NH_3 capacity and desorption performance is ascribed to the moderate Lewis acid-base interaction and cooperative hydrogen bonding between the metal center ligands and NH_3. Further, novel dual-functionalized MILs were designed by simultaneously introducing acidic protons and the Li^+ ion into ILs. Compared with the PIL [2-mim][NTf_2] with 3.04 mol NH_3/mol IL at 313 K and atmospheric pressure, the dual-functionalized MIL [2-mim][Li ($NTf_2)_2$] exhibited the highest NH_3 capacity with 7.01 mol NH_3/mol IL, which is the couple contribution of hydrogen bonding between the protons and NH_3 and the formation of coordination complexes between the Li^+-based anion and NH_3. Moreover, the dual-functionalized MILs also exhibited excellent recyclability, indicating great potentials for NH_3 absorption applications from NH_3-containing industrial gases [13]. Based on the above ideas, a series of lithium (Li)-triethylene glycol (TEG)-chelated ILs with crown –ether-like cation and different anions were designed for multiple-site reversible chemical absorption of NH_3 by combining hydroxyl sites and Li^+ ion with the smallest radius and lightest weight. It was found that Li-TEG-chelated ILs exhibit NH_3 solubilities up to 3.36 mol NH_3/mol IL at 313 K and 102.5 kPa and great reversibility [26].

4.2.1.4 Ionic liquid-based hybrid solvents

Although functionalized ILs showed great NH_3 absorption performance, the higher viscosity and complex synthesis steps of several functionalized ILs than common absorbents hinder further applications. Therefore, in order to overcome such problems and enhance NH_3 absorption performance, ILs have been applied through combining with other molecular solvents to form IL hybrid solvents, especially IL-based deep eutectic solvents (DESs).

Huang et al. [27] prepared a series of 1-ethyl-3-methylimidazolium ([Emim]Cl)-based DESs with different weak acids with the pK_a values in DMSO ranging from 8.2 to 18.6. [Emim]Cl was the HBA, and different weak acids were the HBDs, including imidazole (Im), 1,2,4-triazole (Triz), 1H-tetrazole (Tetz), phenol (Ph), benzimidazole (BenIm), 1H-benzotriazole (BenTriz), and succinimide (Si). The performance of NH_3 absorption and desorption in DESs was examined and correlated with the acidities of HBDs. The result showed that the DESs with stronger acidities exhibit higher NH_3 solubilities but is difficult to strip out due to the strong acidity of Tetz. However, the NH_3 capacities of [Emim]Cl/BenTriz and [Emim]Cl/Triz were totally reversible after 6 times cycles. Deng et al. [28] also designed a kind of DESs with viscosity less than 20 mPa·s by making NH_4SCN paired with Im. NH_3 absorption capacity of NH_4SCN/Im (1:2) DES was 9.65 mol NH_3/kg absorbent due to the strong hydrogen bonding between protic ammonium cation and NH_3.

Deng et al. [29] used ChCl as HBA and selected three dihydric alcohols as HBD to form low viscous DESs. The dihydric alcohol-based DESs absorbed 3.0 mol NH_3/kg DES, and the highest NH_3/CO_2 selectivity was 98 at 303.15 K and 100 kPa. The dissolution Gibbs free energy, enthalpy, and entropy changes of NH_3 solvation were further calculated according to the temperature dependence of Henry's constants. Because 2,3-BD with hydroxyl groups at middle positions exhibits a better steric effect than 1,4-BD with terminal hydroxyl groups, the enthalpy changes ChCl-1,4-BD (1:4) and ChCl-1,3-PD (1:4) were -27.98 and -26.07 kJ/mol, which were more negative than that in ChCl-2,3-BD (1:4) (-23.91 kJ/mol) for NH_3 dissolution. Ren et al. [30] reported the design of novel hybrid DESs with a flexible hydrogen-bonded supramolecular network; the solubility of choline chloride/resorcinol/glycerol (ChCl/Res/Gly) with molar ratio of 1:3:5 is 0.130 g NH_3/g DESs at 313 K and 101 kPa. MD simulations and spectroscopic analysis elucidated that the abundant hydrogen-bonding sites interacted with NH_3 to create a flexible hydrogen-bonded supramolecular network to enable a strong reversible solvation.

Besides, Cao et al. [31] synthesized three PIL absorbents that combined PILs with EG, involving imidazolium nitrate([Im][NO_3]), 1-methylimidazolium nitrate ([mim][NO_3]) and 1,2-dimethylimidazolium nitrate ([Mmim][NO_3]). The viscosity of PIL-based DESs ranged from 14.43 to 4.10 mPa·s at the temperature from 30°C to 70°C. Among the three PIL absorbents, [Im][NO_3]/EG DES with the molar ratio of 1:3 exhibited a faster absorption rate than pure PILs due to the inclusion of EG. The highest mass capacity of [Im][NO_3]/EG with a molar ratio of 1:3 was 0.211 g

NH_3/g DES at 30°C and 100 kPa and was higher than all the reported ILs and IL-based DESs. That was originated from multiple hydrogen bonding between acidic H and hydroxyl groups of the DESs and NH_3 through NMR and Fourier transform infrared (FT-IR) spectra. Most of the DESs currently reported for NH_3 separation had the weak-acidic group but with high viscosities. Jiang et al. [32] designed a new class of DESs with low viscosities by pairing N,N,N',N'-tetramethyl-1,3-propanediamine dihydrochloride ([TMPDA]Cl$_2$) with phenol (PhOH) at the molar ratios of 1:3-7. The viscosity of [TMPDA]Cl$_2$/PhOH DESs was 48.1 mPa·s at 298.2 K, and NH_3 solubility in [TMPDA]Cl$_2$/PhOH DESs was 4.49 mol NH_3/kg DESs at 298.2 K and 13.3 kPa. The efficient and reversible absorption of NH_3 in DESs was attributed to the protonated amine groups of [TMPDA]Cl$_2$ and the phenolic hydroxyl group of PhOH to form a strong interaction with NH_3.

Due to hydrochloride (EaCl) with the protic ionic nature and glycerol (Gly) with multiple hydroxyl groups, Huang et al. [33] designed DESs comprising ethylamine EaCl and Gly for NH_3 absorption that had the strong hydrogen bonding with NH_3. The highest capacity of NH_3 in EaCl/Gly with a molar ratio of 1:2 was 9.631 mol/kg at 298.2 K and 106.7 kPa. The enthalpy changes for NH_3 absorption in EaCl/Gly mixtures were -25.9- -28.0 kJ/mol. In addition, Cheng et al. [34] designed and prepared a series of metal-based DESs by mixing EaCl, metal chloride (SnCl$_2$, ZnCl$_2$, FeCl$_3$, or CoCl$_2$), and Gly. It was found that NH_3 absorption in these DESs is all quite fast and reaches equilibrium in less than 200 s. The maximum value of NH_3 solubilities in metal-based DESs was 17.55 mol/kg at 298.2 K and 103 kPa, and the minimum was 10.24 mol/kg at 298.2 K and 6.8 kPa. Quantum chemistry calculations verified that NH_3 absorption in metal-based DESs is the combination of multiple coordination and hydrogen bonding.

4.2.1.5 Ionic liquid-based NH$_3$ separation technology and applications

Because of extremely low volatility and good affinity with NH_3 of ILs, a new strategy of selective separation and efficient recovery of NH_3 from melamine tail gas containing about 70% NH_3 and 29% CO_2 using the nonaqueous IL system was proposed by Ionic Liquid Research Team from Institute of Process Engineering, Chinese Academy of Sciences. Comprehensively considering multiple factors such as NH_3 capacity, NH_3/CO_2 selectivity, stability, viscosity, and cost of ILs, the targeted IL with multiple hydrogen bonding sites was designed and produced in a large scale. Further, the new process of adjusting NH_3/CO_2 ratios of feed gas and two-stage absorption−desorption was developed, and the high purity of the NH_3 product was obtained, along with low energy consumption and no wastewater. The industrial test pilot plant with a capacity of 50 m^3/h of IL-based NH_3 recovery from melamine tail gas was built, with 3500 hours stable running. This technology passed the appraisal of scientific and technological achievements organized by China Petroleum and Chemical Industry Federation and is regarded as a great breakthrough with the international leading level.

This technology is further extended to other applications. The Ionic Liquids Research Team cooperated with other enterprises to build an industrial demonstration unit in Shaanxi with a capacity of 130 million m^3 of NH_3-containing tail gas ($10000-40000$ mg NH_3/m^3) per year in 2018. It is a two-stage absorption desorption process with IL loading capacity of 45 tons. The plant has been running steadily over 55 months, and the recycled NH_3 can be returned to the ammonium molybdate production line to realize reutilization. The performance of IL absorbent remains consistent, which indicates the stability of the IL absorbent. Compared with the traditional water washing process, the IL-based NH_3 recovery process not only meets the exhaust gas emission standards and reduces environmental pollution, but also obtains the NH_3 product to achieve its reutilization. This technology can be extended to ammonia, metallurgy, organic synthesis, and other industries, leading to broad application prospects.

4.2.2 CO₂ capture with ionic liquids

4.2.2.1 Conventional ionic liquids

Since CO_2 solubility in 1-butyl-3-methylimidazolium hexafluorophosphate ([Bmim][PF$_6$]) at 25°C and pressure up to 40 MPa was reported [35], the extensive studies on CO_2 absorption with conventional ILs with physisorption and functionalized ILs with chemisorption have been carried out in succession [36−39]. For physisorption ILs, compared with cations of ILs, the anions are considered to play a primary role in CO_2 absorption. The presence of long alkyl chains, the fluorination of cations, and ester groups on cations is favorable for improving marginally CO_2 solubility [40−47]. However, some functional groups, such as the hydrogen of the C_2 position on the imidazolium ring substitution with methyl groups, ether groups, and nitrile groups on cations, have a slightly negative impact on CO_2 solubility [48−51]. Even with a little improvement on CO_2 capture, the physisorption ILs still cannot compete with current commercially available solvents due to the lower CO_2 capacity, especially for flue gas with low CO_2 concentration. For example, CO_2 solubility is lower than 0.05 mol CO_2/mol IL when the partial pressure of CO_2 is about 0.15 bar [52]. Therefore, functionalized ILs with chemisorption for highly efficient capture of CO_2 were designed and developed [53−57].

4.2.2.2 Amino ionic liquids

Bates et al. first introduced a primary amine into the imidazolium cation and synthesized the amino-functionalized IL 1-butyl-2-bromopropylamine imidazolium tetrafluoroborate ([NH$_2$p-bim][BF$_4$]). The IL exhibits high capacity of nearly 0.50 mol CO_2/mol IL under ambient pressure [58]. Compared with ILs with amino-functionalized cations, more favorable stoichiometry than one CO_2 to two amines is achieved by tethering the amine to anions. Gurkan et al. reported two phosphonium-based amino acid ILs, trihexyl(tetradecyl) phosphonium prolinate ([P$_{66614}$][Pro]) and trihexyl(tetradecyl)phosphonium methioninate ([P$_{66614}$][Met]), and CO_2 absorption capacity of both ILs reached 0.90 mol CO_2/mol IL, which nearly approached the 1:1 stoichiometry. Zhang et al. [59] also synthesized a dual amino-functionalized cation-tethered IL, 1, 3-di(2-aminoethyl)- 2-methylimidazolium bromide (DAIL). CO_2 absorption capacity of the aqueous solution of DAIL (10%) could be up to 1.05 mol CO_2/mol IL at 30°C and 0.10 MPa, and its CO_2 gravimetric capacity can reach 18.5%(mass fraction) [60]. In addition, a series of dual amino-functionalized ILs were developed by tethering amino groups on both cations and anions, respectively. Lv et al. [61] and Zhou et al. [62] developed two highly efficient absorbents of amine-based amino acid-functionalized ILs 1-aminopropyl-3-methylimidazolium glycinate ([APmim] [Gly]) and 1-aminopropyl-3-methylimidazolium lysine ([APmim][Lys]), respectively. The CO_2 capacities of [APmim][Gly] and [APmim][Lys] aqueous solutions were 1.23 and 1.80 mol CO_2/mol IL, respectively, which were much higher than those of the most existing dual-functionalized ILs due to their low molecular weight.

4.2.2.3 Non-amino ionic liquids

Besides the amino-based ILs, a series of non-amino ILs with chemisorption were also designed and developed for CO_2 capture with both high capacity and low enthalpy [63−66]. A kind of non-amino phosphonium ILs with different azole anions including [P_{66614}][Im], trihexyl (tetradecyl)phosphonium pyrazole ([P_{66614}][Pyr]), trihexyl(tetradecyl)phosphonium trizole ([P_{66614}][Triz]), trihexyl(tetradecyl)phosphonium tetrazole ([P_{66614}][Tetz]), trihexyl (tetradecyl) phosphonium oxazolidinone ([P_{66614}][Oxa]), and trihexyl(tetradecyl)phosphonium phenol ([P_{66614}][PhO]), not only reach equimolar absorption of CO_2 due to the tunable anion basicity, but also only takes about 10 min to complete the absorption, substantially faster than amine-functionalized ILs. The rapid absorption rate is mainly due to a little change of the viscosities of the non-amino ILs during CO_2 absorption because of the absence of hydrogen-bond networks. Wang et al. [64] further designed a series of phenolic ILs to achieve highly efficient and energy-saving absorption of CO_2 through tuning the substituent of the IL anions. CO_2 absorption capacity of trihexyl(tetradecyl)phosphonium p-methylphenolate ([P_{66614}] [4-Me-PhO]), trihexyl(tetradecyl)phosphonium p-methoxyphenolate ([P_{66614}][4-MeO-PhO]), trihexyl(tetradecyl)phosphonium phenolate ([P_{66614}][4-H-PhO]), trihexyl (tetradecyl) phosphonium p-chlorophenolate ([P_{66614}][4-Cl-PhO]), trihexyl (tetradecyl) phosphonium p-trifluorophenolate ([P_{66614}][4-CF$_3$-PhO]), and trihexyl(tetradecyl) phosphonium p-nitrophenolate ([P_{66614}][4-NO$_2$-PhO]) is 0.91, 0.92, 0.85, 0.82, 0.61, and 0.30 mol CO_2/mol IL, respectively. The results indicated that the stronger the electron-withdrawing group (Cl, CF$_3$,and NO$_2$), the lower the CO_2 absorption capacity. The changes in the enthalpy of CO_2 absorption agreed well with the results of CO_2 capacity. The reason is related to a quantitative relationship between CO_2 capacity, absorption enthalpy, and the basicity of phenolic ILs. CO_2 capacity decreased with the decreasing pK_a value of the anion, and the absorption enthalpy of the phenolic ILs increased from -17.1 to -49.2 kJ/mol with the increasing of Mulliken charge on the oxygen atom, which offers a promising method for CO_2 capture with both high capacity and excellent reversibility. Further, Wang et al. [65] synthesized pyridine-containing ILs by introducing a nitrogen-based interacting site on the phenolate anion, and an extremely high capacity of up to 1.60 mol CO_2/mol IL was achieved for trihexyl (tetradecyl) phosphonium 3-methoxy-2-hydroxypyridine ($[P_{66614}][3\text{-}OMe_3\text{-}2\text{-}Op]$) through two site interactions between electronegative oxygen and nitrogen atoms in the anion and CO_2. The use of cooperative interactions in gas separation also provides a gen-eral strategy for enhancing the capacity of other acid gases.

4.2.2.4 Ionic liquid hybrid solvents

Although highly efficient functionalized ILs have been developed for CO_2 capture, their higher viscosities than those of molecular solvents seriously limit their industrial applications. Especially for amino-based ILs, their viscosities are relatively high and increase after CO_2 absorption due to the chemical reaction. For example, the viscosity of 3-aminopropyl tributylphosphonium amino acid ($[aP_{4443}][AA]$) increases nearly threefold after absorption of CO_2 [56], and the viscosity of trihexyl(tetradecyl)-phosphonium isoleucinate ($[P_{66614}][Iso]$) increases over 200-fold when exposed to CO_2 [66]. Mixing ILs with water or organic solvents to form novel IL-based solvents for CO_2 separation is considered as a useful way to solve the inherently high viscosities of ILs [67,68]. For instance, Lv et al. [69] studied the CO_2 absorption performances in 1-hydroxyethyl-3-methylimidazolium glycinate ($[EOHmim][Gly]$)-H_2O systems. CO_2 capacity of the $[EOHmim][Gly]$-H_2O system (0.40 mol/L) is 0.575 mol CO_2/mol absorbent, which is higher than that of the MEA solution (0.457 mol CO_2/mol absorbent) under the same conditions. Besides, the viscosity of $[EOHmim][Gly]$ solution (0.40 mol/L) is found to be 1.01 mPa·s at 30°C, which is very close to that of the MEA solution (0.92 mPa·s). Wang et al. [70] developed the integrated sorption systems consisting of 1:1 mixtures of an alcohol-functionalized IL and a superbase instead of the DBU-alcohol system. The DBU-1-(2-hydroxyethyl)-3-methylimidazolium bis (trifluoromethylsulfonyl) imide ($[Im_{21}OH][NTf_2]$) system has the high CO_2 absorption capacity of 1.04 mol CO_2/mol absorbent at 20°C and 0.10 MPa. Subsequently, considering the weak acidity of C-2 proton of imidazolium-based ILs, Wang et al. [71] investigated the CO_2 absorption performances in the mixtures of $[Bmim][NTf_2]$-DBU and $[Bmim][NTf_2]$-MTBD (1, 3, 4, 6, 7, 8-hexahydro-1-methyl-2H-pyrimido[1,2-a] pyrimidine), respectively. The results suggested that CO_2 absorption capacity in two kinds of IL-based solvents is about 1 mol CO_2/mol absorbent at 23°C and 0.1 MPa, which is superior to conventional ILs. Furthermore, the captured CO_2 is easy to release and recycle with a slight loss of CO_2 capacity. Therefore, the combination of imidazolium-based ILs and superbases will be a good choice for the efficient and reversible capture of CO_2, but more switchable ILs composed of different ILs and solvents need to be explored.

4.3 Application of ionic liquids in liquid—liquid extraction

Liquid—liquid extraction is also known as solvent extraction or extraction (Fig. 4.1). It is the process of separating and extracting components of a liquid mixture with a solvent. In the liquid mixture, an immiscible (or slightly miscible) selected solvent is added, and components in the liquid mixture are separated based on the different distribution ratios between the extract and the raffinate phases.

Liquid—liquid extraction is a chemical separation technology with simple operation, continuous process, large processing capacity, and easy amplification. However, the molecular recognition ability of traditional extractants is weak, making it difficult to efficiently recognize the slight differences in structure and properties of structurally similar compounds [72]. It is usually only applicable to separating substances with significant differences in structural properties. Developing new extractants with more vital molecular recognition ability is a critical scientific problem that must be solved for applying liquid—liquid extraction technology to structurally similar compounds' separation process and realizing energy-saving and emission reduction in the separation process.

ILs provide a new platform for the development of liquid—liquid extraction technology due to their virtually nonvolatile, designable structure and properties, and easy formation of liquid—liquid biphase and have made extensive research progress in the separation of metal ions, organic phase molecules, and biological macromolecules, among other substances. Compared with traditional organic solvents, ionic liquids can be designed with the advantages of good physical and chemical stability, no odor, being nonflammable, and nonexplosive [73], and special hydrogen bonding and π-π interactions, which can enhance the ability of medium-specific recognition of structurally similar compounds and improve the efficiency of extraction and separation.

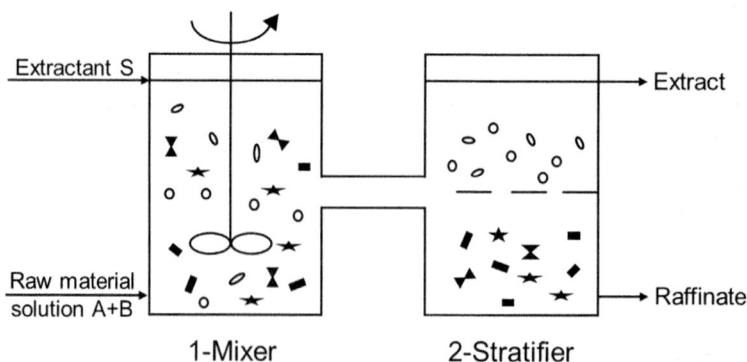

Figure 4.1 Principle of liquid—liquid extraction.

4.3.1 Liquid−liquid extraction with hydrophobic ionic liquids

The earliest report on liquid−liquid extraction with ionic liquids was the extraction of substituted benzene derivatives such as alkylbenzene, aminobenzene, carboxybenzene, and halogenated benzene from water using the hydrophobic IL 1-butyl-3-methylimidazole hexafluorophosphate as the medium. It was found that the less dissociable and more hydrophobic substituted benzene derivatives were more easily extracted into the IL phase [74]. Yu [75] investigated the effects of temperature, pH, inorganic salt type, and concentration on the extraction of ferulic acid (FA) and caffeic acid (CA) by using [Bmim][BF$_4$] and [Hmim][PF$_6$] as the extractants. It was found that the aqueous phase pH had a significant effect on the extraction rate. [Bmim][PF$_6$] and [Hmim][PF$_6$] had high extraction efficiencies for FA and CA. Both had much higher extraction efficiencies for FA than for CA. The extraction of FA and CA was higher when [Bmim][PF$_6$] was the extractant. After stripping with aqueous sodium hydroxide solution, the recoveries of FA in the aqueous phase of the stripping agent were 94%−98% (C$_4$) and 96%−101% (C$_6$), respectively, and the residues of IL in FA were detected to be 1.8% (mass fraction) (C$_4$) and 0.74% (mass fraction) (C$_6$), respectively. The results showed that the extraction efficiency of ionic liquids increased with the decrease of their alkyl groups, while increasing the length of alkyl groups reduced the residual amount of ionic liquids.

Fan et al. [76] utilized 1-methyl-3-alkylimidazolium hexafluorophosphate [C$_n$mim][PF$_6$] and 1-methyl-3-alkylimidazolium tetrafluoroborate [C$_n$mim][BF$_4$] for the extraction of endocrine-disrupting phenols from aqueous solutions and explored the extraction effect of ILs on phenol and the extraction driving force. The extraction effects of different ionic liquids on phenol under the same conditions are shown in Fig. 4.2. The results suggested that increasing the alkyl chain length on the ionic liquid cation increased the

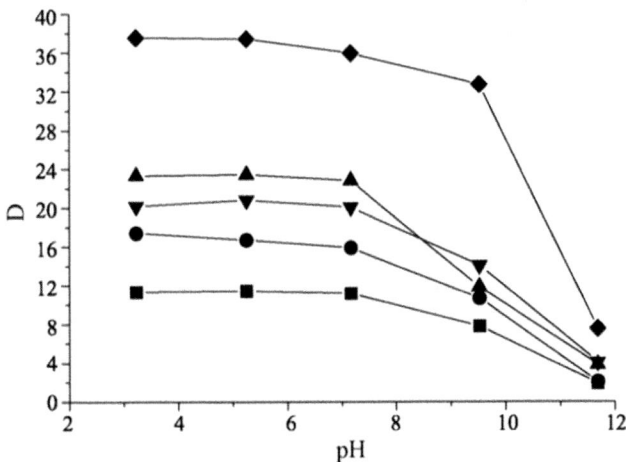

Figure 4.2 The pH dependence of distribution ratios of phenol between ionic liquids and water: (◆), [Omim][BF$_4$]; (▲), [Omim][PF$_6$]; (▼)[Hmim][BF$_4$]; (●), [Hmim][PF$_6$], and (■)[Bmim][PF$_6$] [5].

partition ratio of phenol in the ionic liquid; the extraction efficiency of the ILs containing the $[BF_4]^-$ anion was much higher than that of the ILs containing the $[PF_6]^-$ anion. This result may be due to the combination of hydrogen bonding between the IL anion and phenol hydroxyl H and hydrophobic interactions between the IL imidazole cation and the phenol, which made it easier to transfer phenol from the aqueous solution to the IL extraction phase and increased the extraction rate.

Kumar et al. [77] predicted the removal of cresols (o-cresol, m-cresol, p-cresol) from aqueous solutions by 360 ILs formed by combinations of 15 cations and 24 anions using quantum chemistry-based COSMO-RS theoretical modeling. They compared the data with those reported in the literature to check the accuracy of the predictions and to screen the best methanol extractant. The data showed that trihexyltetradecylphosphinium [THTDP] and 1-octyl-3-methylimidazolium $[Omim]^+$-based ILs gave the highest selectivity among the phosphorus and imidazolium-based cations [78]. The selectivity of different cresols followed the order of: m-cresol > p-cresol > o-cresol. Similarly, for cations based on the pyridine and quinoline, the 1-ethylpyridine key [EPY] and 1-octyl quinoline keys [OQU] gave the highest selectivity. Among the two cations studied against cations based on pyrrolidine, 1-hexyl-1-methylpyrrolidine [HMPL] was the best. Anions such as $[CH_3SO_3]^-$, $[CH_3COO]^-$, $[Cl]^-$, and $[Br]^-$ gave high selectivity due to the spatial shielding effect around their uncharged centers. Among the screened ionic liquids, [THTDP][SAL] and [EMPL][PF_6] gave the highest (662) and lowest (0.01) selectivity values, respectively. This study provided research ideas for treating organic-rich water and wastewater with ionic liquids.

4.3.2 Ionic liquid–molecular solvent complex liquid–liquid extraction

Fluorinated hydrophobic ILs such as $[PF_6]^-$, $[NTf_2]^-$, $[BF_4]^-$, and so forth are weakly hydrogen-bonded, resulting in a lower extraction capacity than conventional hydrophobic media. ILs complexed with molecular solvents to form hybrid solvents can significantly reduce viscosity of the ILs.

Yang et al. [79] designed a series of quaternary phosphonium bromide IL–molecule solvent complex extractants with hydrophobicity and strong hydrogen bond recognition ability to realize the efficient separation of the strongly hydrophilic structurally similar compounds ascorbic acid (AA) and ascorbyl glucoside (AA-2G). When the trihexyltetradecylphosphonium bromide ($[P_{66614}]Br$)-ethyl acetate (molar ratio 1:9) composite extractant was used as the medium, the partition coefficient of AA was as high as 1.36, which was 60–680 times higher than that of the conventional hydrophobic ionic liquids and ethyl acetate extraction. At the same time, the selectivity of AA to AA-2G was greater than 60. By five-stage countercurrent extraction, the purity of AA-2G could be increased from 50% to 96.2%, and the yield was higher than 98%.

Bai et al. [80] constructed a composite hydrophobic extractant consisting of a tridecylmethylammonium-carboxylate-type IL and 1-chlorooctane and realized the selective separation of methacrylic acid from acetic acid in aqueous solution, with a selectivity factor of up to 54.7 under the optimal conditions. The stronger hydrogen bonding between the IL and methacrylic acid was essential in obtaining the above selectivity.

4.3.3 Ionic liquid-based aqueous biphasic system extraction

The small variety of hydrophobic ILs, their poor specific interaction capabilities, and the high cost and poor stability of fluorinated anions severely limit their use in extractive separations. IL-based aqueous biphasic system (IL-based ABS) extraction uses hydrophilic ionic liquids, which are widely available and have adjustable properties. At the same time, inorganic salts, polymers, carbohydrates, amino acids, and other types of substances can be used as the ionic liquid biphasic system of the phase-forming substances, further improving the biphasic system of the physical and chemical properties of the adjustable ability. In addition, compared with conventional biphasic extraction with polymerized inorganic salts, biphasic extraction with ILs has a short phase separation time, easy recycling, and better phase interface and it is not easy to emulsify. Nowadays, IL biphasic extraction has been used to extract, separate, and purify amino acids, proteins, antibiotics, enzymes, and other single solute enrichment or mixtures with significant structural differences.

Chen et al. [81] used [Bmim]Cl and K_2HPO_4 to extract rapeseed meal protein, an active substance in natural products, and the extraction percentage of rapeseed protein was up to 99.1% with the IL concentration of 350 mg/mL and K_2HPO_4 concentration of 150 mg/mL. Wang et al. [82] extracted papain at 303.15 K by applying [Bmim]Br and K_2HPO_4 and the extraction rate of papaya protease was up to 99.1% with the IL concentration of 0.3 g/mL and K_2HPO_4 concentration of 0.3 g/mL. The extraction percentage of the target product was up to 91.2%.

Dreyer et al. [83] constructed a biphasic aqueous system based on quaternary ammonium chloride ILs and phosphate buffer for selective separation of protein mixtures by taking advantage of the differences in the charge states of different types of proteins in solution. The extraction percentage of albumin was as high as 100% in all pH ranges examined. In contrast, other proteins, such as lysozyme and trypsin, were powerfully extracted only when the pH was higher. Capela et al. [84] took advantage of the ability of aliphatic amino acids to act as phase-forming substances to form a biphasic system with quaternary phosphonium ILs and skillfully realized the selective separation of aromatic amino acids and aliphatic amino acids, with the selectivity coefficients between different amino acids reaching 1.5−121, which was much larger than that obtained for the imidazolium-based ILs biphasic system (0.01−0.07). Zhang [85] synthesized a series of poly(ionic liquid)s with specific molecular weights by reversible addition-fragmentation chain transfer polymerization and explored the construction of a poly(ionic liquid)s biphasic system and its application in the field of extraction and found that the poly(ionic liquid)s had better phase-forming ability and extraction efficiency than the IL monomers, which provided new possibilities for the selective separation of structurally similar hydrophilic compounds.

4.3.4 Ionic liquid extraction separation with a similar structure compound

One of the most difficult issues in the chemical industry is the separation of molecules that have similar structural characteristics. Common structural analogs include phenolic compounds, terpenoids, fatty acids, and fatty esters. A wide range of phenolic compounds are found in both plants and animals, and many of these compounds have special physiological properties that make them valuable sources for pharmaceuticals and functional foods. However, they frequently take the form of quite similar homologous monomer combinations, and there are notable variations in the activity and functionality of various homologous monomers. To acquire certain monomers, selective separation is required. Yang et al. [86] investigated the use of IL cosolvent combinations as extractants in liquid–liquid extraction for the first time. They discovered that when utilizing an extractant mixture of acetonitrile and 1-butyl-3-methylimidazole, tocopherol was produced. The selectivity of tocopherol was 11.3, making it more effective in separating materials than a pure solvent or pure ionic liquid. Liu et al. [87] demonstrated that nonaqueous solvent-induced ionic liquid crystals (NILCs) were effective in removing phenolic compounds from wastewater. In order to create NILC, phenols were mixed with ILs after being dissolved in nonaqueous solvents. The phenols in the wastewater were then separated from the resulting mixture using a process known as liquid–liquid extraction. In this study, the NILC, which uses 1,4-dioxane and 1-butyl-3-methylimidazolium hexafluorophosphate, was used to extract phenol from wastewater. The use of NILC in phenolic extraction provided several benefits over conventional extraction techniques, including improved selectivity, a quicker extraction rate, and less solvent usage. Strong hydrogen bond associations with phenolic compounds in weakly polar solvents can result in liquid quaternary ammonium cation/phenolic eutectic solvents, which can also be used for the selective separation of phenols with similar structures. These organic chlorinated quaternary ammonium cations, which are solid at room temperature, such as ammonium chloride, can form these associations.

Qin et al. discussed how to extract vitamin E from deodorized distillates using a biphasic system with solvents of both organic chemical and the salt tetrabutylammonium chloride ($[N_{4,4,4,4}]Cl$). This study used a combination of computational modeling called COSMO-RS and experimental methods to optimize the extraction process and explored the chemical interactions between molecules of vitamin E and parts of the biphasic system. The results showed that the biphasic system is effective in selectively extracting vitamin E from the deodorizer distillate. According to the results of COSMO-RS, the molecular interactions between the vitamin E molecule and the biphasic system's components are extremely important to the extraction. The COSMO-RS analysis revealed that the hydrophobic interactions between the vitamin E molecule and the surfactant were what primarily drove the selective extraction of vitamin E. The surfactant molecules in the aqueous phase created micelles, which enclosed the vitamin E molecules and stopped them from solubilizing there. The results further demonstrated the significance of the organic phase's polarity in the extraction procedure. To ensure selective extraction, the organic phase should be less polar than the vitamin E molecule [88]. Liu et al. [87] used choline fatty acid IL-based water-insoluble liquid crystals, introduced anisotropic self-assembled nanostructures, weakened the influence of strong polarity of ionic liquids, and enhanced their molecular recognition ability. The results showed that the system had very high extraction ability and separation selectivity for substances containing hydrogen bond donors, and the partition coefficient of tocopherol was as high as 50–60, and it was easy to recover.

The fruits, petals, and other parts of plants are used to extract natural essential oil, a significant essence and spice ingredient. Terpenes and their oxygen-containing derivatives, such as terpenoids, terpenol esters, and so forth make up the majority of this oil's crude extract. Terpenes must be thoroughly removed in order to refine the oil. Traditional solvents cannot effectively separate terpenes from their oxygen-containing derivatives due to their extremely comparable structural similarities. Arce et al. [89] studied the possibility of using an ionic liquid as a solvent for citrus essential oil deterpenation by liquid–liquid extraction. They stimulated citrus essential oil using a combination of limonene (terpene) and linalool (deterpenation). The equilibrium phases and quantitative methods were used to analyze the three chemicals using a direct analytical method using ^{1}H NMR. At 298.15 K and 318.15 K, the researchers measured the equilibrium data for limonene/linalool/1-ethyl-3-methylimidazolium ethylsulfate ([Emim][EtSO$_4$]). The experimental results were used to compute the selectivity and distribution ratios of linalool. The experimental LLE data were correlated using the nonrandom two-liquid equation, with the nonrandomness value (α) set to 0.1, 0.2, and 0.3. When the value of α was 0.1, the best results were attained. Each set of temperature data was correlated separately and both sets were done concurrently, yielding an appropriate correlation in every instance. According to the result, [Emim][EtSO$_4$] can be employed as a solvent for citrus essential oil deterpenation by liquid–liquid extraction.

The leaves of Ginkgo biloba include a significant class of terpenoid active compounds known as ginkgolides including a number of homologs, with ginkgolide B having the highest physiological action. Ginkgolide homologs, on the other hand, not only have very comparable structural features, but also possess a high cohesive energy that is hydrophobic and oil-repellent. Their selective separation is extremely challenging. Cao et al. [90] demonstrated that separating ginkgolide homologs could be accomplished using the suggested IL-based biphasic system, which consists of IL, water, and ethyl acetate. The system displayed selectivity, a high extraction capacity, and sufficient distribution coefficients. The numerous interactions between ginkgolides and IL, such as hydrogen bonding, electrostatic interactions, and van der Waals forces, which were proven by using quantum chemistry simulations, were primarily blamed for the enhanced ginkgolide distribution coefficients. By assessing the extraction solvent's Kamlet–Taft parameters, the influence of interactions between ginkgolides and the extraction solvent on the selectivity coefficient was investigated, and the results showed that the selectivity coefficient slightly decreased with the increase of temperature, revealing that room temperature would be appropriate for the separation process. The employed IL-based extraction would be a valid and clean method as an alternative to chromatographic methods for separating bioactive compounds in large-scale operations based on the results of fractional extraction. It is noteworthy that the amount of organic solvents consumed with this method was supposed to be less than 1/11 of the most widely used chromatographic method.

To extract the linoleic acid from soybean oil, Manic et al. [91] used room temperature ILs (RTIL: AMMOENG100 and 1-butyl-3-methylimidazolium dicyanamide [Bmim][DCA]) as the alternative solvents. The liquid–liquid-phase equilibrium was measured as a function of temperature and composition for binary and ternary combinations of (RTIL/soybean oil) and (RTIL/soybean oil/linoleic acid). The Peng–Robinson cubic equation of state and the Mathias–Klotz–Prausnitz mixing rule were used to model the experimental results. This modeling approach was used to predict the liquid–liquid-phase equilibrium for binary and ternary mixtures, and the model's findings were discovered to be in strong agreement with the experimental data. Additionally, the effects of temperature, the initial acid content of the oil, and the solvent-to-oil ratio on distribution coefficients and separation factors were investigated. While raising the initial acid content of the oil has the opposite effect, a rise in temperature often results in larger distribution coefficients and separation factors. The extraction efficiency is also influenced by the solvent-to-oil ratio, with greater ratios often producing better outcomes. It was concluded that RTILs were promising alternative solvents for the extraction of linoleic acid from soybean oil. ILs with aromatic rings, including N-dialkylimidazolium and N-alkylpyridinium were discovered by Cheong et al. to be able to selectively extract and concentrate n–3 polyunsaturated fatty acids. The extraction capacity of $[Hmim][PF_6]$ for n–3 polyunsaturated fatty acid (PUFA) rose by 16.2% under ideal circumstances, and the multistep reverse extraction procedure could further improve the purity of n–3 PUFA to 89% [92]. Through MD simulation, Dong et al. investigated the interaction between three soybean isoflavone glycosides and ILs and discovered that anions are primarily distributed around the phenolic hydroxyl of genistein and that this interaction is mediated by hydrogen bonds [93].

For the separation of AA and AA glucoside (AA-2G), Yang et al. [79] presented a liquid—liquid extraction technique based on hydrophobic IL solutions. The relative purity of AA-2G rose to 96.2%, and the recovery rate was nearly 100% when trihexyl tetradecane phosphonium bromide ethyl citrate was used as the extractant. The separation selectivity and capacity would not be impacted, even at high input concentrations. This method may provide excellent partition coefficients and separation selectivity with minimum solvent consumption and does not require strong acids and bases [94]. To extract and purify proteins, Dreyer et al. [83] employed a biphasic aqueous system based on quaternary ammonium chloride IL and phosphate buffer. According to their research, charge and molecular weight were the deciding variables in the various interactions between proteins and ionic liquids, which resulted in their enrichment in the top layer. The primary mechanism for separation was the electrostatic interaction between positively charged cations and charged amino acid residues on the protein surface. For the purpose of selective separation of mixtures combining aliphatic and aromatic amino acids, Capela et al. [84] created an ABS using aliphatic amino acids and ionic liquids. An extraction percentage of more than 85% was achieved and L-neneneba tryptophan can be recovered in a single step. At the same time, lysine was almost completely enriched in the reverse phase.

Liang et al. [95] used 7 organic solvents (such as dimethyl sulfoxide, acetonitrile, sulfolane, 1,3-propylene glycol, etc.) and 11 ILs (such as 1-hexyl-3-methylimidazolium tetrafluoroborate [Hmim][BF$_4$], 1-butyl-3-methylimidazolium hexafluorophosphate [Bmim][PF$_6$], 1-butyl-3-methylimidazolium bis(trifluoromethylsulfonyl)imide [Bmim][NTf$_2$], 1-butyl-1-methylpyrrolidinium bis(trifluoromethylsulfonyl)imide [BMPy][NTf$_2$], etc.) as the solvent extraction to selectively separate vitamin D$_3$ and tachysterol$_3$. The results demonstrated that sulfolane performed the best extraction among the organic solvents, providing only a selectivity of 1.44 for tachysterol$_3$ over vitamin D$_3$. Unsaturated bond ILs showed great selectivity; [BMPy][NTf$_2$], an ionic liquid-based on pyrrolidinium, for instance, had the best selectivity up to 1.77. In the study, the extraction performance for the selective separation of vitamin D$_3$ and tachysterol$_3$ by solvent extraction was examined in relation to the impacts of various anions, cations, and substituents of ILs. As a result of their unique π-π interactions with the two chemicals, the data demonstrated that ILs with unsaturated linkages had better selectivity. They also found that the selectivity increased with the increase in the length of the alkyl chain of the cation, the size of the anion, and the number of substituents on the cation. To evaluate the feasibility of using ILs as extractants, they simulated the purification and recovery of vitamin D$_3$ via continuous multistage extractions using IL-based liquid—liquid extraction. The simulation showed that the combination of organic solvents and ILs as extracting agents provided acceptable selectivity and distribution coefficients. Insights into the viability of utilizing ILs as extractants for the selective separation of vitamin D_3 and tachysterol$_3$ by solvent extraction were provided by the study, which may have applications in the food and pharmaceutical sectors.

Artemisinin is the most effective antimalarial drug and has efficacy in treating lupus erythematosus, anticancer, and so forth. Currently, artemisinin is mainly from *Artemisia annua* L.; however, the plant composition is complex. During the extraction process, structural analog artemisitene (ARE) will be coextracted into the extraction liquid along with artemisinin. ARE has quite a similar structure to that of the antimalarial artemisinin and thus is difficult to be removed from the target drug. Cao et al. [96] synthesized ammonium-functionalized IL *N*-allyl-*N*,*N*,*N*-trimethylammonium chloride ([AA][Cl]) and used it to remove trace amounts of ARE from a mixture. Several methods, including nuclear magnetic resonance (^{1}H NMR) spectroscopy, FT-IR spectroscopy, and high-performance liquid chromatography (HPLC), were used to characterize the created ILs and the extracted artemisinin. Artemisinin and ARE could be effectively separated using the [AA][Cl] aqueous solution. When ARE concentration in artemisinin was only 0.8%(mass fraction), under ideal circumstances, the separation selectivity could reach 12.23. On this basis, a copolymerized IL with dual responsiveness to temperature and salt has been synthesized, and its cloud point temperature can be precisely controlled by changing the concentration and structure of the poly(ionic liquid). In order to assess the selectivity of the copolymerized ionic liquid for the separation of ARE and artemisinin, HPLC was employed. According to the comparison of the separation efficiency and selectivity of the copolymerized IL with other types of ILs and solvents, the aqueous solution containing poly(isopropylacrylamide) did not specifically recognize the functional groups of ARE and exhibited similar interactions with ARE and ART, leading to a separation selectivity of ARE/ART comparable to the separation effect of the aqueous system. Due to the increased recognition of particular functional groups of ionic liquid in ARE, the concentration, partition coefficient, and separation selectivity of ARE in the extraction phase increased when IL content in the copolymer IL increased [97].

He et al. [98] studied the extraction of vanadium (V) using 1–octyl–3–methylimidazolium chloride ([Omim]Cl), 1–octyl–3–methylimidazolium bromide ([Omim]Br), and 1–octyl–3–methylimidazolium tetrafluoroborate ([Omim][BF$_4$]) as extractants. Investigations were performed into the effects of different diluents, equilibrium time, extraction temperature, and anion species. The findings indicated that *n*–pentanol was the best diluent, and under the extraction circumstances of an equilibrium time of 60 s and an extraction temperature of 298.15 K, the extraction percentages of V by [Omim]Cl, [Omim]Br, and [Omim][BF$_4$] were 97.9%, 96.6%, and 87.0%, respectively. The results of the statistical analysis revealed that the anion species significantly influenced the effectiveness of extraction of V, while the effects of equilibrium time and temperature were not significant in the model. The study's findings suggested that the imidazolium ILs are efficient solvents for the extraction of vanadium, and that the extraction efficiency is influenced by the size of the anions and the intensity of the interaction between the anion and imidazolium cation. N-pentanol is the ideal diluent for vanadium extraction. Vanadium extraction from aqueous solutions may be accomplished using effective and ecologically acceptable technologies.

4.4 Ionic liquids for protein and protein complex extraction

4.4.1 Ionic liquids for protein extraction

As the executors of life activities, proteins play important roles in various fields, such as life science, clinical medicine, biopharmaceutics, and bioengineering. Due to the strong interactions between protein molecules via hydrogen bonds, ionic bonds, or hydrophobic interaction, the low solubility of certain proteins in organic and inorganic solvents prevents the further research [99,100]. Contributing to the good biocompatibility, specific ILs are recognized as the excellent solvents or media for proteins [101−103], and recently, remarkable progress has been made in protein extraction, dissolution, stability, refolding, crystallization, and so forth [104]. In this chapter, the extraction of proteins and protein complexes by ILs is emphasized.

4.4.1.1 Ionic liquids for single protein extraction

The extraction of target proteins from complicated mixtures is of great significance for protein research. Proteins extracted by organic solvents might lose the biological activities due to denaturation. However, IL-based extraction systems might facilitate the acquisition of biofunction-maintained target proteins due to the advantages, including favorable protein solubility, low viscosity, rapid phase separation, and high extraction rate [101−105]. Moreover, the diverse organic cations and inorganic anions combinations, together with the structural designability of ILs, enable them to overcome the constrained selectivity of volatile organic solvents, thus allowing for more efficient protein extraction [106], as demonstrated by Mantz's group, to dissolve silk fibroin protein in 1-butyl-3-methylimidazolium chloride (BmimCl) to improve the solubility of the protein [107].

IL-based ABS is an effective platform for extracting and purifying proteins. Li et al. designed and synthesized a series of environmentally friendly choline ILs and found that these ILs can form ABS with the heat-sensitive, nontoxic, and biodegradable polypropylene glycol 400 (PPG400) [108]. The effects of the anionic structure of ILs were measured at 25°C, and the effects of proteins and differences in top and bottom phase concentrations on the partitioning behavior of some typical proteins were systematically investigated. It was found that bovine serum albumin (BSA), trypsin, papain, and lysozyme could be effectively enriched into the IL-enriched phase. Under optimized conditions, the single step extraction efficiency of ABS can reach 86.4%−99.9%. In addition, increased activity was found after 13 months of storage of trypsin enzymes in ILs. In addition, PPG400 only needs to be heated to be recycled and reused during the extraction process. Therefore, the aqueous solution of choline ILs has the potential to be an excellent long-term storage medium for trypsin, whereas choline IL- PPG400 ATPS has the potential to be used for protein purification. Quental et al. also carried out a similar study, using a series of biocompatible ABS composed of cholinyl ILs and polypropylene glycol to evaluate the extraction performance of BSA. The stability of BSA in IL–rich phase was determined by size-exclusion HPLC and FT-IR spectroscopy. BSA extracted from IL can maintain the natural conformation of protein. Therefore, IL-based ABS is an effective and biocompatible method for isolating and purifying proteins.

Presently, IL–based techniques, encompassing solid—liquid extraction methods like microwave-assisted and ultrasound-assisted extraction, as well as liquid—liquid extraction approaches utilizing hydrophobic IL, ILs-based aqueous biphasic or aqueous micellar biphasic systems, IL-modified materials, and IL-based crystallization methods, have gained extensive applications in protein extraction and separation/purification processes. As a result, ILs can be tailored with unique properties to enhance their performance and meet specific separation requirements.

4.4.1.2 Ionic liquids for protein mixture extraction

Due to the characteristics of good water solubility, strong protein solubility, and high thermal and chemical stability, ILs have not only been well applied to the extraction and structure maintenance of single protein, but also play an important role in the extraction of complicated protein systems such as proteomes and protein complexes. In 2010, Hua's group made the first attempt to use IL solutions for direct protein extraction from yeast cells, opening the way for ionic liquids to be used in complicated protein systems [109]. The effects of 21 different IL solutions on the extraction efficiency of protein were compared, and the 3-(dimethylamino)-1-propyl formate amine ([DMAPA] FA) was selected as the extraction reagent. [DMAPA]FA not only destroys the yeast cell wall and efficiently extracts proteins, but also can be easily removed under vacuum, thus effectively reducing pollution caused by chemical noise. In addition, the chemical properties of the target proteins remained during extraction and were compatible with subsequent two-dimensional gel electrophoresis (2-DE), sodium dodecyl sulfate polyacrylamide gel electrophoresis (SDS-PAGE) and western blotting analysis. This study showed that IL [DMAPA]FA was a promising reagent for protein extraction from yeast cells. Zhang's group further applied IL [DMAPA]FA combined with a low-temperature and high-pressure cell breakage method to extract whole proteins from *Chlorella pyrenoidosa* cells with rigid cell walls and found that the extraction efficiency was superior to the three traditional methods (freeze-thawing, ultrasonication, and freeze-thawing combining with ultrasonication) [110]. The above research preliminarily shows the application potential of ILs in the extraction of complex proteins.

4.4.1.2.1 Ionic liquids for membrane proteome extraction

Membrane proteins play a crucial role in the functionality of biofilms and are intricately connected to various biological processes, including cellular communication, ion transport, and signal transduction [111,112]. In the field of drug development, more than 60% of drug targets are membrane proteins [113]. Therefore, deep coverage analysis of membrane proteome is of great significance for revealing the pathogenesis of disease and discovering new drug targets [113,114]. Nevertheless, the extensive hydrophobic nature and limited extraction efficiency of membrane proteins significantly impede the progress of membrane proteome research [115]. Therefore, the development of sample preparation methods with high extraction efficiency, good trypsin digestion compatibility, and no interference for liquid chromatography-mass spectrometry (LC-MS) analysis is the key to membrane proteome analysis. Addressing this issue, SDS has been effectively employed for the extraction of both hydrophilic and hydrophobic proteins [116,117]. However, SDS is difficult to remove, and the low concentration of residual SDS in the sample inhibits the activity of trypsin, diminishing trypsin digestion efficiency and impeding the identification process in LC-MS. Although the filter-assisted sample preparation (FASP) strategies can efficiently remove SDS, the multistep sample processing results in the loss of low-abundance membrane proteins [118]. Additionally, signal interference from high-abundance proteins during identification poses obstacles to achieving comprehensive coverage in membrane proteome analysis.

Zhang's group firstly proposed a membrane proteomic sample preparation strategy based on IL 1-butyl-3-methylimidazole tetrafluoroborate ([Bmim]BF_4) [119]. 1% (volume fraction) [Bmim]BF_4 was used to extract integrated membrane proteins (IMPs) from rat brain tissue and compared with common methods of protein sample preparation. Compared with Rapigest and urea extraction methods, the number of IMPs identified by [Bmim]BF_4 extraction method was increased by 25% and 80%, respectively. Compared with SDS and methanol preparation methods, the number of IMPs identified was more than 3 times higher, which was due to good protein solubilization and easy removal in the desalting process of [Bmim]BF_4, fully demonstrating the application potential of [Bmim]BF_4 in membrane proteomic analysis.

In order to improve the recovery rate of hydrophobic peptide segment of membrane proteins, Zhang's group further optimized the separation conditions of membrane proteins obtained by the [Bmim]BF_4 extraction method [120]. During the desalting process, peptides underwent prefractionation via a reverse-phase trapping column. This step effectively mitigated sample complexity, resulting in a notable increase of over 43% in the identification of hydrophobic peptides and IMPs. In addition, the effects of C_{18} and C_8 stationary phase on the separation of hydrophobic peptides were studied, and C_8 column separation with weak retention was selected. Compared with the C_{18} phase, the number of hydrophobic peptides and IMPs identified by the C_8

phase were significantly increased by 29% and 20%, respectively, indicating that the polarity of the stationary phase has a significant effect on the analysis of membrane protease hydrolysis products. These results indicate that [Bmim]BF$_4$-assisted extraction and integrated sample classification and C$_8$ stationary phase separation can effectively promote the in-depth identification of IMPs.

In order to ensure the coverage of proteomic analysis based on bottom-up analysis strategies, the ideal protein extraction system should have both excellent protein solubility and enzyme activity compatibility. Although [Bmim]BF$_4$ has shown good protein solubilization and trypsin compatibility in IMP analysis, it is still necessary to systematically study the effect of the IL structure on membrane protein solubilization to reveal the key factors affecting the solubilization ability of ILs. Zhao et al. systematically studied the effect of the IL structure on membrane protein dissolution. By evaluating the ability of different imidazole ILs to dissolve bacteriorhodopsin (BR, a model membrane protein including seven transmembrane regions), it was found that 1-dodecyl-3-methylimidazole chloride ([C$_{12}$im]Cl) had the best solubility and its relative solubility to BR was 7 times that of [Bmim]BF$_4$ [121]. Subsequent MD simulation analysis revealed that in the BR-C$_{12}$im-Cl system, the alkyl chain was adsorbed onto the BR surface rather than the [C$_{12}$im]Cl cationic ring (Fig. 4.3) [122]. The results illustrated that [C$_{12}$im]Cl primarily bound to hydrophobic amino acid residues exposed in membrane proteins, and the hydrophobic interaction of alkyl substituents based on imidazole ILs played a central role in membrane protein solubilization. By examining the enzyme activity compatibility of [C$_{12}$im]Cl, it was found that compared with common extraction systems such as SDS and methanol, the trypsin activity of [C$_{12}$im]Cl extraction system was the highest, indicating that the [C$_{12}$im]Cl extraction system had both excellent protein solubility and enzyme activity compatibility (Fig. 4.4). It is expected to be a new

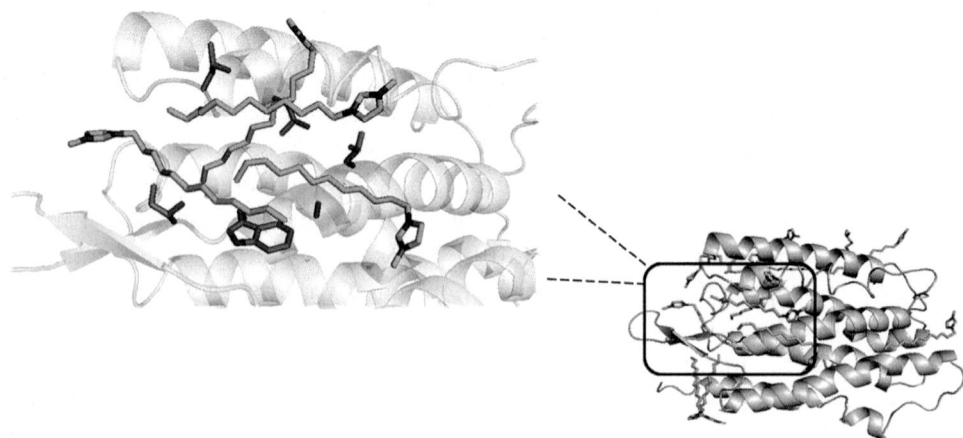

Figure 4.3 Conformation of bacteriorhodopsin and [C$_{12}$im]Cl simulation system [122].

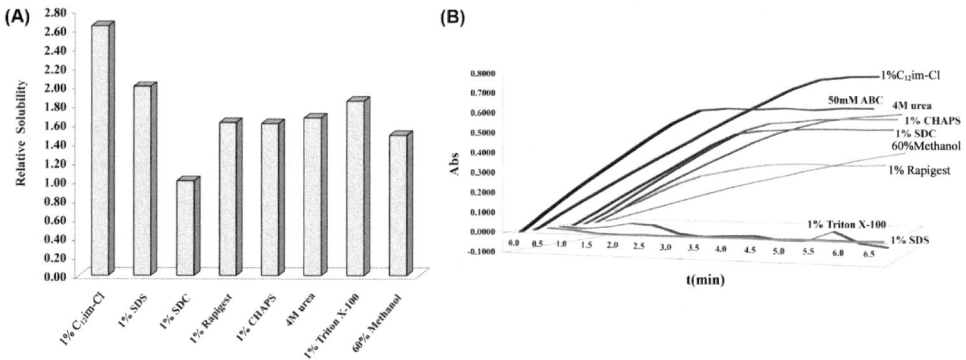

Figure 4.4 Comparison of the effects of the [C$_{12}$im]Cl system and common protein extraction systems on membrane protein solubility (A) and trypsin activity (B) [121].

extraction system for proteome sample preparation. The removal of [C$_{12}$im]Cl after trypsin digestion with strong cation exchange trapping column not only avoided protein precipitation during digestion, but also did not interfere with subsequent LC-MS analysis. The [C$_{12}$im]Cl-assisted sample preparation method was applied to the analysis of rat brain tissues. Compared with the SDS-assisted extraction method, the number of IMPs and hydrophobic peptides identified increased by 1.4 times and 3.5 times, respectively, which fully demonstrated the application prospect of [C$_{12}$im]Cl-assisted sample preparation method in the large-scale analysis of membrane proteins. Sui et al. used [C$_{12}$im]Cl for the extraction of velvet antler cartilage protein in response to the serious interference of high abundance proteoglycan in the analysis of cartilage proteome [123]. The results showed that compared with the cetylpyridine chloride (CPC) extraction method developed by Vincourt et al., the identification quantity identified by [C$_{12}$im]Cl extraction method was 2.4 times that of the CPC extraction method (663 vs.279) [124]. To mitigate the inhibitory impact of high-abundance proteins on low-abundance membrane proteins and enhance the identification scope of membrane proteins, Fang et al. introduced an in-situ classification method for membrane proteins involving sequential extraction assisted by [C$_{12}$im]Cl [125]. Through the sequential use of extraction reagents with distinct dissolution capabilities, the complexity of the extracted protein components was systematically reduced. This approach significantly enhanced the identification capabilities of low-abundance membrane proteins, leading to the high-coverage identification of 5553 membrane proteins and 2573 integral membrane proteins in HeLa cells.

In summary, the IL-based membrane protein extraction system not only has strong protein dissolution ability, but also has compatibility with subsequent enzymatic hydrolysis and mass spectrometry analysis, which can realize large-scale analysis of membrane proteins with strong hydrophobicity.

4.4.1.2.2 Ionic liquids for whole proteome extraction

Considering the benefits of ILs in membrane proteome extraction, their extended utilization in comprehensive whole proteome analysis holds paramount importance. This application is crucial for mapping the complete proteome and unraveling the intricacies of life processes. Zhao et al. combined [C_{12}im]Cl with the conventional urea extraction system for in-depth analysis of the whole proteome of HeLa cells [126]. The [C_{12}im]Cl extraction step can recover the strong hydrophobic and low-abundance membrane proteins lost during the extraction of 8 M urea, reduce the inhibition of high-abundance proteins, and improve the coverage of proteome identification. Previous research results showed that the C_8 column was beneficial to improve the separation efficiency of hydrophobic peptides and was more suitable for the identification of membrane proteins. Therefore, peptides extracted from 8 M urea and [C_{12}im]Cl were separated on C_{18} and C_8 capillary columns, respectively. This method successfully identified 11313 proteins, spanning over 8 orders of magnitude in abundance. It generated a comprehensive dataset for the HeLa cell proteome, surpassing the identification results achieved by Mann's SDS-based FASP method (10255 proteins) [118].

Zhao et al. further developed a [C_{12}im]Cl IL-based FASP method (i-FASP), which realized the efficient identification of 3337 proteins in trace HeLa cells (1000 cells) [122]. Because the low concentration of [C_{12}im]Cl has excellent compatibility with trypsin, and no additional desalting steps are required after enzymatic hydrolysis, the sample loss caused by multiple elution is reduced. As a result, the sample preparation time of i-FASP was about 4 h shorter than that of FASP method, and the number of proteins identified was increased by 53%, providing technical support for achieving high-throughput proteomic scale analysis (Fig. 4.5). Furthermore, this method has

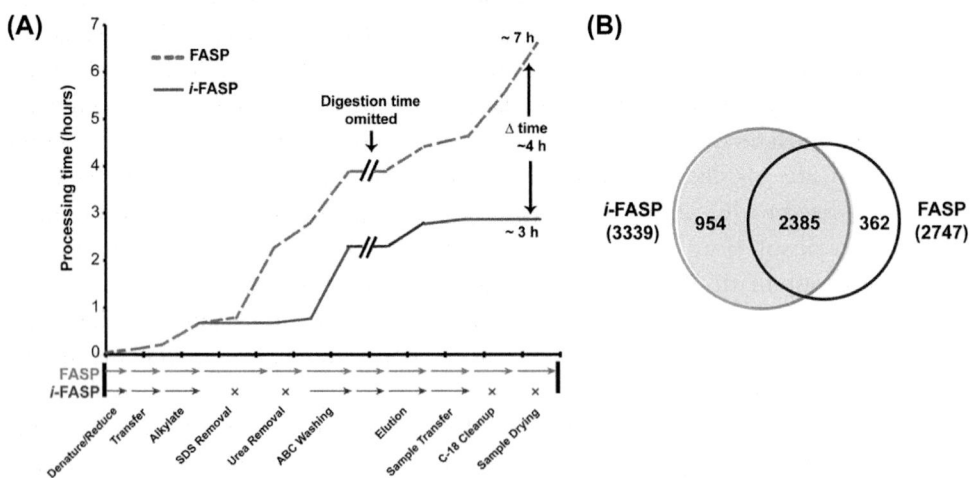

Figure 4.5 Comparison of i-FASP and FASP method workflows (A) and commonality in the proteins identified by the i-FASP and FASP methods (B) [122].

been successfully applied to the nonstandard quantitative analysis of human liver cancer and adjacent tissue samples, and its accuracy, reproducibility, and coverage are significantly improved compared with the FASP method, and key target molecules such as NES and CAP1 are closely related to liver cancer metastasis.

Isoenzymes are enzymes with different molecular structures but catalyze the same reaction. The accurate identification of isoenzymes is of great significance for the deep coverage analysis of the whole proteome. Cytochrome P450 enzymes (CYPs) and uridine diphosphoglucuronosyl transferases (UGTs) are two common isoenzymes in mammals. SDS-PAGE separation and enzymolysis in gel are common pretreatment strategies. However, the enzymolysis in gel is not conducive to the recovery of peptides, affecting the identification accuracy of CYPs and UGTs, and the enzymolysis in gel is time-consuming and difficult to automate. In response to this problem, Sun et al. used the sample preparation method based on [Bmim][BF$_4$] extraction system to analyze the proteome of human liver microsomes and identified a total of 27 CYPs and 12 UGTs [127]. The high average sequence coverage and the number of specific peptide matches demonstrated in this result enabled the successful differentiation of isoenzymes CYP2C9 and CYP2C19 with sequence similarity of 91%, ensuring reliable identification between CYPs and UGTs isoenzymes. Compared with Peng et al. for cleaning and enrichment with sodium carbonate (pH = 11.5), the proteins and IMPs identified by this method were increased 4-fold and 2.7-fold.

Posttranslational modifications of proteins occur all the time in vivo and are diverse in variety, but limited by their low abundance, they are susceptible to interference from high-abundance proteins during sample analysis, resulting in low identification coverage. Zhang's group used the developed IL extraction system for posttranslational modification proteome analysis [128−130]. As one of the most common posttranslational modifications, protein glycosylation is related to many biological processes such as molecular recognition, receptor activation, and signal transduction. Qiao et al. established a strategy for identification of mouse brain N-glycosylated proteome based on [Bmim]BF$_4$ [128]. In the analysis of glycoproteins in insoluble and soluble components, 462 nonredundant N-glycoprotein groups, including 316 transmembrane glycoproteins, were successfully identified, and accordingly 849 characteristic N-glycoaglycone types were identified, 267 of which had not been previously reported. This method provides technical support for further study of N-glycosylation modification in the brain, especially for the discovery of candidate drug targets and biomarkers.

The use of ILs in protein sample preparation technology has found widespread application in phosphorylated proteomics research. In the pursuit of understanding the regulatory role of protein phosphatase 2 A (PP2A) in liver disease development, Chen et al. employed a [C$_{12}$im]Cl-assisted sequential extraction method for phosphorylated proteomic analysis on mouse liver tissues with a specific deletion of the Ppp2r1a gene (encoding the PP2A Aα subunit) in hepatocytes [129]. A total of 351 phosphorylated proteins associated with fibrosis were identified, including 108 not previously reported, indicating their potential involvement in the initiation of liver fibrosis. Further proteomic and functional analysis revealed that the dysregulation of the epidermal growth factor receptor (EGFR) signaling pathway following PP2A deletion is pivotal in mediating liver fibrosis. Li et al. developed a large number of plant sample preparation methods based on $[C_{12}im]Cl$ in view of the difficulty of wall breaking of plant samples and the low efficiency of protein extraction, and developed an efficient $[C_{12}im]Cl$ precipitation removal strategy based on ion exchange mediated by bis(trifluoromethane)sulfonimide lithium $[(CF_3SO_2)_2NLi]$ [130]. The peptides obtained were compatible with subsequent liquid-phase separation mass spectrometry identification. Compared with the commonly used 8 M urea extraction method, the method increased the protein extraction ability of tobacco tissue by 1.9 times. Combined with immobilized metal ion affinity chromatography enrichment methods, 14441 phosphopeptides (5153 phosphorylated proteins) were identified from 10 mg tobacco leaf samples, which is the most comprehensive dataset of phosphorylated proteins in tobacco and has greatly promoted the systematic interpretation of the function of plant phosphorylated proteins.

In recent years, proteome analysis has greatly promoted the research process in the field of biomedicine. Through the comparison of proteomic information between organisms in normal and pathological states, it offers crucial theoretical and technical support for comprehensively understanding the mechanisms underlying disease occurrence, identifying drug targets, and investigating mechanisms of drug resistance. Sun et al. introduced a novel method for deep coverage quantitative analysis of membrane proteins using the $[C_{12}im]Cl$-based extraction system [131]. Employing dimethylation labeling for quantitative analysis, the technique was utilized for the proteomic analysis of the liver cancer cell lines Bel/5Fu (5-fluorouracil-resistant) and Bel7402 (5-fluorouracil-sensitive). The strategy identified 103 differentially expressed proteins, highlighting the close association between the membrane protein EPHX1 and drug resistance in Bel/5Fu cells. Further studies showed that downregulating the expression of EPHX1 protein could reduce the chemotherapy resistance of liver cancer. To gain deeper insights to the potential mechanism of sorafenib resistance in hepatocellular carcinoma cells, Chu et al. used the i-FASP method to conduct quantitative proteomic analysis of hepatocellular carcinoma cell line Huh7 and its drug-resistant cell line Huh7-R [132]. A total of 89 differentially expressed proteins were found, among which the expression of membrane protein folate receptor alpha (FOLR1) was significantly upregulated in drug-resistant cells. Results from biological experiments showed that FOLR1 drives drug resistance in hepatocellular carcinoma cells by inducing autophagy, providing new insights into the mechanism of sorafenib resistance in hepatocellular carcinoma. Liu et al. utilized the $[C_{12}im]Cl$ extraction system to study the dynamic changes in proteins during the epithelial-mesenchymal transition (EMT) induced by transforming growth factor β in lung adenocarcinoma cells (A549 cells) [133]. In three biological replicates, they quantified a total of 7206 proteins, with 368 proteins exhibiting changes greater than 1.5-fold in expression. Notably, the fibronectin and vimentin (mesenchymal markers) showed upregulation, while E-cadherin and desmoplakin (epithelial markers) exhibited downregulation. Functional analysis revealed that the differentially expressed proteins were primarily enriched in various biological processes related to EMT, such as cytoskeleton reorganization, cell adhesion, and migration. Hence, the IL-based extraction system facilitates dynamic protein analysis, offering new perspectives on comprehending the mechanisms in tumor development. Yang et al. used $[C_{12}im]Cl$-assisted protein sample preparation method to pretreat hemodialysis membranes. The qualitative and quantitative analysis of the adsorbed proteins showed that the proteins adsorbed on the dialysis membranes were related to complement activation, blood coagulation, and leukocyte-related biological processes, and it was speculated that the adsorbed proteins on the dialysis membranes were one of the key factors leading to hemodialysis complications [134]. Han et al. extended the application of the $[C_{12}im]Cl$-assisted protein sample preparation method to examine the proteome adsorbed on two commercially available dialysis membranes of the same material but with different fluxes [135]. They compared the variations in isoelectric point, molecular weight, and protein categories between the two groups of proteins. This study offers an effective approach for evaluating the biocompatibility of dialysis membranes. Given the advantages of ILs in protein extraction, it is anticipated that, as research progresses, the proteomic analysis method based on IL extraction will hold promising applications in the biomedical field.

In addition, Liu et al. developed a solution rapid digestion method based on $[C_{12}im]Cl$ and successfully applied it to the extraction and rapid enzymatic hydrolysis of proteins from formalin fixed paraffin-embedded (FFPE) liver cancer tissue samples [136]. By optimizing the amount of different enzymes and the digestion time and selecting the best digestion conditions (trypsin to protein ratio 1:10, Lys-C to protein ratio 1:20, mass ratio), the digestion time of protein was shortened from 16 h to 1 h. In contrast to the conventional FASP method, this approach significantly reduced protein digestion time while maintaining a comparable number of identified proteins and peptides, indicating the potential application value of this method in the proteomic analysis of FFPE tissues. For precipitated insoluble proteins, Ichimura et al. developed a polymeric proteome sample preparation method based on the extraction of IL 1-butyl-3-methylimidazolium thiocyanate ([Bmim][SCN]) [137]. Compared with traditional systems such as formic acid, SDS, and guanidine hydrochloride, the IL [Bmim][SCN]- 0.5 mol/L NaOH solution [40% (volume fraction) IL] (i-soln) system had a stronger extraction capacity for egg-aggregating proteins. In addition, [Bmim][SCN] was applied to the analysis of insoluble polymeric proteins deposited in aged fer-15 mutant nematoides, and 176 proteins were successfully identified, including 12 transmembrane proteins, which was higher than the identification result of 70% formic acid extractants (2 transmembrane proteins). It shows the advantage of [Bmim][SCN] in hydrophobic protein extraction.

The remarkable advantages of ILs as a novel protein extraction system are evident in achieving extensive coverage in whole proteome analysis. Nevertheless, the current utilization of ILs-based extraction systems for whole proteome analysis remains in its early stages, requiring ongoing expansion into various fields such as food safety, environmental toxicology, and forensic identification.

4.4.2 Ionic liquids for protein complex extraction

Membrane protein complexes (MPCs) play a key role in intercellular communication, signal transduction, and molecular recognition [111]. The abnormal function of MPCs can lead to a variety of diseases and drug resistance in clinical treatment [114]. Therefore, it is of great significance to analyze the composition and function of MPCs. However, the abundance of membrane proteins in MPCs is low and the hydrophobicity is strong, so it is difficult for the existing extraction solvent systems to take into account the strong hydrophobicity of membrane proteins and the easy dissociation of MPCs, which seriously restricts the analytical coverage of MPCs. Therefore, there is an urgent need to develop sample extraction systems that can simultaneously ensure the extraction efficiency of MPCs and their structural stability.

In the early stage, Zhang's group screened IL [C_{12}im]Cl, which is suitable for efficient extraction of proteomics and compatible with subsequent trypsin enzyme activity, through MD simulation and experimental evaluation of ILs with different structures. However, due to the strong interaction force between [C_{12}im]Cl and proteins, the structural stability of the protein complex will be destroyed in the process of extraction. Therefore, Zhang's group further screened ILs that could maintain the stability of protein interaction through experiments and, combined with molecular docking analysis, explained the mechanism of their maintenance of PPI stability. On this basis, an IL-assisted MPCs extraction system, i-TAN [containing 1% NP-40 triethylamine acetate (TEAA)], was developed [138]. The i-TAN system improved the extraction ability of MPCs. Compared with the traditional NP-40 extraction system, the number of EGFR-interacting proteins identified from HeLa cells was increased from 73 to 90 by using i-TAN system (Fig. 4.6).

In order to investigate the stabilizing effect of TEAA on protein—protein interactions, the classical interacting protein model protein A — IgG was selected for molecular docking simulation (Fig. 4.7). The results show that [N222]$^{+}$ can occupy the anticompetitive binding sites between protein A-IgG interfaces and exert salt bridge and hydrophobic interactions. [Ac]$^{-}$ can also occupy the anticompetitive binding site between the protein A-IgG interface, exerting salt bridge and hydrophobic interactions. Furthermore, cations and anions occupy different binding sites, which can synergically promote PPI stability. The i-TAN system was further applied to the difference analysis of EGFR complex

(A)

NP-40   *i*-TAN

(B)

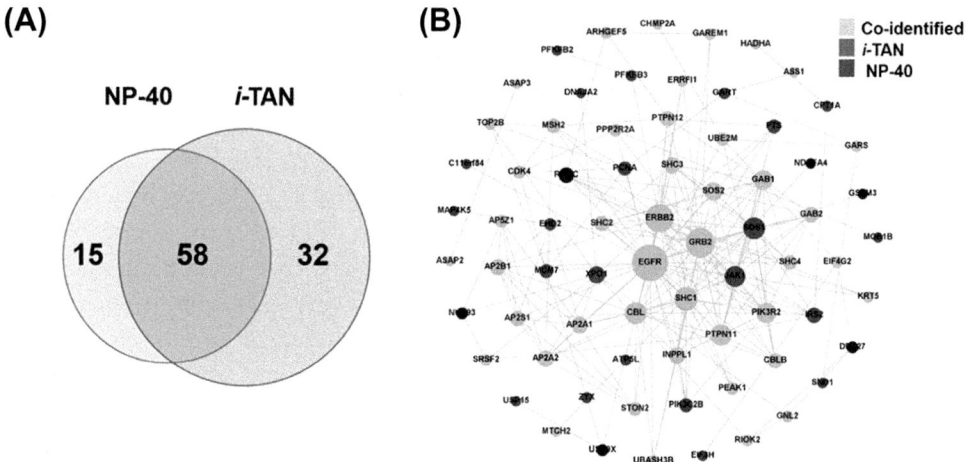

Figure 4.6 (A) Overlap of EGFR-interacting proteins identified by the NP-40 and *i*-TAN. (B) PPI network from STRING database formed by the EGFR-interacting proteins identified by the NP-40 and *i*-TAN [138].

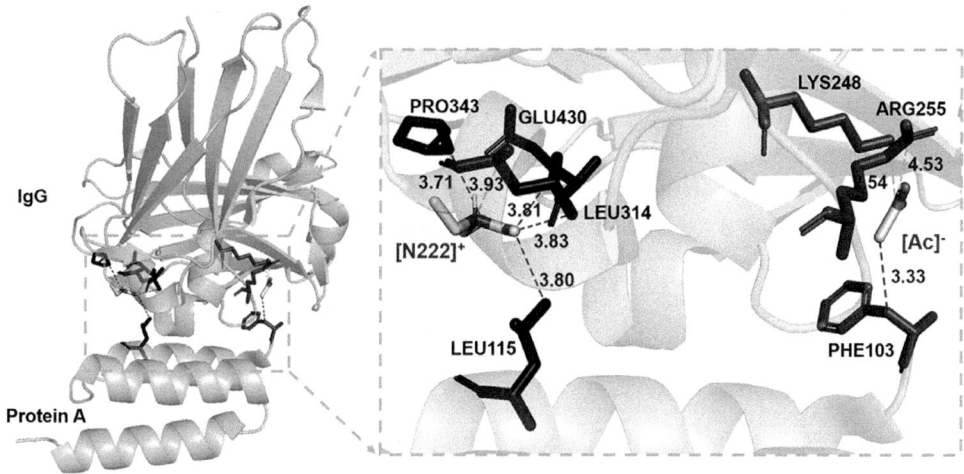

Figure 4.7 The molecular docking simulation results for protein A-IgG and *i*-TAN system. Note: The crystal structure of protein A-IgG (PDB file 5U4Y) was derived from the PDB database. Numbers indicate interaction distances (Å) [138].

composition in trastuzumab-sensitive and –resistant breast cancer cells, and a total of 74 EGFR interacting proteins were identified, of which 13 proteins were more abundant in sensitive cells and 41 proteins were more abundant in resistant cells. Bioinformatics analysis showed that in addition to the activation of the ERBB signaling pathway and its downstream PI3K pathway affecting trastuzumab treatment response [139], the upregulation of AP-2 junction complex protein also promoted clathrin–mediated endocytosis, resulting in the downregulation of ERBB2 expression and downstream signaling pathway activity.

This results in poor response of trastuzumab-resistant cells to trastuzumab, which provides a new way to understand the mechanism of trastuzumab resistance in breast cancer. In addition, based on the previous results that membrane protein FOLR1 was significantly high expression in Huh7-resistant cells [132], Chu et al. further extracted MPCs in hepatocellular carcinoma cells using the *i*-TAN system. 124 FOLR1 interacting proteins were identified, and subsequent biological verification showed that FOLR1 is closely related to the autophagy signaling pathway and the overexpression of FOLR1 may be involved in abnormal folic acid transport, thus promoting autophagy and then causing sorafenib resistance. The association between FOLR1 and sorafenib resistance was again demonstrated.

Compared with the traditional surfactant extraction system, the micellar environment formed by the IL-assisted extraction method is more uniform and can occupy the binding sites between proteins to promote the stability and extraction of MPCs. Based on the designability of cationic and anion structures of ionic liquids, combined with theoretical simulation techniques such as molecular docking, it is expected to screen the optimal ILs and establish an MPC extraction system based on ILs (without surfactants).

4.5 Membrane separation process with ionic liquids

4.5.1 Introduction

Considering the fascinating features of ILs, a series of ILMs have been developed by incorporating functionalized ILs with support membranes [140,141]. Due to the benign stability and versatility of ILs, the ILMs exhibit superiority in membrane performance and stability compared with traditional organic solvent-based membranes. During the last decade, ILMs have been reported to be used in the fields of gas separation (carbon capture, utilization and storage, CCUS, for example), water treatment, resources recovery, fuel cells, solar cells, and electrochemical devices [142−144]. The publication concerning the fabrication and application of ILMs also follows an exponential growth. Among them, the topic on separation of gases, such as CCUS and VOCs, which make up the majority, is discussed, and there are comprehensive reviews on relative progress, which is also discussed in Section 4.2. In addition to this, the applications of ILMs on liquid separation including the recovery or removal of metal ions and organic substances from aqueous solutions and seawater desalination are of great significance in view of resources recycling and environmental remediation. Targeting at different types of ILMs, including supported IL membranes (SILMs), IL−polymer membranes (ILPMs), poly (ionic liquid) membranes (PILMs), and IL polymer inclusion membranes (ILPIMs), the preparation strategy, applicability, stability, transport mechanism, and further development tendency are summarized. This section intends to provide a preliminary database for ILMs and their performance in liquid separation. It is expected to guide the exploration and design of more ILMs with excellent separation performance in respective fields by providing insights into existing structures and functionality.

4.5.2 Ionic liquid membranes and preparation strategy

4.5.2.1 Supported ionic liquid membranes

SILMs are a kind of membranes with ILs impregnating in the membrane pores [145]. The basement of the membranes could be inorganic or organic with porous structures. As reported, there are mainly three physical methods for the preparation of SILMs, that is, impregnation, pressure-induced, and vacuum-induced methods. In brief, the key issue is how to immobilize ILs on the pores of support membrane through a physical method.

(1) *Impregnation method*

Impregnation is one of the most convenient and cost-effective methods, in which the immobilization of IL is achieved by soaking the support membrane in specific IL under ambient conditions. The residual ILs on the surface could be further removed by washing of wiping up with a filter paper.

(2) *Pressure-induced method*

For this method, the ILs are introduced into the pores of a membrane by applying nitrogen pressure, and the pressure is released until the thin layer of IL is apparent on the surface of the membrane. After that, the excess ILs are removed and the SILMs are finally obtained. Recently, SILMs with [Bmim]Cl, [Bmim][BF$_4$], or [Bmim][NTf$_2$] supported in the nylon organic membrane were prepared using the pressure method and the resulting SILMs confirmed that the pores of the support membrane were completely filled with each of the ILs.

(3) *Vacuum-induced method*

The immobilization of ILs by the vacuum-induced method includes several steps: ① placing the support membrane in a vacuum chamber to remove the air in the pores; ② spreading out specific ILs on the surface of the membrane under vacuum; and ③ removing the excess ILs on the membrane surface.

The SILMs hold the advantages of high selectivity and low solvent holding, but are also limited by the insufficient membrane stability in large scale and long-time operations. The membranes' performance and selectivity could suffer from marked decline due to the loss of functionalized ILs from the support during continuous operation. In this case, the research concerning the improvement of the process stability of SILMs should be strengthened, which would be discussed in following section.

4.5.2.2 Ionic liquid–polymer membranes

ILPMs are synthesized by the incorporation of ILs into a polymer matrix, which is considered as a successful approach to overcome the drawbacks of SILMs, especially the insufficient stability. The ILPMs could be also mentioned as IL gel membranes (ILGMs) [146]. The ILs are stabilized into a polymeric membrane through a physical or chemical process. For ILPMs, the polymer membranes serve as mechanically stable support and the functionalized ILs exhibit certain trade-off between permeability and selectivity, resulting in enhanced membrane performance. In general, the main preparation strategies cover physical blending and chemical grafting, including surface-grafting and filler-grafting.

(1) Physical blending

For this method, the ILs and polymer matrix are mixed and dissolved in solvents to form a homogeneous polymer solution. After that, the casting solution is poured into a glass plate and forms mechanically stable dense membranes, in which the thickness of the ILPMs is controlled by the casting knife. Then the membrane is placed in an oven for solvent evaporation and stored for further use. Since the facile blending process, this method has been extensively used in the preparation of ILPMs.

(2) Chemical grafting

Besides physical blending, chemical grafting is another effective method for ILPMs owing to the abundant active functional groups of membrane components. Due to the electrostatic nature of ILs, they are able to interact strongly with the polymers via pronounced hydrogen bonding, coulombic forces, and van der Waals interactions. Moreover, ILs could be chemically grafted to the membrane surface or filler, serving as the membrane additives. For the former, the support membrane is first synthesized using solvent casting or interfacial polymerization. After that, the ILs are poured onto the membrane surface and react with the active group of membrane under specific conditions or activation by radiation and UV light process [8]. Through the chemical bonding of ILs to the membrane surface after membrane formation, the surface properties of membrane surface could be tuned by choosing different ILs, and the separation performance, antifouling properties, and stability are expected to be improved. This method has been extensively used in the fabrication of ILPMs, which is especially appropriate for the modification of polyamide membranes; polyamide chains could be split by the hydrogen bonds with imidazolium ILs or react with amine-based ILs via the acyl chloride-amine esterification reaction.

Due to the diversity of ILs and polymer matrices, a series of ILPMs have been fabricated and applied in the fields of gas or liquid separation. The stability of chemical grafting ILPMs is more satisfactory with those of the SILMs in virtue of the chemical interaction of ILs and membranes. Moreover, the synergistic effects of ILs, membranes and fillers provide more opportunities for the application of ILPMs. The selection of suitable fillers and polymer matrices, the filler dispersion, and the compatibility between fillers and membrane should be considered for different application scenarios in further research.

4.5.2.3 Poly(ionic liquid) membranes

Benefiting from the preeminent properties of PILs, poly(ionic liquid) membranes are further developed to resolve the mechanical stability issue. Due to their polymer macrostructure, PILMs not only have enhanced processability and durability, but also improved mechanical stability. PILMs largely consist of the poly(acrylate) or poly(styrene) backbone with imidazolium cations tethered as side chains where the anion is not chemically bonded to the main polymer chain. The fabrication strategy of PILMs is similar to that of ILGMs subject to the synthesis of PILs. On the one hand, the PILs are primarily synthesized through various ways, such as radical polymerization, photopolymerization of IL monomers, and rapid counterion exchange.

With the advantage of PILs and membrane, PILMs have found themselves in numerous applications, such as metal ion separation, solid polymer electrolyte, and gas separation. Regarding the diversity of PILs and basic polymers for PILMs synthesis, further research is required to find new chemical structures of PILs and suitable combination of PILs and carriers. Moreover, more attention should be given in the preparation cost to increase the economic feasibility.

4.5.2.4 Ionic liquid polymer inclusion membranes

ILPIMs are a kind of membrane using a polymer binder to create a membrane structure and a liquid membrane as the active sites, whose principle is similar to that of SILMs. The first application of PIMs in the area of separation was to eliminate metal species from wastewater, and electrically driven ILPIMs were also developed [147]. ILPIMs are formed by casting a solution containing a base polymer, an extractant (carrier), and a plasticizer, in which the base polymers ensure the mechanical strength; extractants transport and bind the substances passing through the membrane, and plasticizers increase the plasticity or fluidity. ILs could be considered as a carrier and a plasticizer at the same time, and ILs could provide efficient and selective transport with improved stability, and the resulting ILPIMs would be more chemically resistant and thermally stable.

4.5.3 Application in liquid separation

4.5.3.1 Metal separation

The recovery of scarce metals including rare metal (Li for example) and the removal of heavy metals (Hg, Cr, Cd, etc.) from aqueous solutions have been universal in the industrial process. These metal separation processes are critical in view of sustainable development and environment issues. Compared with the traditional solvent extraction and adsorption method, the membrane process is recognized for its fewer energy consumptions, smaller carbon footprints, and convenient operations [148]. Moreover, the ILMs have been used in such a metal separation process for the unique superiority, such as high selectivity and stability.

A series of ILs with cationic functional groups, such as imidazoliums, quaternary ammonium, or quaternized bipyridine groups, were adopted for the grafting and modification of NF membranes, and the representative ILPMs were prepared.

Wu and coworkers prepared a positively charged NF membrane by grafting amine-functionalized ionic liquid 1-(3-aminopropyl)-3-methylimidazolium bis(trifluoro-methanesulfonyl)imide ([APmim][NTf$_2$]) onto the surface of the polyamide/polyacrylo-nitrile (PA/PAN) membrane [Fig. 4.8A] [150]. The ([APmim][NTf$_2$]) modification not only narrowed the surface pore size, but also changed the surface charges from negative to positive. The resulting membrane presented high water permeance [37.8 L/(m^2·h)] and Li/Mg selectivity ($S_{Li/Mg}$, 8.12). A strong electrolyte monomer (diaminoethimida-zole bromide salt, DAIB) was further introduced for the modification of the nascent polyamide membrane, and a nano-heterogeneous membrane (DAIB/PEI-TMC) was synthesized [Fig. 4.8C] [151]. DAIB modification occurred both on top and inside the

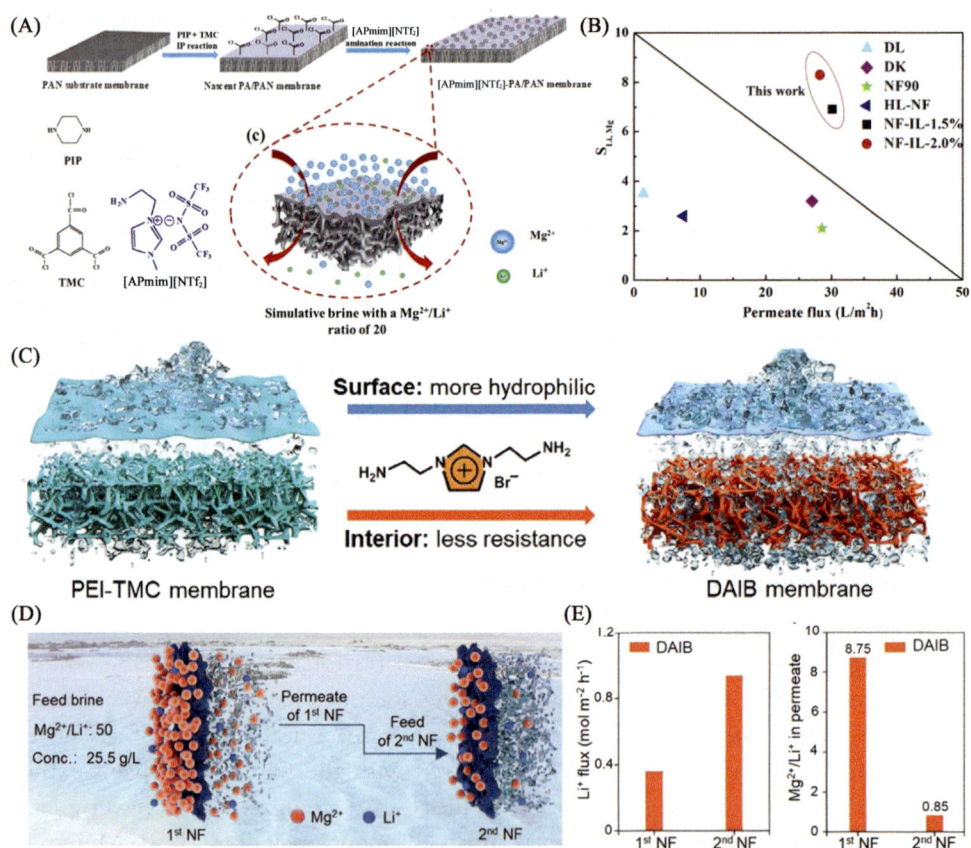

Figure 4.8 (A) Schematic preparation of the [APmim][NTf$_2$]-PA/PAN NF membrane for separation of Li$^+$ from a simulative brine. (B) Comparison of performance of $S_{Li,Mg}$ vs permeate flux, commercial NF membranes, and NF-IL membranes for a simulative brine [12]. (C) Schematic illustration of absorption and transport of water and hydrated ions through DAIB membranes. (D) Schematic of two-step nanofiltration to process the concentrated salt mixture and (E) Li$^+$ flux and Mg^{2+}/Li$^+$ in permeate of the first and second nanofiltration [149].

PEI-TMC membrane, leading to loose membranes featuring more positive charge and higher hydrophilicity. As a result, the water permeance of the DAIB/PEI-TMC membrane was improved by 5 times with good stability in a 200-h continuous Li/Mg nanofiltration process. The IL modification and generated nanoscale structural heterogeneity led to enhanced surface hydrophilicity and reduced internal resistance through the membrane selective layer.

Heavy metal removal from water bodies is a serious concern for the environment protection and human health. Heavy metals (Hg, As, Pb, Cr, Cd, Zn, Co, Ni, Ag, and U) have been considered toxic substances due to their harmful effects on the ecosystem, which could also lead to terrible human disease. Thus far, membrane processes have been regarded as a distinguished invention in the wastewater treatment process. Imdad has given a critical review on heavy metals removal using ILMs from the industrial wastewater, with focus on the SILMs [152]. In addition, ILPMs have been also used in heavy metal removal. Zheng et al. prepared 1-vinyl-3-butyl imidazolium tetrafluoroborate (VBImBF$_4$) grafted PES nanofibrous membranes (PES-g-IL), which exhibited excellent Cd(III) removal ability [153]. Besides heavy metals removal, Wang and coworkers developed the Gemini IL-modified nacre-like reduced graphene oxide click membranes for ReO_4^-/TcO_4^- removal [154].

4.5.3.2 Organic separation

As of now, ILMs have been successfully applied for the highly selective separation of organic substances, for example, acetone, butanol, and ethanol (ABE), aromatic hydrocarbons, organic acids, and dyes [145]. Among them, the SILMs make up the majority in the previous reports. For example, Mai et al. fabricated 1-octyl-3-methylimidazolium bis(trifluoromethylsulfonyl)imide ([Omim][NTf$_2$]) /polydimethylsiloxane (PDMS) SILM through vacuum-induced method and used for ABE recovery from aqueous solution [155]. Rdzanek et al. also blended trihexyl (tetradecyl) phosphonium tetracyanoborate (P$_{6,6,6,14}$ tcb) and 1-hexyl-3-methylimidazolium tetracyanoborate (Im$_{6,1}$ tcb), with polyether block amide and immobilized them into the pores of PSf or polypropylene membranes for ABE recovery from water [156]. In addition, three kinds of trihexyl (tetradecyl) phosphonium ionic liquids (i.e., [THTDP][Br], [THTDP][N(CN)$_2$] and [THTDP][NTf$_2$]) were immobilized on PVDF membranes and used for ethanol recovery from dilute aqueous solution. At a feed concentration of 2%(mass fraction) ethanol, the selected SILM was able to maintain its functionality for ~240 h with high ethanol flux [$> 2.2 \, kg/(m^2 \cdot h)$] and selectivity (> 320).

4.5.3.3 Water desalination

Water desalination has attracted considerable interest in order to tackle water scarcity, which remains a grand challenge threatening human health and sustainability [157]. Limited by the unsatisfactory performance of commercial NF membranes, functionalized ILs were introduced for surface modification and different ILPMs were further developed. Peng et al. reported the quaternized diaminoethylpiperzine (QAEP)-based polyamide nanofiltration membranes (Fig. 4.9). The roughness of the PIP-TMC-QAEP membrane (RMS = 29.2 nm) almost doubled compared to the pristine membrane (RMS = 15.3 nm). The PIP-TMC-QAEP membrane exhibited high salt rejection coupled with high permeate flux that was 3 times as high as that of the pristine membranes without IL [158]. This group also explored the 1-aminoethyl-3-methylimidazolium bromide–grafted polyamide membrane, and the water permeance was improved by ∼4 times with high salt rejection (Na_2SO_4, 95%). The amino acid IL (AAIL) was also employed to functionalize the nanofiltration membrane. The introduction of AAIL increased the water permeability, pore size, and Na_2SO_4/NaCl selectivity [94].

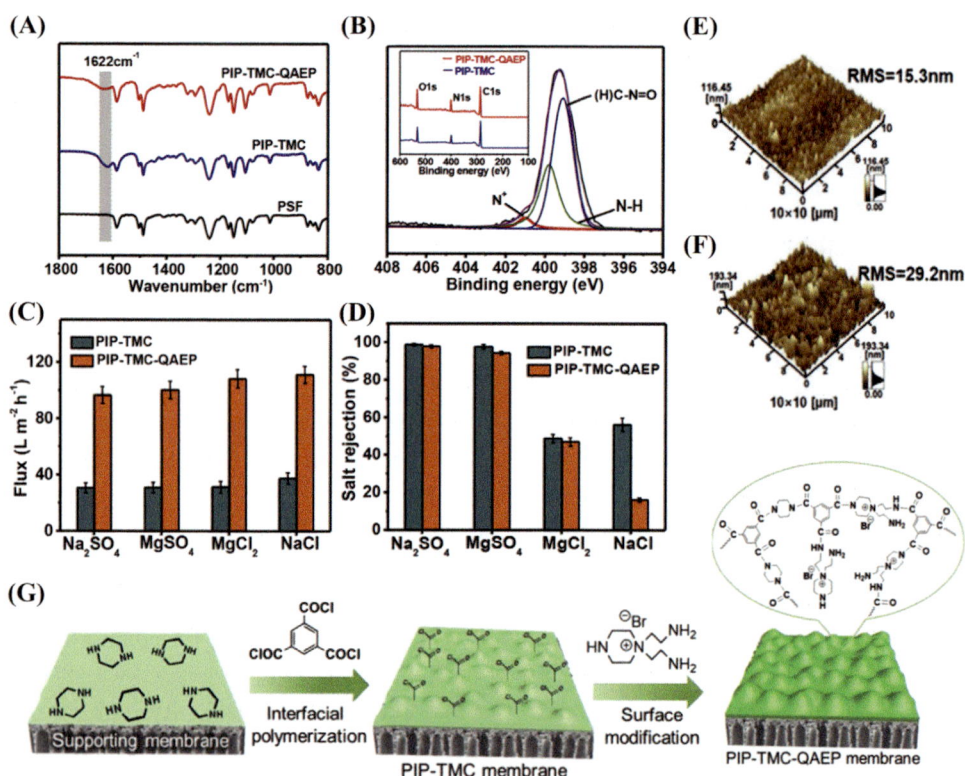

Figure 4.9 (A) ATR-FT-IR spectra of polysulfone, PIP-TMC, and PIP-TMC-QAEP membranes. (B) N1s spectrum of the PIP-TMC-QAEP membrane (inset is the surface XPS spectra of the PIP-TMC-QAEP and PIP-TMC membranes). (C) Water flux and (D) salt rejection of PIP-TMC-QAEP and PIP-TMC membrane. (E, F) AFM images of PIP-TMC and PIP-TMC-QAEP [158]. (G) The surface modification of PIP-TMC polyamide membranes with QAEP monomer [94].

4.5.4 Transport mechanism

In the traditional membrane separation process, the Donnan−Steric pore model with dielectric exclusion is currently a better match for the study. Solute transport is divided into four main steps: ① partitioning at the solution−membrane interface in contact with the feed solution; ② transport process within the nanopore described by the extended Nernst−Planck equation; ③ partitioning process at the solution−membrane interface in contact with the permeate; and ④ diffusion away from the membrane on the permeate side.

For IL-based membranes, the addition of ionic liquids can change the structure and pore size, chargeability, hydrophobicity, and other properties of the membrane. In the case of ILPMs and PILMs, the functionalized ILs exhibit strong chemical interaction with the target component in the aqueous solutions, resulting in an improved rejection ratio for specific metals or organic substances. Moreover, the surface charge, hydrophobicity, and pore size are different with the pristine support membrane. In general, the mass transport process and mechanism of the substances for traditional membranes and ILMs undergo similar steps; however, the addition of ILs facilitates the mass transfer process, either the interaction between the target component with the membrane surface (step ①) or the transport in the membrane pores (step ②). In this case, further work will focus on the transport model or development of more accurate models for the description and prediction of the mass transport process in specific ILMs.

References

[1] Palomar J, Miquel MG, Bedia J, et al. Task-specific inoic liquids for efficient ammonia absorption. Separation and Purification Technology 2011;82:43−52.
[2] Ruiz E, Ferro VR, de Riva J, et al. Evaluation of ionic liquids as absorbents for ammonia absorption refrigeration cycles using COSMO-based process simulations. Applied Energy 2014;123:281−91.
[3] Bedia J, Palomar J, Gonzalez-Miquel M, et al. Screening ionic liquids as suitable ammonia absorbents on the basis of thermodynamic and kinetic analysis. Separation and Purification Technology 2012;95 (12):188−95.
[4] Holbrey JD, Reichert WM, Swatloski RP, et al. Efficient, halide free synthesis of new, low cost ionic liquids: 1, 3-dialkylimidazolium salts containing methyl-and ethyl-sulfate anions. Green Chemistry 2002;4(5):407.
[5] Li ZJ, Zhang XP, Dong HF, et al. Efficient absorption of ammonia with hydroxyl-functionalized ionic liquids. Rsc Advances 2015;5(99):81362−70.
[6] Shang D, Zhang X, Zeng S, et al. Protic ionic liquid [Bim][NTf$_2$] with strong hydrogen bond donating ability for highly efficient ammonia absorption. Green Chemistry 2017;19(4):937−45.
[7] Shang D, Bai L, Zeng S, et al. Enhanced NH$_3$ capture by imidazolium-based protic ionic liquids with different anions and cation substituents. Journal of Chemical Technology & Biotechnology 2018;93(5):1228−36.
[8] Zhao T, Zeng S, Li Y, et al. Molecular insight into the effect of ion structure and interface behavior on the ammonia absorption by ionic liquids. AIChE Journal 2022;68(11):e17860.
[9] Yuan L, Gao H, Jiang H, et al. Experimental and thermodynamic analysis of NH$_3$ absorption in dual-functionalized pyridinium-based ionic liquids. Journal of Molecular Liquids 2021;323:114601.

[10] Yuan L, Zhang X, Ren B, et al. Dual-functionalized protic ionic liquids for efficient absorption of NH$_3$ through synergistically physicochemical interaction. Journal of Chemical Technology & Biotechnology 2020;95(6):1815−24.

[11] Luo X, Qiu R, Chen X, et al. Reversible construction of ionic networks through cooperative hydrogen bonds for efficient ammonia absorption. ACS Sustainable Chemistry & Engineering 2019;7(11):9888−95.

[12] Deng D, Deng X, Li K, et al. Protic ionic liquid ethanolamine thiocyanate with multiple sites for highly efficient NH$_3$ uptake and NH$_3$/CO$_2$ separation. Separation and Purification Technology 2021;276(119298):2−8.

[13] Shang DW, Zeng SJ, Zhang XP, et al. Highly efficient and reversible absorption of NH$_3$ by dual functionalised ionic liquids with protic and Lewis acidic sites. Journal of Molecular Liquids 2020;312.

[14] Li J, Luo L, Yang L, et al. Ionic framework constructed with protic ionic liquid units for improving ammonia uptake. Chemical Communications 2021;57(36):4384−7.

[15] Sun X, Li G, Zeng S, et al. Ultra-high NH$_3$ absorption by triazole cation-functionalized ionic liquids through multiple hydrogen bonding. Separation and Purification Technology 2023;307 (122825):2−9.

[16] Zhang J, Zheng L, Ma Y, et al. A mini-review on NH$_3$ separation technologies: recent advances and future directions. Energy & Fuels: An American Chemical Society Journal 2022;36 (24):14516−33.

[17] Zhang L, Dong H, Zeng S, et al. An overview of ammonia separation by ionic liquids. Industrial & Engineering Chemistry Research 2021;60(19):6908−24.

[18] Zeng S, Cao Y, Li P, et al. Ionic liquid−based green processes for ammonia separation and recovery. Current Opinion in Green and Sustainable Chemistry 2020;25:100354.

[19] Wang JL, Zeng SJ, Chen N, et al. Research progress of ammonia adsorption materials (in Chinese). Chin. J. Process Eng. 2019;19(1):14−24.

[20] Zeng SJ, Shang DW, Yu M, et al. Applications and perspectives of NH$_3$ separation and recovery-with ionic liquids. CIESC Journal 2019;70(3):791−800.

[21] Wang RZ, Wang LW. Adsorption refrigeration green cooling driven by low grade thermal energy. Chinese Science Bulletin 2005;50(3):193−204.

[22] Chen W, Liang SQ, Guo YX, et al. Investigation on vapor−liquid equilibria for binary systems of metal ion-containing ionic liquid [Bmim]Zn$_2$Cl$_5$/NH$_3$ by experiment and modified UNIFAC model. Fluid Phase Equilibria 2013;360(25):1−6.

[23] Kohler FTU, Popp S, Klefer H, et al. Supported ionic liquid phase (SILP) materials for removal of hazardous gas compounds—efficient and irreversible NH$_3$ adsorption. Green Chemistry 2014;16:3560.

[24] Zeng S, Liu L, Shang D, et al. Efficient and reversible absorption of ammonia by cobalt ionic liquids through Lewis acid-base and cooperative hydrogen bond interactions. Green Chemistry 2018;20 (9):2075−83.

[25] Wang J, Zeng S, Huo F, et al. Metal chloride anion-based ionic liquids for efficient separation of NH$_3$. Journal of Cleaner Production 2019;206:661−9.

[26] Cai Z, Zhang J, Ma Y, et al. Chelation-activated multiple-site reversible chemical absorption of ammonia in ionic liquids. AIChE Journal 2022;68(5):17632 1−8.

[27] Jiang WJ, Zhang JB, Zou YT, et al. Manufacturing acidities of hydrogen-bond donors in deep eutectic solvents for effective and reversible NH$_3$ capture. ACS Sustainable Chemistry & Engineering 2020;8(35):13408−17.

[28] Li K, Fang H, Duan X, et al. Efficient uptake of NH$_3$ by dual active sites NH$_4$SCN-imidazole deep eutectic solvents with low viscosity. Journal of Molecular Liquids 2021;339:116724−30.

[29] Deng X, Duan X, Gong L, et al. Ammonia solubility, density, and viscosity of choline chloride−dihydric alcohol deep eutectic solvents. Journal of Chemical and Engineering Data: the ACS Journal for Data 2020;65(10):4845−54.

[30] Li Y, Ali MC, Yang Q, et al. Hybrid deep eutectic solvents with flexible hydrogen-bonded supramolecular networks for highly efficient uptake of NH$_3$. ChemSusChem 2017;10(17):3283−3283.

[31] Cao Y, Zhang X, Zeng S, et al. Protic ionic liquid-based deep eutectic solvents with multiple hydrogen bonding sites for efficient absorption of NH_3. AIChE Journal 2020;66(8):16253.

[32] Cao Y, Zhang J, Ma Y, et al. Designing low-viscosity deep eutectic solvents with multiple weak-acidic groups for ammonia separation. ACS Sustainable Chemistry & Engineering 2021;9(21):7352−60.

[33] Jiang WJ, Zhong FY, Liu Y, et al. Effective and Reversible Capture of NH_3 by ethylamine hydrochloride plus glycerol deep eutectic solvents. ACS Sustainable Chemistry & Engineering 2019;7 (12):10552−60.

[34] Cheng NN, Li ZL, Lan HC, et al. Remarkable NH_3 absorption in metal-based deep eutectic solvents by multiple coordination and hydrogen-bond interaction. AIChE Journal 2022;68(6):17660 1−12.

[35] Blanchard LA, Hancu D, Beckman EJ, et al. Green processing using ionic liquids and CO_2. Nature 1999;399(6731):28−9.

[36] Zhang XP, Zhang XC, Dong HF, et al. Carbon capture with ionic liquids: overview and progress. Energy & Environmental Science: EES 2012;5(5):6668−81.

[37] Zhao YS, Zhang XP, Zhen YP, et al. Novel alcamines ionic liquids based solvents: preparation, characterization and applications in carbon dioxide capture. International Journal of Greenhouse Gas Control 2011;5(2):367−73.

[38] Wappel D, Gronald G, Kalb R, et al. Ionic liquids for post-combustion CO_2 absorption. International Journal of Greenhouse Gas Control 2010;4(3):486−94.

[39] Vega LF, Vilaseca O, Llovell F, et al. Modeling ionic liquids and the solubility of gases in them: recent advances and perspectives. Fluid Phase Equilibria 2010;294(1−2):15−30.

[40] Muldoon MJ, Aki SNVK, Anderson JL, et al. Improving carbon dioxide solubility in ionic liquids. The Journal of Physical Chemistry, B. Condensed Matter, Materials, Surfaces, Interfaces & Biophysical 2007;111(30):9001−9.

[41] Gonzalez-Miquel M, Talreja M, Ethier AL, et al. COSMO-RS studies: structure-property relationships for CO_2 capture by reversible ionic liquids. Industrial & Engineering Chemistry Research 2012;51(49):16066−73.

[42] Yokozeki A, Shiflett MB, Junk CP, et al. Physical and chemical absorptions of carbon dioxide in room-temperature ionic liquids. The journal of Physical Chemistry, B. Condensed Matter, Materials, Surfaces, Interfaces & Biophysical 2008;112(51):16654−63.

[43] Blanchard LA, Gu ZY, Brennecke JF. High-pressure phase behavior of ionic liquid/CO_2 systems. The Journal of Physical Chemistry, B. Condensed Matter, Materials, Surfaces, Interfaces & Biophysical 2001;105(12):2437−44.

[44] Shariati A, Peters CJ. High-pressure phase behavior of systems with ionic liquids: II. The binary system carbon dioxide + 1-ethyl-3-methylimidazolium hexafluorophosphate. Journal of Supercritical Fluids 2004;29(1−2):43−8.

[45] Schilderman AM, Raeissi S, Peters CJ. Solubility of carbon dioxide in the ionic liquid 1-ethyl-3-methylimidazolium bis(trifluoromethylsulfonyl)imide. Fluid Phase Equilibria 2007;260(1):19−22.

[46] Jacquemin J, Husson P, Majer V, et al. Influence of the cation on the solubility of CO_2 and H_2 in ionic liquids based on the bis(trifluoromethylsulfonyl)imide anion. Journal of Solution Chemistry 2007;36(8):967.

[47] Shin EK, Lee BC, Lim JS. High-pressure solubilities of carbon dioxide in ionic liquids: 1-Alkyl-3-methylimidazolium bis (trifluoromethylsulfonyl)imide. Journal of Supercritical Fluids 2008;45(3):282−92.

[48] Cadena C, Anthony JL, Shah JK, et al. Why is CO_2 so soluble in imidazolium-based ionic liquids? Journal of the American Chemical Society 2004;126(16):5300−8.

[49] Bara JE, Gabriel CJ, Lessmann S, et al. Enhanced CO_2 separation selectivity in oligo(ethylene glycol) functionalized room-temperature ionic liquids. Industrial & Engineering Chemistry Research 2007;46(16):5380−6.

[50] Carlisle TK, Bara JE, Gabriel CJ, et al. Interpretation of CO_2 solubility and selectivity in nitrile-functionalized room-temperature ionic liquids using a group contribution approach. Industrial & Engineering Chemistry Research 2008;47(18):7005−12.

[51] Jiang B, Huang ZH, Zhang LH, et al. Highly efficient and reversible CO_2 capture by imidazolate-based ether-functionalized ionic liquids with a capture transforming process. Journal of the Taiwan Institute of Chemical Engineers 2016;69:85−92.

[52] Cui GK, Wang JJ, Zhang SJ. Active chemisorption sites in functionalized ionic liquids for carbon capture. Chemical Society Reviews 2016;45(15):4307−39.

[53] Zhang JM, Zhang SJ, Dong K, et al. Supported absorption of CO_2 by tetrabutylphosphonium amino acid ionic liquids. Chemistry: A European Journal 2006;12(15):4021−6.

[54] Zhang YQ, Zhang SJ, Lu XM, et al. Dual amino-functionalised phosphonium ionic liquids for CO_2 capture. Chemistry: A European Journal 2009;15(12):3003−11.

[55] Xue ZM, Zhang ZF, Han J, et al. Carbon dioxide capture by a dual amino ionic liquid with amino-functionalized imidazolium cation and taurine anion. International Journal of Greenhouse Gas Control 2011;5(4):628−33.

[56] Liu XM, Zhou GH, Zhang SJ, et al. Molecular dynamics simulation of dual amino-functionalized imidazolium-based ionic liquids. Fluid Phase Equilibria 2009;284(1):44−9.

[57] Zeng SJ, Wang J, Bai L, et al. Highly selective capture of CO_2 by ether-functionalized pyridinium ionic liquids with low viscosity. Energy & Fuels 2015;29(9):6039−48.

[58] Bates ED, Mayton RD, Ntai I, et al. CO_2 capture by a task-specific ionic liquid. Journal of the American Chemical Society 2002;124(6):926−7.

[59] Gurkan BE, de la Fuente JC, Mindrup EM, et al. Equimolar CO_2 absorption by anion-functionalized ionic liquids. Journal of the American Chemical Society 2010;132(7):2116−17.

[60] Zhang JZ, Jia C, Dong HF, et al. A novel dual amino-functionalized cation-tethered ionic liquid for CO_2 capture. Industrial & Engineering Chemistry Research 2013;52(17):5835−41.

[61] Lv BH, Jing GH, Qian YH, et al. An efficient absorbent of amine-based amino acid-functionalized ionic liquids for CO_2 capture: High capacity and regeneration ability. Chemical Engineering Journal 2016;289:212−18.

[62] Zhou ZM, Zhou XB, Jing GH, et al. Evaluation of the multi-amine functionalized ionic liquid for efficient postcombustion CO_2 capture. Energy & Fuels: An American Chemical Society Journal 2016;30(9):7489−95.

[63] Bhattacharyya S, Shah FU. Ether functionalized choline tethered amino acid ionic liquids for enhanced CO_2 capture. ACS Sustainable Chemistry & Engineering 2016;4(10):5441−9.

[64] Wang CM, Luo HM, Li HR, et al. Tuning the physicochemical properties of diverse phenolic ionic liquids for equimolar CO_2 capture by the substituent on the anion. Chemistry: A European Journal 2012;18(7):2153−60.

[65] Luo XY, Guo Y, Ding F, et al. Significant Improvements in CO_2 capture by pyridine-containing anion-functionalized ionic liquids through multiple-site cooperative interactions. Angewandte Chemie-international Edition 2014;53(27):7053−7.

[66] Gurkan B, Goodrich BF, Mindrup EM, et al. Molecular design of high capacity, low viscosity, chemically tunable ionic liquids for CO_2 capture. Journal of Physical Chemistry Letters 2010;1(24):3494−9.

[67] Cao LD, Gao JB, Zeng SJ, et al. Feasible ionic liquid-amine hybrid solvents for carbon dioxide capture. International Journal of Greenhouse Gas Control 2017;66:120−8.

[68] Cao LD, Huang JH, Zhang XP, et al. Imidazole tailored deep eutectic solvents for CO_2 capture enhanced by hydrogen bonds. Physical Chemistry Chemical Physics 2015;17(41):27306−16.

[69] Lv BH, Xia YF, Shi Y, et al. A novel hydrophilic amino acid ionic liquid C_2OHmim Gly as aqueous sorbent for CO_2 capture. International Journal of Greenhouse Gas Control 2016;46:1−6.

[70] Wang CM, Mahurin SM, Luo HM, et al. Reversible and robust CO_2 capture by equimolar task-specific ionic liquid-superbase mixtures. Green Chemistry 2010;12(5):870−4.

[71] Wang CM, Luo HM, Luo XY, et al. Equimolar CO_2 capture by imidazolium-based ionic liquids and superbase systems. Green Chemistry 2010;12(11):2019−23.

[72] Yang QW, Bao ZB, Xing HB, et al. Progress in the separation of structurally similar compounds by ionic liquid extraction. Chemical Progress 2019;38(1):91−9.

[73] Feng J, Peng XM, Li CQ, et al. Application of ionic liquids in the extraction of natural product actives. Applied Chemical Engineering 2019;48(4):945−9.

[74] Huddleston JG, Willauer HD, Swatloski RP, et al. Room temperature ionic liquids as novel media for 'clean' liquid−liquid extraction. Chemical Communications 1998;16:1765−6.

[75] Yu YY, Zhang W, Cao SW. Extraction of ferulic acid and caffeic acid with ionic liquids. Chinese Journal of Analytical Chemistry 2007;35(12):1726−30.

[76] Fan J, Fan Y, Pei Y, et al. Solvent extraction of selected endocrine-disrupting phenols using ionic liquids. Separation and Purification Technology 2008;61(3):324−31.

[77] Kumar L, Banerjee T, Mohanty K. Prediction of selective extraction of cresols from aqueous solutions by ionic liquids using theoretical approach. Separation Science and Technology 2011;46(13):2075−87.

[78] Miyano Y, Kobashi T, Shinjo H, et al. Henry's law constants and infinite dilution activity coefficients of cis-2-butene, dimethylether, chloroethane, and 1,1-difluoroethane in methanol, 1-propanol, 2-propanol, 1-butanol, 2-butanol, isobutanol, tert-butanol, 1-pentanol, 2-pentanol, 3-pentanol, 2-methyl-1-butanol, 3-methyl-1-butanol, and 2-methyl-2-butanol. The Journal of Chemical Thermodynamics 2006;38(6):724−31.

[79] Yang Q, Guo S, Liu X, et al. Highly efficient separation of strongly hydrophilic structurally related compounds by hydrophobic ionic solutions. AIChE Journal 2017;64(4):1373−82.

[80] Bai Y, Yan R, Tu W, et al. Selective separation of methacrylic acid and acetic acid from aqueous solution using carboxyl-functionalized ionic liquids. ACS Sustainable Chemistry & Engineering 2017;6(1):1215−24.

[81] Chen MM, Yuan L, Gao M, et al. Study on biphasic extraction of rapeseed meal protein by ionic liquid and its phase behavior. Chinese Journal of Cereals and Oils 2013;28(6):56−61.

[82] Wang WT, Zhang HD, Jiang ZG, et al. Ionic liquid biphasic extraction of papain and optimization of. Modern Food Science and Technology 2014;30(9):210−16.

[83] Dreyer S, Salim P, Kragl U. Driving forces of protein partitioning in an ionic liquid-based aqueous two-phase system. Biochemical Engineering Journal 2009;46(2):176−85.

[84] Capela EV, Quental MV, Domingues P, et al. Effective separation of aromatic and aliphatic amino acid mixtures using ionic-liquid-based aqueous biphasic systems. Green Chemistry 2017;19(8):1850−4.

[85] Zhang JZ. Construction and extraction performance of poly(ionic liquid) biphasic system [D/OL]. Zhejiang University 2016.

[86] Yang Q, Xing H, Su B, et al. Improved separation efficiency using ionic liquid−cosolvent mixtures as the extractant in liquid−liquid extraction: A multiple adjustment and synergistic effect. Chemical Engineering Journal 2012;181−182:334−42.

[87] Liu XX, Yang QW, Bao ZB, et al. Nonaqueous lyotropic ionic liquid crystals: preparation, characterization, and application in extraction. Chemistry-a European Journal 2015;21(25):9150−6.

[88] Cheng H, Li J, Wang J, et al. Enhanced vitamin E extraction selectivity from deodorizer distillate by a biphasic system: a COSMO-RS and experimental study. ACS Sustainable Chemistry & Engineering 2018;6(4):5547−54.

[89] Arce A, Pobudkowska A, Rodríguez O, et al. Citrus essential oil terpenless by extraction using 1-ethyl-3-methylimidazolium ethylsulfate ionic liquid: effect of the temperature. Chemical Engineering Journal 2007;133(1):213−18.

[90] Cao Y, Xing H, Yang Q, et al. High performance separation of sparingly aqua-/lipo-soluble bioactive compounds with an ionic liquid-based biphasic system. Green Chemistry 2012;14(9):2617−25.

[91] Manic MS, Najdanovic-visak V, Da Ponte MN, et al. Extraction of free fatty acids from soybean oil using ionic liquids or poly (ethyleneglycol)s. AIChE Journal 2011;57(5):1344−55.

[92] Cheong LZ, Guo Z, Yang Z, et al. Extraction and enrichment of n-3 polyunsaturated fatty acids and ethyl esters through reversible π−π complexation with aromatic rings containing ionic liquids. Journal of Agricultural and Food Chemistry 2011;59(16):8961−7.

[93] Dong K, Cao Y, Yang Q, et al. Role of hydrogen bonds in ionic-liquid-mediated extraction of natural bioactive homologues. Industrial & Engineering Chemistry Research 2012;51(14):5299−308.

[94] Xiao HF, Chu CH, Xu WT. Amphibian-inspired amino acid ionic liquid functionalized nanofiltration membranes with high water permeability and ion selectivity for pigment wastewater treatment. Journal of Membrane Science 2019;586:44−52.

[95] Liang R, Bao Z, Su B, et al. Feasibility of ionic liquids as extractants for selective separation of vitamin D_3 and tachysterol$_3$ by solvent extraction. Journal of Agricultural and Food Chemistry 2013;61(14):3479−87.

[96] Cao Y, Tan X, Zhan G, et al. Novel process for selective separation of trace artemisitene from artemisinin by ammonium functional ionic. AIChE Journal 2022;68(8):17711 1−13.

[97] Cao Y, Wang Y, Chen B, et al. Design of dual stimuli-responsive copolymerized ionic liquid with flexible phase transition temperature and its application in selective separation of artemisitene/artemisinin. ACS Sustainable Chemistry & Engineering 2023;11(11):4463−72.

[98] He J, Tao W, Dong G. Study on extraction performance of vanadium (V) from aqueous solution by octyl-imidazole. Ionic Liquids Extractants 2022;12(5):854.

[99] Han Q, Ryan TM, Rosado CJ, et al. Effect of ionic liquids on the fluorescence properties and aggregation of superfolder green fluorescence protein. Journal of Colloid and Interface Science 2021;591:96−105.

[100] Shukla SK, Mikkola JP. Use of ionic liquids in protein and DNA chemistry. Frontiers in Chemistry 2020;8.

[101] Egorova KS, Gordeev EG, Ananikov VP. Biological activity of ionic liquids and their application in pharmaceutics and medicine. Chemical Reviews 2017;117(10):7132−89.

[102] Cevasco G, Chiappe C. Are ionic liquids a proper solution to current environmental challenges? Green Chemistry 2014;16(5):2375−85.

[103] Fujita K, MacFarlane DR, Forsyth M. Protein solubilising and stabilising ionic liquids. Chemical Communications 2005;38:4804−6.

[104] Zhao Q, Chu H, Zhao B, et al. Advances of ionic liquids-based methods for protein analysis. TrAC Trends in Analytical Chemistry 2018;108:239−46.

[105] Ventura SPM, e Silva FA, Quental MV, et al. Ionic-liquid-mediated extraction and separation processes for bioactive compounds: past, present, and future trends. Chemical Reviews 2017;117 (10):6984−7052.

[106] Trujillo-Rodriguez MJ, Nan H, Varona M, et al. Advances of ionic liquids in analytical chemistry. Analytical Chemistry 2019;91(1):505−31.

[107] Phillips DM, Drummy LF, Conrady DG, et al. Dissolution and regeneration of bombyx mori silk fibroin using ionic liquids. Journal of the American Chemical Society 2004;126(44):14350−1.

[108] Li Z, Liu X, Pei Y, et al. Design of environmentally friendly ionic liquid aqueous two-phase systems for the efficient and high activity extraction of proteins. Green Chemistry 2012;14(10):2941−50.

[109] Ge L, Wang XT, Tan SN, et al. A novel method of protein extraction from yeast using ionic liquid solution. Talanta 2010;81(4−5):1861−4.

[110] Wang X, Zhang X. Optimal extraction and hydrolysis of Chlorella pyrenoidosa proteins. Bioresource Technology 2012;126:307−13.

[111] Cao S, Peterson SM, Muller S, et al. A membrane protein display platform for receptor interactome discovery. Proceedings of the National Academy of Sciences of the United States of America 2021;118(39).

[112] Freed DM, Bessman NJ, Kiyatkin A, et al. EGFR ligands differentially stabilize receptor dimers to specify signaling kinetics. Cell 2017;171(3):683−95 e18.

[113] Hu Z, Yuan J, Long M, et al. The cancer surfaceome atlas integrates genomic, functional and drug response data to identify actionable targets. Nature Cancer 2021;2(12):1406−22.

[114] Santos R, Ursu O, Gaulton A, et al. A comprehensive map of molecular drug targets. Nature Reviews. Drug Discovery 2017;16(1):19−34.

[115] Kitata RB, Dimayacyac-Esleta BR, Choong WK, et al. Mining missing membrane proteins by high-pH reverse-phase stagetip fractionation and multiple reaction monitoring mass spectrometry. Journal of Proteome Research 2015;14(9):3658−69.

[116] Vuckovic D, Dagley LF, Purcell AW, et al. Membrane proteomics by high performance liquid chromatography-tandem mass spectrometry: Analytical approaches and challenges. Proteomics 2013;13(3−4):404−23.

[117] Chang YH, Gregorich ZR, Chen AJ, et al. New mass-spectrometry-compatible degradable surfactant for tissue proteomics. Journal of Proteome Research 2015;14(3):1587−99.

[118] Wisniewski JR, Zougman A, Nagaraj N, et al. Universal sample preparation method for proteome analysis. Nature Methods 2009;6(5):359−62.

[119] Sun L, Tao D, Han B, et al. Ionic liquid 1-butyl-3-methyl imidazolium tetrafluoroborate for shotgun membrane proteomics. Analytical and Bioanalytical Chemistry 2011;399(10):3387−97.

[120] Zhao Q, Sun L, Liang Y, et al. Prefractionation and separation by C8 stationary phase: effective strategies for integral membrane proteins. Talanta 2012;88:567−72.

[121] Zhao Q, Fang F, Liang Y, et al. 1-Dodecyl-3-methylimidazolium chloride-assisted sample preparation method for efficient integral membrane. Analytical Chemistry 2014;86(15):7544−50.

[122] Fang F, Zhao Q, Chu H, et al. Molecular dynamics simulation-assisted ionic liquid screening for deep coverage proteome analysis. Molecular & Cellular Proteomics 2020;19(10):1724−37.

[123] Sui Z, Weng Y, Zhao Q, et al. Ionic liquid-based method for direct proteome characterization of velvet antler cartilage. Talanta 2016;161:541−6.

[124] Vincourt JB, Lionneton F, Kratassiouk G, et al. Establishment of a reliable method for direct proteome characterization of human articular. Molecular & Cellular Proteomics 2006;5(10):1984−95.

[125] Fang F, Zhao Q, Li X, et al. Dissolving capability difference based sequential extraction: a versatile tool for in-depth membrane proteome analysis. Analytica Chimica Acta 2016;945:39−46.

[126] Zhao Q, Fang F, Shan Y, et al. In-depth proteome coverage by improving efficiency for membrane proteome analysis. Analytical Chemistry 2017;89(10):5179−85.

[127] Sun L, Zhang Y, Tao D, et al. SDS-PAGE-free protocol for comprehensive identification of cytochrome P450 enzymes and uridine diphosphoglucuronosyl transferases in human liver microsomes. Proteomics 2012;12(23−24):3464−9.

[128] Qiao X, Tao D, Qu Y, et al. Large-scale N-glycoproteome map of rat brain tissue: simultaneous characterization of insoluble and soluble protein fractions. Proteomics 2011;11(21):4274−8.

[129] Chen L, Guo P, Li W, et al. Perturbation of specific signaling pathways is involved in initiation of mouse liver fibrosis. Hepatology (Baltimore, Md.) 2021;73(4):1551−69.

[130] Li Y, Fang F, Sun M, et al. Ionic liquid-assisted protein extraction method for plant phosphoproteome analysis. Talanta: The International Journal of Pure and Applied Analytical Chemistry 2020;213:120848−55.

[131] Sun R, Dong C, Li R, et al. Proteomic analysis reveals that EPHX1 Contributes to 5-fluorouracil resistance in a human hepatocellular carcinoma cell line. Proteomics. Clinical Applications 2020;14 (4):e1900080.

[132] Chu H, Wu C, Zhao Q, et al. Quantitative proteomics identifies FOLR1 to drive sorafenib resistance via activating autophagy in hepatocellular carcinoma cells. Carcinogenesis 2021;42 (5):753−61.

[133] Liu J, Zhou Y, Shan Y, et al. A multiplex fragment-ion-based method for accurate proteome quantification. Analytical Chemistry 2019;91(6):3921−8.

[134] Yang K, Liu J, Sun J, et al. Proteomic study provides new clues for complications of hemodialysis caused by dialysis membrane. Science Bulletin (Beijing) 2017;62(18):1251−5.

[135] Han S, Yang K, Sun J, et al. Proteomics investigations into serum proteins adsorbed by high-flux and low-flux dialysis membranes. PROTEOMICS—Clinical Applications 2017;11(11−12):1700079.

[136] Liu C, Si X, Yan S, et al. Development of the C12Im-Cl-assisted method for rapid sample preparation in proteomic. Analytical Methods 2021;13(6):776−81.

[137] Taoka M, Horita K, Takekiyo T, et al. An ionic liquid-based sample preparation method for next-stage aggregate proteomic analysis. Analytical Chemistry 2019;91(21):13494−500.

[138] Chu H, Zhao Q, Liu J, et al. Ionic system liquid-based extraction for in-depth analysis of membrane protein complexes. Analytical Chemistry 2022;94(2):758−67.

[139] Harari D, Yarden Y. Molecular mechanisms underlying ErbB2/HER2 action in breast cancer. Oncogene 2000;19(53):6102−14.

[140] Tome LC, Marrucho IM. Ionic liquid-based materials: a platform to design engineered CO_2 separation membranes. Chemical Society Reviews 2016;45(10):2785−824.

[141] Qian W, Texter J, Yan F. Frontiers in poly(ionic liquid)s: syntheses and applications. Chemical Society Reviews 2017;46(4):1124−59.

[142] Lei LF, Pan FJ, Lindbråthen A, et al. Carbon hollow fiber membranes for a molecular sieve with precisecutoff ultramicropores for superior hydrogen separation. Nature Communications 2021;12:268.

[143] Ahmad MG, Chanda K. Ionic liquid coordinated metal-catalyzed organic transformations: a comprehensive review. Coordination Chemistry Reviews 2022;472:1−31.

[144] Cowan MG, Gin DL, Noble RD. Poly(ionic liquid)/Ionic liquid ion-gels with high "free" ionic liquid content: platform membrane materials for CO_2/light gas separations. Accounts of Chemical Research 2016;49(4):724−32.

[145] Wang JF, Luo JQ, Feng SC, et al. Recent development of ionic liquid membranes. Green Energy & Environ 2016;1:43−61.
[146] Sasikumar B, Arthanareeswaran G, Ismail AF. Recent progress in ionic liquid membranes for gas separation. Journal of Molecular Liquids 2018;266:330−41.
[147] Elias G, Díez S, Fontàs C. System for mercury preconcentration in natural waters based on a polymer inclusion membrane incorporating an ionic liquid. Journal of Hazardous Materials 2019;371:316−22.
[148] Li X, Liu YX, Wang J, et al. Metal−organic frameworks based membranes for liquid separation. Chemical Society Reviews 2017;46:7124−44.
[149] Soyekwo F, Wen H, Liao D, et al. Nanofiltration membranes modified with a clustered multiquaternary ammonium-based ionic liquid for improved magnesium/lithium separation. ACS Applied Materials & Interfaces 2022;14:32420−32.
[150] Wu HH, Lin YK, Feng WY, et al. A novel nanofiltration membrane with [MimAP][Tf$_2$N] ionic liquid for utilization of lithium from brines with high Mg^{2+}/Li$^+$ ratio. Journal of Membrane Science 2020;603:117997.
[151] Peng HW, Zhao Q. A Nano-heterogeneous membrane for efficient separation of lithium from high magnesium/lithium ratio brine. Advanced Functional Materials 2021;31:2009430.
[152] Imdad S, Dohare RK. A critical review on heavy metals removal using ionic liquid membranes from the industrial wastewater. Chemical Engineering and Processing 2022;173(108812):1−13.
[153] Zheng X, Ni CJ, Xiao WW, et al. Ionic liquid grafted polyethersulfone nanofibrous membrane as recyclable adsorbent with simultaneous dye, heavy metal removal and antibacterial property. Chemical Engineering Journal 2022;428(3):132111.2−11.
[154] Wang KC, Yan ZY, Fu LL, et al. Gemini ionic liquid modified nacre-like reduced graphene oxide click membranes for ReO$_4^-$/TcO$_4^-$ removal. Separation and Purification Technology 2022;302 (122073):2−9.
[155] Mai NL, Kim SH, Ha HS, et al. Selective recovery of acetone-butanol-ethanol from aqueous mixture by pervaporation using immobilized ionic liquid polydimethylsiloxane membrane. Korean Journal of Chemical Engineering 2013;30:1804−9.
[156] Rdzanek P, Heitmann S, Górak A, et al. Application of supported ionic liquid membranes (SILMs) for biobutanol pervaporation. Separation and Purification Technology 2015;155:83−8.
[157] Elimelech M, Phillip WA. The future of seawater desalination: energy, technology, and the enviroment. Science (New York, N.Y.) 2011;333:712−17.
[158] Peng HW, Tang QQ, Tang SH, et al. Surface modified polyamide nanofiltration membranes with high permeability and stability. Journal of Membrane Science 2019;592:117386.

CHAPTER 5

Photo- and electrochemical process with ionic liquids

Contents

5.1 Overview   143
5.2 Photocatalytic and photoelectrocatalytic process with ionic liquids   144
    5.2.1 Application of ionic liquids in photocatalytic systems   144
    5.2.2 Application of ionic liquids in photoelectrocatalytic systems   148
5.3 Application of ionic liquids in solar cells   150
    5.3.1 Ionic liquids as additives in perovskite precursor solutions   151
    5.3.2 Ionic liquids as solvent for perovskite solar cells   153
    5.3.3 Ionic liquid-modified charge transport layers for perovskite solar cells   155
    5.3.4 Ionic liquids for interface modification   157
5.4 Ionic liquids intensify the electrochemistry process   159
    5.4.1 Oxygen reduction reaction behaviors in different ionic liquids electrolyte systems   161
    5.4.2 Study on electrochemical lignin depolymerization using p-benzyloxyl phenol as model compounds   164
    5.4.3 Depolymerization of lignin by electrochemical method in ionic liquids   179
5.5 Application of ionic liquids in new energy batteries   184
    5.5.1 Application of ionic liquids in lithium-ion batteries   184
    5.5.2 Application of ionic liquids in supercapacitors   189
    5.5.3 Application of ionic liquids in flow batteries   192
References   197

5.1 Overview

Ionic liquids (ILs) have been successfully applied in various photochemical and photo-electrochemical processes due to their unique properties. ILs are typically used as solvents or catalysts in photochemical reactions. Their high thermal stability makes them excellent solvents for photochemical reactions, which require high temperatures. Many studies demonstrated that the ability of ILs to stabilize charged species can enhance the selectivity and efficiency of the reactions. A prime example of their usage is in solar cells, where the viscosity and nonvolatility of ILs make them ideal candidates for use in dye-sensitized solar cells and perovskite solar cells (PSCs).

In electrochemical reactions, ILs often serve as electrolytes due to their high ionic conductivity and broad electrochemical stability window. For instance, they are used in lithium–ion batteries and fuel cells to improve their performance and safety.

Furthermore, as a nonaqueous supported electrolyte with good conductivity, ILs provide possibilities to design certain special electrochemical synthesis and electroplating process. It has been found that the oxygen reduction reaction (ORR) process could be regulated to produce different sorts of reactive oxygen species in a supporting electrolyte system with the controlled number of protons, thereby triggering different electrolysis processes. For example, certain lignin model compounds might undergo different depolymerization processes and form different depolymerization products just using aprotic or protonic ILs as the supporting electrolyte system.

This chapter focuses on the comprehensive understanding of the functions of ILs in photocatalytic and electrocatalytic processes and reviews the applications of ILs in lignin depolymerization, solar cells, lithium–ion batteries, and supercapacitors. The key advantages and challenges associated with using ILs in photocatalytic and electrocatalytic reactions, along with recent developments and promising future directions in this field are presented. It is worth noting that due to space limitations, this chapter does not cover some important topics on the broader application of ILs in photochemical and electrochemical processes, like their recovery and reuse, potential toxicity and environmental impact, and their high cost compared to conventional solvents or electrolytes.

5.2 Photocatalytic and photoelectrocatalytic process with ionic liquids

ILs are salts composed of larger organic cations and inorganic or organic anions in a liquid state at room temperature, which are stable and act as both solvent and catalyst. In photocatalytic processes involving ILs, most studies have loaded ILs onto photocatalysts, which can promote charge separation and introduce catalytic sites to improve photocatalytic efficiency. In the photoelectrochemical process, ILs are mainly present as reaction medium to promote the solvation diffusion and charge transport of molecules, which is conducive to the adsorption and activation of molecules and improves the performance of the catalytic system.

5.2.1 Application of ionic liquids in photocatalytic systems

For photocatalysis, ILs can be used for photocatalytic reduction of CO_2 [1–4], photocatalytic hydrogen production [5], photocatalytic degradation [6–8], and photocatalytic organic reactions [9–11].

An IL ([Bmim][NTf$_2$])-loaded self-doped TiO_2 nanotube (ILs-RTNT) composite photocatalyst was developed for the photoreduction of CO_2 to acetic acid [1]. The experiments demonstrated that acetic acid yield of ILs-RTNTs reached as high as 88.1 $\mu mol/g_{cat}/h$, and there was no decrease after six cycles. It is found that the defects constructed by self-doping and the surface modification by ILs introduced more active sites, strengthened the light utilization ability of the photocatalyst and the adsorption of CO_2, changed the electronic structure, enhanced the separation of photoinduced electron—hole pairs, as well as lowered the energy barrier for electron transfer of the photocatalyst.

Wang et al. developed a new protocol to photochemically reduce CO_2 to CO at mild conditions using ILs for capture and activation of CO_2 [2]. Various ILs were applied in the visible light catalytic system combining [Ru(bpy)$_3$]Cl$_2$(bpy = 2,2'-bipyridine) and CoCl$_2 \cdot 6H_2O$ as a light sensitizer and an electron mediator. The efficiency of CO_2 photoreduction is strongly related to the chemical properties of counterions (NTf$_2{}^-$, L-L$^-$, TfO$^-$, Ac$^-$, DCA$^-$, and BF$_4{}^-$) and the organic functional groups on the imidazolium cation (Emim, Bmim, Hmim, Omim, and BDimim). Also, the mutual action of IL and H_2O plays a critical role in the improvement of photocatalytic activity toward CO_2 reduction. The best promotional effect of the ILs is the NTf$_2{}^-$ anion due to the structural symmetry and low viscosity, compared to other anions such as L-L$^-$, TfO$^-$, Ac$^-$, DCA$^-$, and BF$_4{}^-$.

Jing and coworkers constructed a nanocomposite photocatalytic system (IL/Co-bCN) including g-C$_3$N$_4$ nanosheets coloaded by IL and borate-anchored Co single atoms for CO_2 and water photoreduction to CO and CH$_4$ under UV–vis light irradiation [3]. The loading of typical ILs like [Emim][BF$_4$], [Bmim][BF$_4$], [Emim][NTf$_2$], and [Bmim][NTf$_2$] demonstrated different catalytic performances. Among them, the best one is [Emim][BF$_4$], which greatly enhanced the photoactivity for CO_2 photoreduction. It is found that the introduction of [Emim][BF$_4$] could improve CO_2 reduction selectivity and inhibit H$_2$ evolution. IL has been verified to function as a favorable electron acceptor and also a catalyst for CO_2 reduction, while Co single atoms have been proved to trap holes and catalyze water oxidation. The possible photocatalytic mechanism of CO_2 photoconversion proposed that carboxylation at the C$_2$ position of imidazolium cations would occur, which is triggered by the reaction with CO_2 [3].

A Z-scheme heterojunction (CoO$_x$-BVO/CN-IL) based on BiVO$_4$/g-C$_3$N$_4$ constructed by spatially separated dual sites with CoO$_x$ clusters and imidazolium IL proposes a new avenue for CO_2 photoreduction to CO [4]. DFT calculations demonstrate that the H atom of IL is connected to CO_2 with a weak H-bonding character, and the CO_2 molecule could bend at the interface between IL and CN. The charge transfer of BVO/CN is significantly promoted after IL modification. μs-TAS spectra indicate that the modified IL could effectively trap electrons and activate CO_2, accelerating the electron kinetics for CO_2 reduction in such a CO_2-enriched environment. The synergistic effects of modified CoO$_x$ and IL facilitate the modulation of the Z-scheme charge transfer, and activating water and CO_2 boosts overall CO_2 reduction reactions. The conversion pathway shows that CO_2 molecules tend to complex with the hydrogen atom of the imidazolium cation to form carboxylate intermediates and then get attacked by the photogenerated electrons to form COOH* species to yield CO.

Yildirim and coworkers reported a photocatalyst system (IL-Dye:Pt/TiO$_2$) composed of 1 wt% Pt/TiO$_2$ particulates sensitized with N719 dye (di-tetrabutylammonium cis-bis-(isothiocyanato)bis(2,2'-bipyridyl-4,4'-dicarboxylato) ruthenium(II)) and coated by a thin layer of IL [Bmim][BF$_4$] [5]. The UV-visible absorbance spectrum shows that dye sensitization and IL coating do not change the bandgap of Pt/TiO$_2$, while Pt impregnation lowers it slightly. The IL can stabilize the dye on the photocatalyst surface, while simultaneously enhancing the charge transfer between the dye and TiO$_2$ and preventing the recombination of photogenerated electron−hole pairs. Thus the performance of photocatalytic hydrogen production was increased from 21 μmol/h · g$_{cat}$ to 70 μmol/h · g$_{cat}$ by coating the dye-sensitized particulates with an IL over Pt/TiO$_2$.

The introduction of IL is a good way to improve the photocatalytic activity of the MOF material. Four kinds of IL/MIL-68(In)-NH$_2$ photocatalysts were synthesized for visible light catalytic degradation of doxycycline hydrochloride (DOXH), including diethylenetriamine acetate ([DETA][OAc]), diethylenetriamine hexafluorophosphate ([DETA][PF$_6$]), 1-ethyl-3-methylimidazole acetate ([Emim][OAc]), and 1-ethyl-3-methylimidazole hexafluorophosphate ([Emim][PF$_6$]) [12].

A possible photocatalytic mechanism shows that the introduction of ILs enhanced the migration of photoexcited electrons and effectively promoted the separation of photoexcited electron−hole pairs (Fig. 5.1). IL$_{DAc}$/MIL-68(In)-NH$_2$ can be stimulated to generate photogenerated holes (h$^+$) and electrons (e$^−$) in the HOMO and LUMO

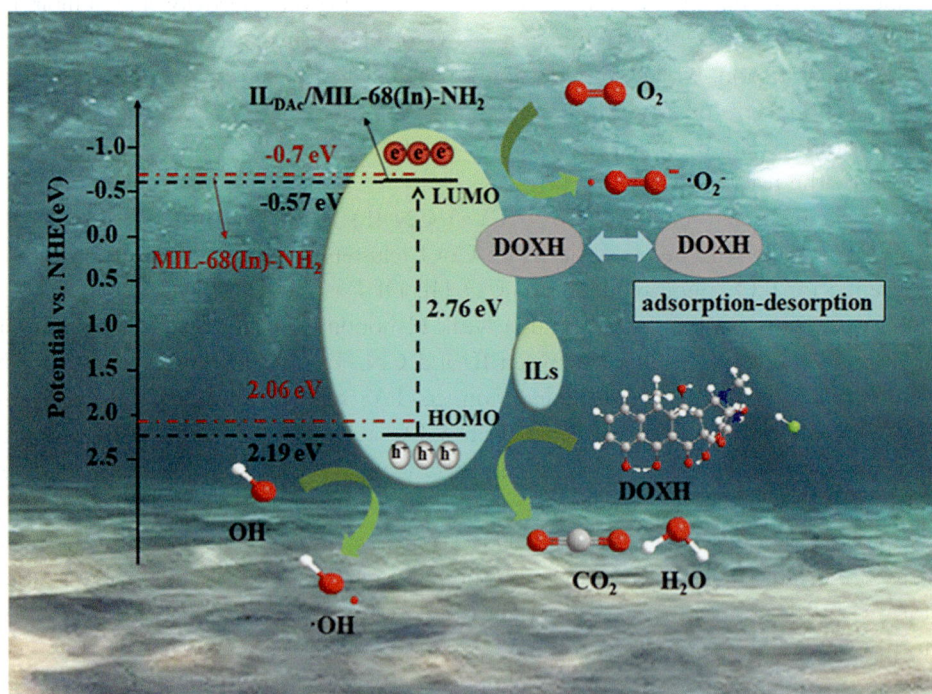

Figure 5.1 Photocatalytic mechanism for the degradation of DOXH by IL$_{DAc}$/MIL-68(In)-NH$_2$ [12]. *DOXH*, Doxycycline hydrochloride.

under visible light irradiation. DOXH can be directly oxidized and decomposed into CO_2 and H_2O by photogenic holes with strong oxidation capacity. Among them, IL_{DAc}/MIL-68(In)–NH_2 can provide more catalytic active centers and increased active sites, and show the highest photocatalytic activity with a 92% removal rate of DOXH. Cycle experiments indicate that IL_{DAc}/MIL-68(In)–NH_2 composites have excellent stability and reusability.

Based on green chemistry principles, supported IL phase (SILP) catalysis has stimulated interest in grafting IL-like units onto porous materials with a high surface area. A magnetically retrievable ferrocene-appended supported IL phase photocatalyst containing a molybdate anion ([FemIL@SiO_2@Mag]$_2$$MoO_4$) has been synthesized for degradation of methyl orange (MO) under UV light (365 nm) irradiation [6]. A plausible photodegradation mechanism of MO shows that in the UV light irradiation, excitation of the electron from the valence band (VB) to conduction band of [FemIL@SiO_2@Mag]$_2$$MoO_4$ leads to the separation of electrons and holes. Then photoinduced holes left in VB accept electrons from the hydroxyl group forming highly oxidative ·OH radicals, which are responsible for degradation of MO dye molecules. This study establishes a new heterogeneous ferrocene-based Fenton system, and the degradation percentage of MO as a pollutant was calculated about 99% within 30 min. The photocatalyst exhibited remarkable recyclability maintained in the range of 99%–93% up to six consecutive cycles.

Poly(ionic liquid)s (PILs) have gained great interest in material research due to their favorable properties of both IL and conventional polyelectrolyte, especially for serving as a supporter of functional materials. Meng et al. developed a PIL nanofiber membrane with phosphotungstic acid (PW_{12}) as photocatalysts (PW-PIL) to degrade dye wastewater under visible light [7]. The honeycomb-like cavity may increase the surface area of the membrane, and the bandgap energy of PW-PIL is 3.09 eV. As the imidazolinium groups with the positive charge can seize the negatively charged PW_{12} via coulombic interaction, a built-in field could be formed between PW_{12} and PIL chain, which would enhance the transfer of photogenerated electrons. The photogenerated holes of the VB level react with OH^- to form ·OH, which further degrades MO. The degradation rate was found to be around 98% and can be well maintained after five cycles. A sustainable process can be obtained when the photocatalyst PW-PIL utilizes the sun as a cheap and clean source of light on a large scale.

A polymer-supported photocatalyst of Rose Bengal (RB) immobilized on supported IL-like (RB-SILLPs) phases was developed for the photooxygenation of furoic acid to butanolide [8]. The presence of these IL-like units facilitates the immobilization and stabilization of quantum dots, which works as light antennas to increase the light-harvesting capacity at wavelengths not corresponding to the optimal absorption of RB. Due to the easy separation of products and the continuous reutilization of the photocatalyst in consecutive cycles, avoiding downtime and the consequent loss of productivity, the photocatalytic reactions can be used for continuous-flow photochemical processes. This RB-SILLP system was very stable and could be reused for eight reaction cycles without any decay in its photocatalytic activity [9].

A heterogeneous photocatalyst, metal-free Brønsted acid-functionalized porphyrin grafted with benzimidazolium-based IL (BAPBIL), was synthesized for esterification of the oleic acid with a 5 W LED as a visible light source [10]. In the system, hydrogen peroxide (H_2O_2) was a hole and electron scavenger, sodium nitrate ($NaNO_3$) was an electron scavenger, and EDTA was a hole scavenger. The catalyst can be effectively recycled up to the fifth cycle without affecting its activity for the synthesis of ethyl oleate and has a wide range with other alcohols (methanol, propanol, and butan-1-ol) and other acid (levulinic acid with methanol, ethanol, and octanol).

A catalytic (E)- to (Z)-isomerization of olefins using a photoredox catalyst in IL is reported, and the reaction can be conducted in a continuous flow system to scale up [11]. The reaction was applied with a two-phase system consisting of toluene/[Bmim][BF$_4$]. The photosensitizer should be, due to its charge, well absorbable in an IL phase, whereas the olefin should be readily separable using a corresponding apolar solvent phase (toluene). Separation was readily achieved using this protocol, and the photocatalyst@IL system can be carried out eight cycles without loss of reactivity.

5.2.2 Application of ionic liquids in photoelectrocatalytic systems

Compared with single photocatalysis, photoelectrochemical technology uses solar and electrical energy as energy sources. By applying a certain bias voltage, photoelectrocatalysis triggers the bending of the semiconductor energy band and the directional movement of photogenerated carriers, which inhibits the electron−hole combination and jointly excites the generation and transfer of photogenerated carriers to improve the catalytic efficiency.

IL organic solution, acting as an effective electrolyte or a cocatalyst, can be used for photoelectrochemical reduction of CO_2 [13−15] or water oxidation [16] and in photoelectrochemical (PEC) cells [17].

Cronin and workers reported the photoelectrochemical reduction of CO_2 to CO by using a III−V compound semiconductor photocatalyst in an IL solution firstly [13]. TiO$_2$-passivated InP was used as the working electrode in a three-terminal potentiostat. Photoelectrochemical reduction of CO_2 with a faradaic efficiency (FE) of 99% at an underpotential of 0.78 V in the [Emim][BF$_4$] electrolyte was observed. NMR spectra showed that the [Emim]$^+$ ions in solution form an intermediate complex with $CO_2{}^-$, thus lowering the energy barrier of this reaction.

The results indicate that the nonaqueous IL electrolyte is important for suppressing hydrogen evolution and enables CO_2 reduction by forming the intermediate complex

[Emim-CO$_2$]*, which lowers the energy of the reaction pathway. Thus IL [Emim][BF$_4$] plays the role of not only an electrolyte but also a homogeneous catalyst for CO$_2$.

A series of molecular chromophores and catalysts assembling on Ru + Re@CuGaO$_2$ electrodes are used for solar-driven CO$_2$ photoelectrochemical reduction to formic acid in an IL [Bmim][TfO], acting as a CO$_2$ absorbent and electrolyte [14]. The Ru and Re complexes not only enhanced the FE of the multielectron processes but also showed higher stability.

In aqueous medium, H$_2$ was the major product with the FE of 80%, while in organic media (acetonitrile solutions of [Bmim][TfO]), formate was obtained as the major product, with a higher selectivity to CO$_2$ reduction. C$_{2+}$ products like ethanol and propanol were also promoted in the organic solvents, in particular with the IL electrolyte. Thus the imidazolium-based electrolytes as cocatalysts may influence the CO$_2$RR mechanism. The beneficial role of the IL is in the high CO$_2$ solubility, in promoting the conversion of CO$_2$ into formic acid and C$_{2+}$ alcohols, as well as in improving the stability of the system and suppressing the production of H$_2$.

Wang et al. reported the photoelectrochemical reduction of CO$_2$ into formic acid in an IL (1-aminopropyl-3-methylimidazolium bromide, [NH$_2$C$_3$mim]Br) aqueous solution, which functions as an absorbent and electrolyte at ambient temperature and pressure [15]. The onset potential of CO$_2$-purged aqueous [NH$_2$C$_3$mim]Br was anodically shifted due to the fact that the IL had the better ability to lower the overpotential through the formation of reaction intermediates, thereby overcoming the initial formation of a CO$_2^{\cdot-}$ anionic radical.

Quantum chemical calculations show that the C2$\cdots$CO$_2$ bond length in [NH$_2$C$_3$mimCO$_2$]$_{(ad)}$ (1.75 Å) is much shorter than that in [BmimCO$_2$]$_{(ad)}$ (1.90 Å), which means that the interaction between the C2 atom of [NH$_2$C$_3$mim]$^+$$_{(ad)}$ and CO$_2$ is much stronger than that between the C2 atom of [Bmim]$^+$$_{(ad)}$ and CO$_2$. This can also be supported by the calculated binding energy between CO$_2$ and the cation of each molecule of IL (–23.6 and –3.4 kJ/mol for [NH$_2$C$_3$mim]$^+$$_{(ad)}$ and [Bmim]$^+$$_{(ad)}$, respectively). Compared with other ILs, [NH$_2$C$_3$mim]Br-assisted capture and conversion of CO$_2$ exhibits a high FE (94.1%) and high electro-to-chemical efficiency (η_{ECE} = 86.2%) for HCOOH at an applied voltage of 1.7 V.

Manganese oxide (MnO$_x$) were used for photoelectrochemical water oxidation in buffered aqueous electrolytes containing amine ILs (n-butylammonium nitrate [BAN]) [16]. Photocurrents as high as 4.5 mA cm^{-2} were obtained at 1.0 V versus SCE (η = 540 mV) in the Bi-buffered BAN electrolyte at pH 9. Thus the use of buffered electrolytes improved photocurrent and significantly enhanced the photoelectrochemical activity by producing H$_2$O$_2$ and O$_2$ at the same time. The MnO$_x$ as the photoanode is not only photoactive, but also a highly active and earth–abundant water oxidation catalyst.

Based on the ILs as the electrolyte, a new type of silicon microhole array PEC cells were fabricated and displayed better stability than that containing conventional aqueous solution [17]. 1-Propyl-3-methylimidazolium iodide (PMII) and 1-ethyl-3-methylimidazolium thiocyanate were prepared and mixed as electrolytes, which also contained 0.05 M I_2 and 0.1 M LiI. In the electrolyte system, ILs act as an excellent solvent media for the I^-/I_3^- pair, and PMII and LiI also can afford partial I^- ions for the redox pair. The experimental results validate the concept of using interpenetrating networks and ILs to produce stable and efficient PEC cells and emphasize the importance of the effect of semiconductor/liquid junction contact on device performance.

As a novel material and green solvent, ILs have a broad application in the photo-/photoelectrocatalytic process with the roles as cocatalysts, in improving solubility and conductivity of electrolyte, and in enhancing the selectivity of products. However, it is still in the primary stage of basic research and industrialization, and the following problems still need to be solved: (1) exploring the mechanism of cocatalytic reaction of ILs; (2) developing structural stable ILs is the key to achieve its large-scale promotion; (3) exploring the enhancement technology of the new photo/photoelectrocatalytic reduction system.

5.3 Application of ionic liquids in solar cells

Since the last decade, there has been a consistent and significant ongoing global endeavor to enhance the efficiency and stability of PSCs. These efforts have been driven by the fascinating array of characteristics displayed by perovskite materials, which include notably high absorption coefficients, low exciton binding energies, cost-effective manufacturing, and extensive charge-carrier diffusion lengths [18]. Presently, the power conversion efficiency (PCE) has reached a significant milestone of 26.1%, a remarkable advance from its initial value of 3.8% in 2009, thereby rivaling that of commercial crystalline silicon cells [19]. However, challenges related to the degradation induced by light, moisture, and long-term photo-instability have hampered the widespread applications of PSCs. In response to these challenges, several strategies have been adopted by different groups and researchers such as optimizing film crystallinity, doping with anions, cations, or ILs, engineering the charge transport layers (CTLs), and use of an efficient passivation layer [20]. Among these approaches, the utilization of ILs holds the potential for significant advancements.

ILs are a new class of materials containing liquid molten salts having a melting point of typically less than 100°C; they stand out as a promising group of substances suitable for incorporation into PSCs. These substances are characterized by their compositional makeup comprising diverse cations (such as ammonium and imidazolium, phosphonium, guanidium, and pyridinium) along with anions (such as halide and carboxylic acid, acetate, tetrafluoroborate, and formate among others) [21]. The distinctive dualistic organic-inorganic composition of these ions confers upon ILs a range of captivating chemical and physical attributes, including notable conductivity, thermal resilience, proficient solvating ability, and minimal vapor pressure. These attributes are amenable to meticulous modulation through purposeful design of their chemical and structural attributes. As a result of these inherent qualities, ILs find practical utility across a spectrum of applications that contribute to the enhancement of PSCs [22]. The ILs usually possess high ionic conductivity, excellent thermal stability, low vapor pressure, and solvate capability. Reported literature substantiates that the incorporation of ILs may fulfill diverse roles in the fabrication and functioning of perovskite PSCs [23].

This specific review presents a methodical examination of the role of ILs in PSCs. The roles of ILs are categorized into four distinct groupings based on their functions: ILs serving as solvents, the impacts of IL additives on perovskite properties, ILs as interface modifiers, and ILs acting as CTLs or IL-modified CTLs. The impact of ILs on various aspects such as perovskite crystallization, nucleation, surface defect mitigation, energy level alignment, and charge transport is succinctly summarized. Additionally, the influence of ILs on device performance is deliberated upon in this review, followed by a future perspective on the potential applications of ILs in PSC module technology.

5.3.1 Ionic liquids as additives in perovskite precursor solutions

Hunting of green additives is part of an ongoing global effort to enhance the stability and efficiency of PSCs. Empirical evidence substantiates that the incorporation of ILs as an additive yield substantially boosts both the operational efficiency and long-lasting robustness of the devices. This phenomenon can be attributed to the noteworthy capacity of ILs to profoundly impact the growth of perovskite crystals and the morphology of the films. Controlling morphology and crystallization during the fabrication of perovskite-based thin films are primary considerations because morphology and crystallinity affect the quality of the films [24]. In the realm of kinetics, the swift evaporation of the solvent coupled with rapid crystallization typically yields perovskite films exhibiting suboptimal crystalline characteristics. This manifests as elevated defect concentrations, the presence of pinholes, and the formation of diminutive grains. These combined factors collectively undermine the efficiency and stability of PSCs [25]. Several IL additives have been employed previously such as imidazolium [26], pyridinium [27], ammonium [28], phosphonium [29], and guanidinium [30]. Wei et al. employed 1,3-dimethylimidazolium iodide IL as an additive to triple-cation-based perovskite, which successfully passivates the defects and regulates the crystallization kinetics, resulting in a high-quality perovskite film with enlarged grain size being achieved [31]. Likewise, Yin et al. introduced ethylpyridinium chloride as an additive. This addition not only facilitated the uniform formation of perovskite through nucleation but also contributed electrons to counterbalance trap states present on the surfaces of the perovskite [27].

The incorporation of ILs can restrict fast crystallization, elevate the boiling point, and reduce the vapor pressure during the fabrication of perovskite films. These effects collectively contribute to the augmentation of the quality of PSCs [32]. Liu et al. [33] investigated the effects of multifunctional potassium hexafluorophosphate KPF$_6$ additive on crystallization kinetics and stability of the PSCs. The corresponding mechanism indicates that KPF$_6$ promotes crystallization kinetics and passivates grain boundaries and internal ionic defects in the perovskite film, as shown in Fig. 5.2A. As a result, the efficiency of the PSCs improved to 22.04%. This improvement persisted even after subjecting the cells to 1000 hours of light soaking and aging under one-sun illumination, during which they retained 80% of their initial PCE. ILs can interact with perovskites via different bonding approaches such as hydrogen bonding, chelate bonding, or intermediate interactions, thereby retarding the rate of perovskite crystal growth. Seo et al. [36] incorporated methylammonium formate (MAFa) as an additive to the perovskite precursor

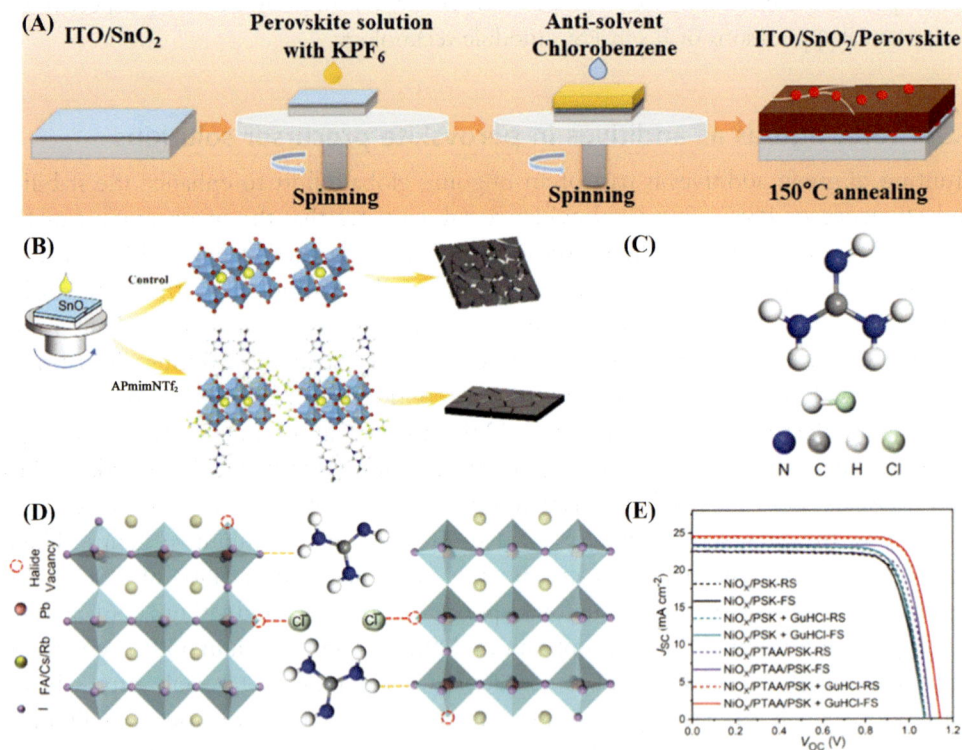

Figure 5.2 (A) Schematic illustration of the fabrication process of KPF$_6$ additive-based perovskite films [33]. (B) Graphical representation of the passivation mechanism of [APmim][NTf$_2$] in perovskite crystallization [34]. (C) Chemical structure of the GuHCl additive. (D) Schematic illustration of the guanidine additive interaction with the perovskite crystal structure. (E) Corresponding J−V curves [35].

solution. The mechanism study indicates that during the process of crystal growth, initially, $HCOO^-$ and Pb^{2+} led to intermediate products. Subsequently, through the annealing phase, there was a progressive substitution of $HCOO^-$ by I^-, ultimately resulting in the complete replacement of $HCOO^-$ anions by I^- ions, thereby delaying the progression of crystal growth. Similarly, Zhou and coworkers reported that within the PbI_6^{4-} framework, the ammonium cations existing in IL $[PF_6]^-$ can engage in interactions with I^- ions, to establish an $N-H\cdots I^-$ hydrogen bonding. This interaction then leads to the retardation of perovskite crystals growth [37]. Recently, we employed amine-functionalized IL (1-aminopropylimidazolium bis(trifluoromethylsulfonyl) imine, [APmim][NTf$_2$]), as a multifunctional additive to the perovskite to control the crystallization kinetics of the perovskite film [34]. The microcosmic mechanism investigation discloses that [APmim][NTf$_2$] has a strong interaction with the Pb^{2+} and X^- in perovskite crystals, consequently improving the crystallization and hydrophobicity of the perovskite film (Fig. 5.2B). Tahir et al. [35] employed the additive engineering approach by incorporating guanidinium hydrochloride (GuHCl) additive to an MA-free, Br-free RbCsFAPbI$_3$-based perovskite absorber, as shown in Fig. 5.2C. Results indicate that the introduction of the GuHCl additive could facilitate the establishment of hydrogen bonding between adjacent grains (Fig. 5.2D). Consequently, PSCs demonstrate an efficiency of 22.78%, accompanied by notable values for open-circuit voltage (V_{oc}), short-circuit current (J_{sc}), and fill factor (FF) of 1.14 V, 24.52 mA/cm^2, and 81.5%, respectively (Fig. 5.2E).

5.3.2 Ionic liquids as solvent for perovskite solar cells

Exploration of state-of-the-art solvents for the fabrication of PSCs is another significant aspect of the research endeavor. A variety of solvents have been employed for the fabrication of PSCs. The combination of dimethylformamide (DMF) and dimethyl sulfoxide (DMSO) is frequently utilized to retard the crystallization process, thereby leading to the formation of films that exhibit enhanced device performance due to their smooth and compact characteristics. Nevertheless, a restraint associated with these solvents pertains to the inadvertent creation of plumbate intermediates, exerting an adverse influence on the final film characteristics. Consequently, there is a discernible need to explore alternative solvent systems. ILs are considered a complete alternative to conventional solvents for solvent modification. This is ascribed to the considerable intrinsic attributes of ILs, exhibiting notable qualities like thermal and electrochemical characteristics.

Wang and coworkers introduced ILs based on methylammonium acetate [MA]$^+$[Ac]$^-$ to synthesize CsPbI$_{3-x}$Br$_x$ perovskites [38]. The initial precursor solution, as prepared, exhibits a colorless appearance (refer to Fig. 5.3A), indicative of robust solvated ion interactions that contribute to the stabilization of the precursor solution. A

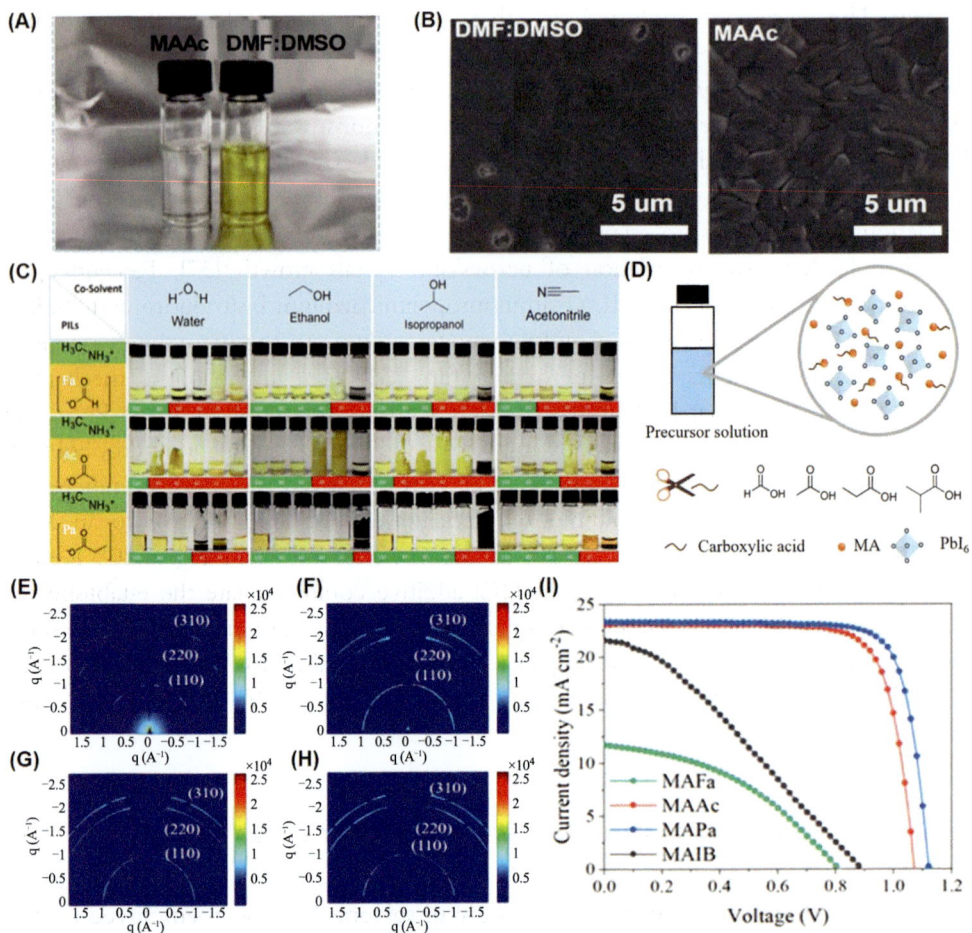

Figure 5.3 (A) Photographs of CsPbI$_2$Br perovskite solutions with [MA]$^+$[Ac]$^-$ IL and DMF/DMSO as solvents. (B) SEM images of CsPbI$_2$Br perovskite films made using DMF/DMSO and [MA]$^+$[Ac]$^-$ IL solvents [38]. (C) The images portray perovskite ink configurations utilizing IL MAFa, MAAc, and MAPa as both individual solvents and cosolvents [39]. (D) Schematic diagram of different IL functional groups. 2D GIWAXS data of the perovskite films manufactured by (E) MAFa, (F) MAAc, (G) MAPa, and (H) MAIB solvents. (I) J–V curves of these IL-based PSCs [40]. DMF, Dimethylformamide; DMSO, dimethyl sulfoxide; MAFa, methylammonium formate.

distinctive observation was the distinct light–green color of the resultant solution, in contrast to the typically observed dark–yellow color for PbI$_2$ solutions in DMF/DMSO (depicted in Fig. 5.3A). In this particular context, perovskite thin films are produced with diminished iodide defects, complemented by morphological benefits (see Fig. 5.3B). Snaith et al. [23] conducted a study on the encapsulation using 1-butyl-3-methylimidazolium tetrafluoroborate ([Bmim][BF$_4$]) IL. Their research focused on the performance of inverted planar PSCs, indicating that the initial PCE of these cells (20%)

experienced a minimal reduction of approximately 5%. This reduction occurred after subjecting the cells to continuous illumination using simulated full-spectrum sunlight for a duration exceeding 1800 hours at a temperature range of $70°C-75°C$. Mathur et al. [39] systematically investigated the solubility of MA-based perovskite precursors, utilizing ILs (MAFa, MAAc, and MAPa) as single solvents and cosolvents, as illustrated in Fig. 5.3C. They successfully formed well-oriented $CH_3NH_3PbI_3$ crystals via PbI_2 dissolution in IL MAPa. However, challenges persist in achieving high-quality perovskite films at ambient temperature due to IL viscosity. Incorporating MAPa as a cosolvent in a MAPa/acetonitrile/DMSO system led to 15.46% PCE in MA-based PSCs. Song et al. [23] investigated the interaction between different functional groups of IL-based solvents and Pb^{2+} in the precursor solution (Fig. 5.3D). Results indicate that these IL-based solvents coordinate with PbI_2 to form intermediate phases and passivate halide vacancies during the annealing process, as illustrated in Fig. 5.3E—H. Consequently, they enhance the efficiency and stability of the cell (Fig. 5.3I).

As previously examined, a multitude of rationales exist for the potential enhancement of perovskite film quality through the utilization of ILs as solvents. First, owing to their covalent bonding characteristics, ILs can facilitate favored crystal orientation. Second, ILs can induce the expansion of micellar dimensions within precursor solutions due to the comparatively diminished chemical interaction strength between ILs and perovskite constituents. This outcome consequently yields perovskite films that are both uniform and densely structured. Third, the notably elevated boiling points of ILs enable a residual quantity to persist within perovskite films, thereby effectuating the passivation of defects.

5.3.3 Ionic liquid-modified charge transport layers for perovskite solar cells

Owing to unique chemical structures, the incorporation of ILs into the other parts of PSCs have garnered enormous attention in the subsequent years. Previous investigation exhibits that ILs display solid interaction toward the electron transport layer (ETL) and hole transport layer (HTL) materials exerting a significant influence on both electrical properties and morphology. This portion of the review will analyze some cutting-edge investigations centered around the incorporation of ILs for ETLs and HTLs.

Huang et al. [41] introduced a low-temperature $(100°C-150°C)$ tetramethylammonium hydroxide (TMAH) additive-modified SnO_2 ETL. The incorporation of TMAH not only facilitates electron extraction but also promotes charge transport at the grain boundaries of perovskite. Similarly, Zhang et al. [42] employed anion engineering through integration of a series of guanidinium salts containing distinct anions (guanidinium thiocyanate [GASCN], guanidinium sulfate [GASO$_4$], guanidine acetate [GAAc], and guanidine chloride [GACl]) at the perovskite/SnO$_2$ interface (Fig. 5.4A). Owing to the presence of coordination and/or ionic bonds between anions and the perovskite film, along with the SnO_2 film, all anion modifiers can proficiently mitigate defects on the surfaces of both the perovskite and SnO_2 layers. Furthermore, this can concurrently

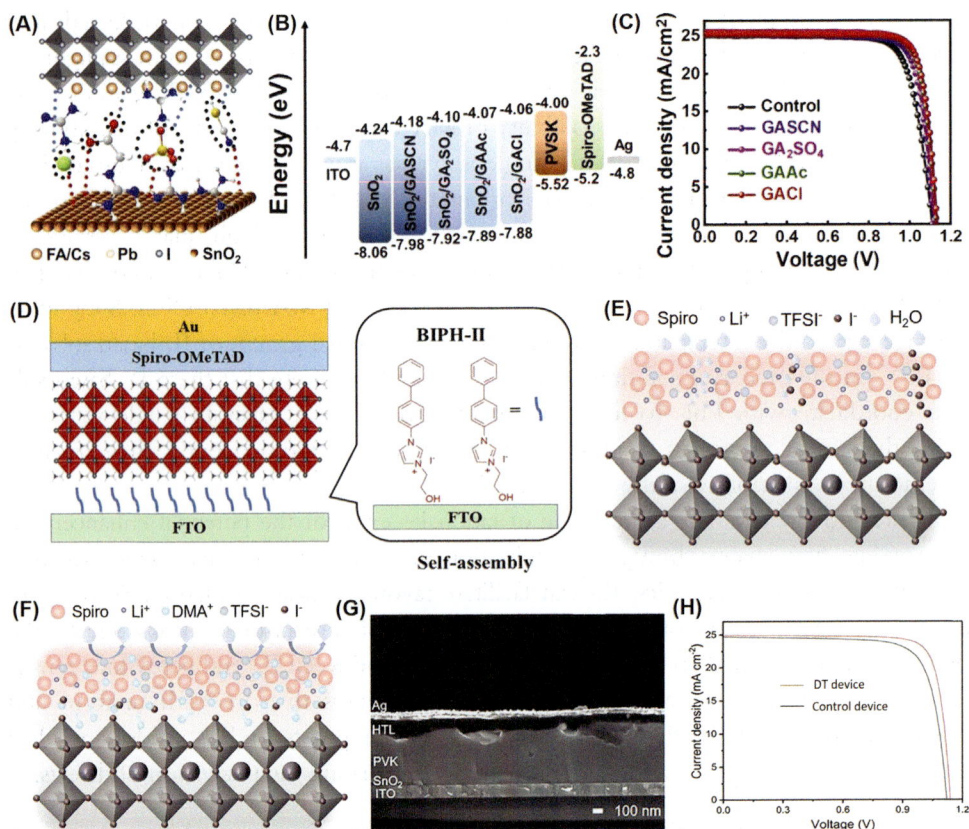

Figure 5.4 (A) Schematic illustration displaying the chemical interaction of different GA IL-based modifiers and perovskite layer as well as ETL layer. (B) The corresponding energy band diagram and (C) $J-V$ curves [42]. (D) The device structure of the PSC modified with BIPH-II [43]. The graphical illustration of aging processes of devices without (E) and with (F) the addition of DMATFSI to spiro-OMeTAD. (G) Cross-section SEM image. (H) Current density–voltage graph of the DMATFSI and LiTFSI codoped device and LiTFSI doped device [44]. DMATFSI, Dimethylammonium bis(trifluoromethanesulfonyl)imide; ETL, electron transport layer.

adjust the alignment of energy bands at the interface and promote the crystallization process of PbI_2 and perovskite materials (Fig. 5.4B). As a result, devices modified with $GASCN^-$, $GA_2SO_4^-$, $GAAc^-$, and $GACl^-$ groups demonstrate PCEs of 22.76%, 23.43%, 23.57%, and 23.74%, respectively. These values significantly surpass the PCE of the control device, which stands at 21.84% (Fig. 5.4C). Wang et al. [43] deposited an ultrathin layer of the hydroxyethyl-functionalized imidazolium iodide IL BIPH-II on the FTO surface via a self-assembly technique (Fig. 5.4D). This strategic modification led to a substantial enhancement in PCE, alongside the notable achievement of maintaining stable performance even under conditions of 10%–15% humidity.

Spiro-OMeTAD, a predominant HTL, has been extensively studied in conjunction with lithium bis(trifluoromethanesulfonyl)imide (LiTFSI) and 4-tert-butylpyridine (t–BP) as an additive, aiming to augment the performance of PSCs. Nevertheless, the inclusion of p-type dopant t–BP and LiTFSI has been found to exert adverse effects on the hysteresis behavior and long-term stability of PSCs, thereby imposing constraints on the viability of large-scale manufacturing and practical implementation of PSC technology. Therefore ILs possessing elevated levels of ionic conductivity and hydrophobic characteristics have been employed as p-type dopants within HTLs to enhance the efficiency and stability of PSCs. Wu et al. [44] employed dimethylammonium bis(trifluoromethanesulfonyl)imide (DMATFSI) IL as an additive to the HTL, which significantly manipulated p-doping and enhanced the conductivity, hole mobility, and hydrophobicity of the HTL (Fig. 5.4E and F). Furthermore, DMATFSI-modified HTL passivates the surface of perovskite PCE, which achieved greater than 23% (Fig. 5.4G and H).

5.3.4 Ionic liquids for interface modification

The enhancement of efficiency, stability, and the optimization of the fabrication process for PSCs necessitate comprehensive evaluation. This is crucial for advancing these materials toward a state suitable for potential industrial production and real-world implementation. For both n-i-p and p-i-n type PSCs, elevated defect densities and inadequate contacts at the buried and top interfaces limit their performance. To address this issue, many ILs were implemented to passivate the defect, enhance the quality of perovskite film, and improve the carriers of charges between CTLs and perovskite absorbers [21]. In this part of the review, we will investigate contemporary passivation approaches applying ILs layer for top and buried interfaces.

Zhang et al. [45] employed the 1-butyl-2,3-dimethylimidazolium chloride ([Bmmim] Cl) IL layer to passivate surface defects in perovskite films. The bifunctional [Bmmim]Cl IL owing to the electron-rich nitrogen of the layer significantly passivated the uncoordinated Pb^{2+} and Cs^+ ions. Furthermore, the hydrophobic alkyl side chain within [Bmmim]Cl augmented the hydrophobic characteristics of the perovskite films. As a result, the PCE of the [Bmmim]Cl-modified device exhibited a remarkable development from 6.15% to 9.92%. Similarly, Zhu et al. [46] introduced a 1-ethyl-3-methylimidazolium bromide ([Emim]Br) IL capping layer on the surface of $FAPbI_3$ to passivate undercoordinated Pb^{2+} and halide ion migration (Fig. 5.5A and B). The ([Emim]Br) IL-modified PSC device exhibits PCE of 24.33% under one-sun illumination with negligible hysteresis, and a large area (10.75 cm^2) integrated module achieves PCE of 20.33%. Huang et al. [47] reported an IL, methyltrioctylammonium trifluoromethanesulfonate (MATS), as a passivation layer at the interface of perovskite film to passivate for the defects. Taking advantage of the low vapor pressure, this nonvolatile IL remained within the grain boundaries and flaw of perovskite

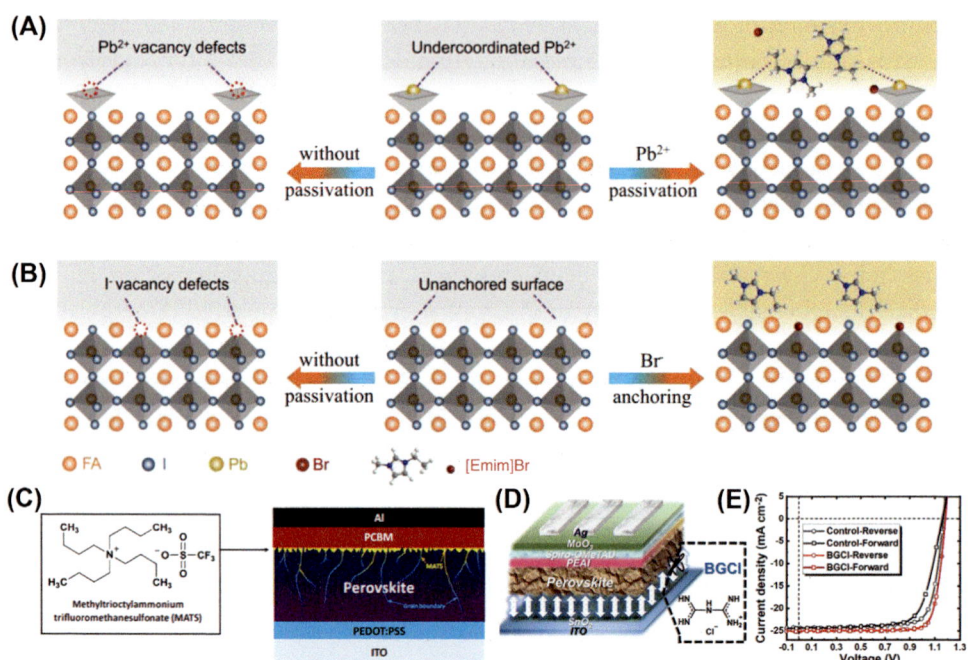

Figure 5.5 (A, B) Schematic illustration of the ionic liquid [Emim]Br anchoring on the perovskite surface [46]. (C) Molecular structure of MATS and schematic diagram of device structure of MATS-treated planar PSCs [47]. (D) Schematic device architecture modified by BGCl (E) and corresponding *J–V* curves [48]. BGCl, Biguanide hydrochloride; MATS, methyltrioctylammonium trifluoromethane-sulfonate; PSCs, perovskite solar cells.

surfaces (Fig. 5.5C). As a result of the effective passivation by MATS, the electron extraction, charge transfer resistance, and related PCE of the devices were significantly improved.

In n–i–p type PSCs, TiO_2 has been usually employed as an ETL. However, the comparatively low electron mobility and elevated electronic trap densities of TiO_2 hamper its widespread application. To counter this issue, Yang et al. introduced the [Bmim][BF_4] layer between the interfaces of TiO_2 and perovskite absorber [20]. Based on the theoretical calculations, the [BF_4]$^-$ group favored to bond to TiO_2, whereas the [Bmim]$^+$ group was more likely to adsorb on the perovskite surface, as a result of which surface dipole will form between the ETL and perovskite. As a result, electron mobility of [Bmim][BF_4]-modified TiO_2 was improved four times. Xiong et al. applied simultaneous interfacial modification and crystallization techniques by incorporating biguanide hydrochloride (BGCl) between SnO_2 and the perovskite layer (Fig. 5.5D) [48]. The BGCl-modified PSC device achieved an impressive certified PCE of 24.4% (Fig. 5.5E) and maintained 95% of its initial PCE after aging for over 500 h at 20°C and 30% relative humidity in ambient conditions.

5.4 Ionic liquids intensify the electrochemistry process

The ORR refers to the electrochemical processes of O_2 molecules on the cathode surface [49,50]. Under various electrolysis conditions, such as in different sorts of supporting electrolyte and/or at different cathode potentials, O_2 molecules might be reduced through different pathways on the cathode, and then different kinds of reduction products can be obtained [51]. The product of the 1e-ORR process, wherein one O_2 molecule accepts one electron, is the superoxide radical $\cdot O_2^-$ or its protonated $\cdot OOH$; the product of the 2e-ORR process is H_2O_2 or its deprotonated HOO^-; and the product of 4e-ORR is H_2O or OH^- [52−54]. Theoretically, H_2O_2 undergoes another proton-coupled electron transfer (PCET) process to obtain hydroxyl radicals, but hydroxyl radicals are difficult to be observed in the ORR process and more usually obtained through the oxidation process of H_2O on the anode [55,56]. The above-mentioned oxygen-containing species ($\cdot O_2^-$, HOO^-/H_2O_2, $\cdot OH$, etc.) are also called active oxygen species due to their short lifespan but high reactivity. The various ORR reaction pathways mentioned above and the interconversion relationships between these active oxygen species are shown in Fig. 5.6 [57].

In the past few years, researchers have found that the properties of the electrolyte, such as aqueous or nonaqueous, might affect the ORR pathways [57−60]. ILs have been used as the electrolyte due to their low volatility, good thermal stability, high ionic conductivity, nonflammability, and wide electrochemical window, and the ORR processes in ILs have been well-studied in the field of Li-O_2 batteries, O_2 gas sensors, and PEMFC [61−64]. More and more researchers are trying to intensify some oxidative chemical processes such as lignin depolymerization and methane conversion by those active oxygen species, in which these active oxygen species play important roles to improve the conversion efficiency and the products selectivity [65−68].

Lignin is the second largest biomass resource in the nature only after cellulose. As shown in Fig. 5.7, the lignin molecule is an unordered polymer, which is composed

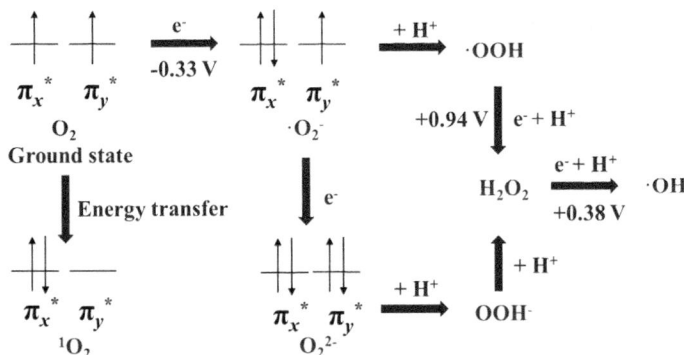

Figure 5.6 Molecular orbitals of O_2 and reactive oxygen species.

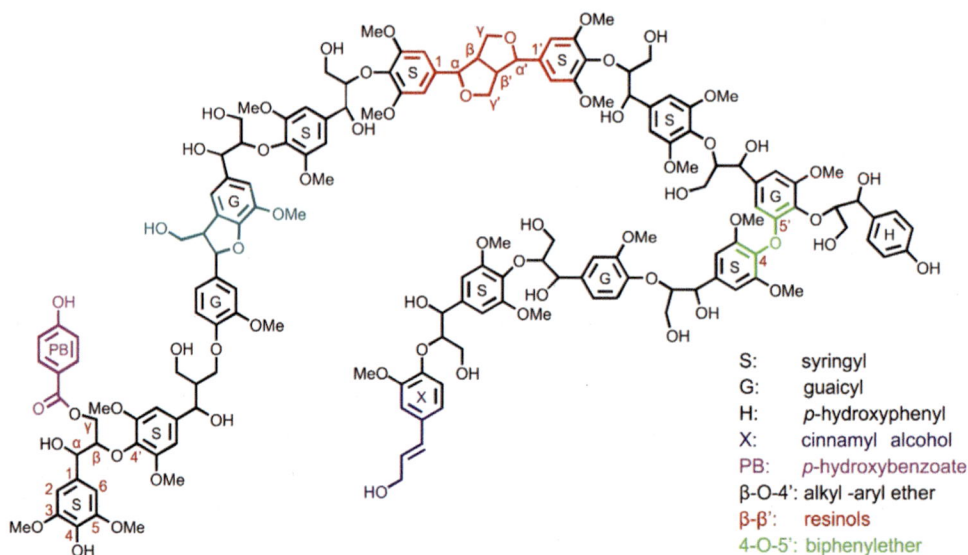

Figure 5.7 Diagrammatic drawing of the molecular structure of lignin and the main sorts of linkage bonds present in lignin. *Reprinted from Li M, Pu Y, Ragauskas AJ. Current understanding of the correlation of lignin structure with biomass recalcitrance. Frontiers in Chemistry, 2016, 4: 45. Open Access.*

of thousands of phenylpropane units linked with each other by C-C bonds or ether bonds. There are mainly three kinds of phenylpropane units in lignin molecules: eugenol unit (S), guaiacyl unit (G), and *p*-hydroxyphenyl unit(H). These structural units are interconnected through a series of linkage bonds, such as β-O-4 bonds (β-aromatic ether), 4-O-5 bonds (diaryl ether), α-O-4 bonds (α-aryl ether), β-5 bonds (phenyl coumarin), β-β bonds (resin alcohol), and so forth [69,70]. Researchers hope to selectively break the linkage bonds among the units in lignin to obtain small-weight aromatic compounds, and the latter could be used as platform materials for various chemicals, so as to alleviate the oil dependence in current chemical industry [71,72].

The oxidative lignin depolymerization refers to the process to break down the linkage bonds in lignin molecules using oxidants such as nitrobenzene and O_2 catalyzed by metal oxides or H_2O_2. The advantage of this scheme is it is easy to scale, while its disadvantages include the need to consume reagents, difficulty in controlling reaction conditions, and susceptibility to peroxidation of the products. One of electrochemical schemes for the oxidative lignin depolymerization is that applying active oxygen species generated in situ through the ORR process in ILs as oxidants for lignin depolymerization. The advantage of this scheme is that the sort of active oxygen is regulated to be $\cdot O_2^-$ or HOO^-/H_2O_2, which has moderate oxidation ability, so as to effectively avoid excessive oxidation. Moreover, ILs provide good solubility for lignin and can effectively improve the depolymerization efficiency.

In short, this section focusses on the studies on the generation of active oxygen species by regulating the ORR process in ILs, and the bond-breaking mechanism of lignin model compounds through the electrochemical indirect oxidation process involving the active oxygen species and the effects on the conversion percent and the product selectivity of an actual lignin sample.

5.4.1 Oxygen reduction reaction behaviors in different ionic liquids electrolyte systems

5.4.1.1 Oxygen reduction reaction behavior in aprotic [Bmim][BF₄] and the effect of ionic liquid viscosity

The ORR behavior in an aprotic IL ([Bmim][BF$_4$]) recorded by the cyclic voltamme-try (CV) method is shown in Fig. 5.8. The significant cathodic current observed under an O$_2$ atmosphere (red line) compared to that under Ar atmosphere (black line) origi-nated from the ORR processes. The transferring electron number during the ORR process on this case was calculated as 1.36 at about -0.9 V by the rotating ring-disk electrode (RRDE) method. The reduction peak **1c** at -1.03 V shown in Fig. 5.8 is attributed to a 1e-ORR process, and the oxidation peak **1a** at -0.94 V is attributed to the oxidation process of $\cdot O_2^-$ to O$_2$.

Furthermore, ORR behaviors in three different aprotic electrolyte systems ([Bmim][BF$_4$], [Bmim][BF$_4$] –MeCN (7:3 v/v), MeCN) were compared with the same other conditions using the CV method [73]. In the mixture of [Bmim][BF$_4$]-MeCN (green line), a significant increase in the current of peak **1c** is observed, which means that the reaction rate of ORR on the electrode surface increases, which is caused by the decrease in viscosity of the electrolyte system enhancing the mass

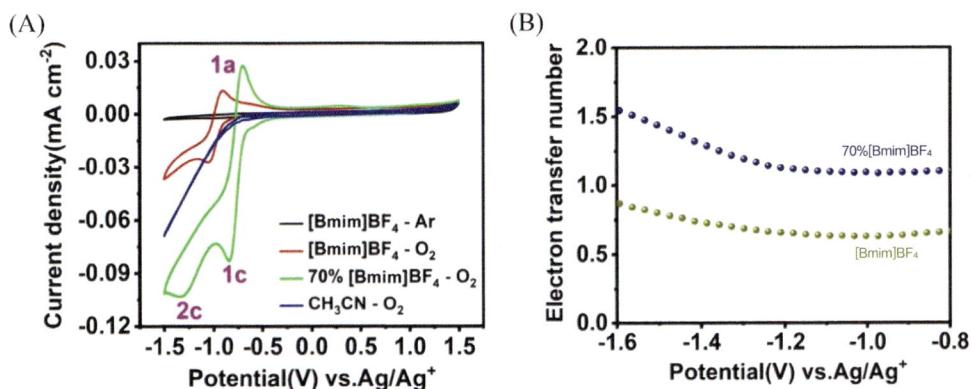

Figure 5.8 (A) CV curves of GCE in Ar and O$_2$ atmosphere in various electrolytes with the scanning rate of 50 mV/s. (B) The transferring electron number during ORR in [Bmim][BF$_4$] calculated by the RRDE method. *CV*, Cyclic voltammetry; *ORR*, oxygen reduction reaction; *RRDE*, rotating ring-disk electrode.

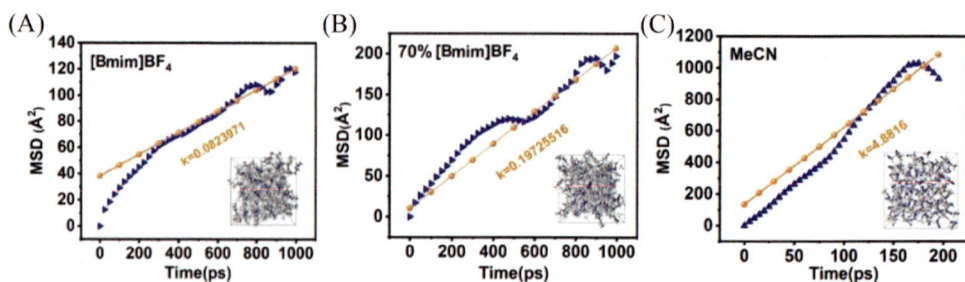

Figure 5.9 The structural models and the fitting of the root-mean-square of displacement versus time in various electrolytes.

transfer of O_2 in the electrolyte, and an increase in the current of peak **1a** sequentially is observed. It is worth noting that an additional peak **2c** corresponding to the generation of HOO^- by the 2e-ORR process is also observed, which confirms that the [Bmim][BF$_4$]-MeCN system (with a viscosity of 21.95 mPa s) is more conducive to the diffusion of O_2 than that in the [Bmim] [BF$_4$] system (with a viscosity of 104.96 mPa s), resulting in a large amount of $\cdot O_2^-$ formed on the electrode surface, thereby indirectly enhancing the 2e-ORR process. Although the current for O_2 reduction can be clearly observed in the pure MeCN (blue line), there is no obvious oxidation peak, indicating that the active species (i.e., $\cdot O_2^-$) generated at this case is little, which might be caused by the low reaction rate constant of O_2 on the surface of the glassy carbon electrode in MeCN due to the heterogeneous reaction system of ORR [74].

The molecular dynamics (MD) of O_2 in the above three electrolyte systems have been simulated, and the structural model and the fitting of the root-mean-square of displacement versus time are shown in Fig. 5.9. The results show that the diffusion coefficients of O_2 in [Bmim][BF$_4$], [Bmim][BF$_4$]-MeCN, and MeCN are 13.7×10^{-6} cm^2/s, 32.9×10^{-6} cm^2/s, and 8.1×10^{-5} cm^2/s, respectively. According to the MD simulation results, the O_2 molecule has a high adsorption energy in [Bmim][BF$_4$]-MeCN (-20.93 kJ/mol), indicating that van der Waals interaction energy, electrostatic interaction, and intermolecular bond energy of the system might change when a small amount of MeCN is present in [Bmim][BF$_4$], which is beneficial for improving the stability of O_2 in the system. That is the reason for improving the ORR reaction rate of the system, increasing the Faraday efficiency of ORR, and accelerating the kinetic rate of the ORR process on the electrode surface.

5.4.1.2 Oxygen reduction reaction behaviors in [Bmim][BF$_4$] containing a small amount of water

In previous studies, the presence of a trace amount of water in an aprotic IL was often regarded as an impurity, focusing on the effect on the electrochemical window and physical properties (such as conductivity and viscosity) of ILs [75,76]. Considering from the electrochemical point of view, all ORR processes that occur involve protons except for the 1e-ORR process. Therefore whether or not free protons are present and how many are they in the electrolyte system must have an impact on the ORR electrode process and so do the type and amount of the generated active oxygen species. Here, a small amount of water was added into [Bmim][BF$_4$] as a proton source, and the effect of the presence of trace amounts of H$_2$O on the sorts of the active oxygen species generated by ORR was investigated.

Fig. 5.10 shows the CV curves in [Bmim][BF$_4$] containing different amounts of water under an O$_2$ atmosphere. As shown in Fig. 5.10A, the electrochemical window of the electrolyte system gradually becomes narrower as the H$_2$O content increases, which is due to the oxygen evolution reaction of H$_2$O molecules, while the significant cathodic peaks around -1.03 V under an O$_2$ atmosphere are assigned to the ORR process. By zooming in the ORR region shown in Fig. 5.10B, the reduction peak **1a** and the oxidation peak **1c** corresponding to 1e-ORR could be clearly observed when the water content is less than 0.6 mol/L. Moreover, the current intensity of peak **1c** increases with the increase of H$_2$O content, while the current of peak **1a** decreases with the increase of H$_2$O content. When the water content reaches 1.8 mol/L, the peak **1a** disappears, and only the peak **1c*** at -0.97 V corresponding to ORR is observed. It was found that the ORR peak current increases and the peak potential shifts positively as the free proton H$^+$ in the system increases, and the change in peak potential is often due to a change in thermodynamic factors (i.e., changes in the ORR paths).

The I-V curves of the RRDE at 1600 rpm in [Bmim][BF$_4$] containing different contents of water are shown in Fig. 5.11A, and the calculated transferring electron numbers during the ORR process are shown in Fig. 5.11B. The results show that about $1.17 \sim 1.81$ electrons were transferred during the ORR process in [Bmim][BF$_4$] with the water content of 0.6 mol/L, which means some of the 1e-ORR products

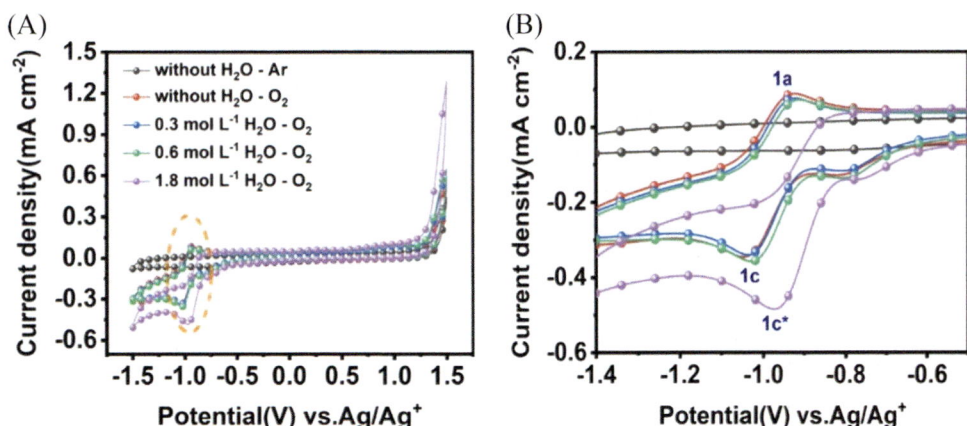

Figure 5.10 (A) Effect of water addition on ORR for [Bmim][BF$_4$] under an O$_2$ atmosphere at a scanning rate of 50 mV/s; (B) partial magnification of the ORR region. *ORR*, Oxygen reduction reaction.

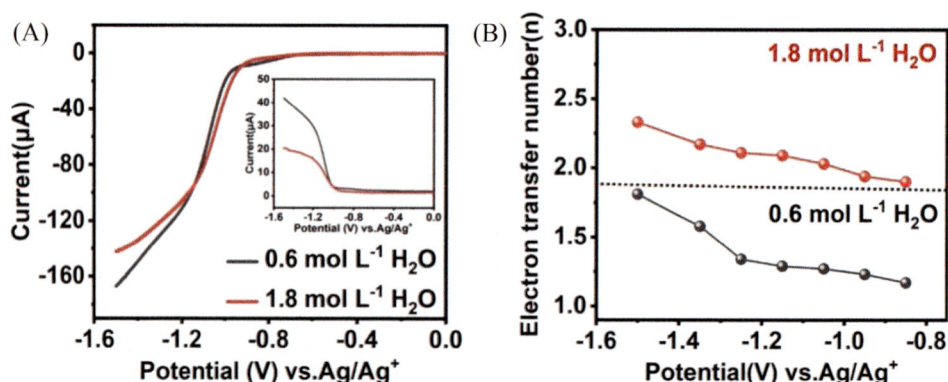

Figure 5.11 (A) *I-V* curves of RRDE under an O_2 atmosphere in electrolyte systems with different water contents at 1600 rpm; (B) the calculated electron transferring number for ORR based on RRDE curves. *ORR*, Oxygen reduction reaction; *RRDE*, rotating ring-disk electrode.

$(\cdot O_2^-)$ further undergo another proton-coupling electron transferring process due to the influence of proton H^+ and result in the mixture of superoxides and peroxides as the ORR products. When the water content increases to 1.8 mol/L, about 1.90−2.33 electrons were transferred during the ORR process, which means mainly 2e-ORR occurred in the system and H_2O_2 was the main ORR products. In summary, studies on ORR behaviors in [Bmim][BF_4] with a small amount of water indicate that the presence and the amounts of free protons H^+ in ILs can change the transferring electron numbers of ORR and the different type of ORR products can be obtained.

5.4.1.3 Oxygen reduction reaction behaviors in the protonic ionic liquid [HNEt₃][HSO₄]

The CV curves in a protonic IL [HNEt$_3$][HSO$_4$] under Ar and O_2 atmospheres are shown in Fig. 5.12A. Unlike the aprotic IL [Bmim][BF_4] system, only one cathode peak (peak **2c**) was observed in [HNEt$_3$][HSO$_4$]. The *I-V* curve of the RRDE at 1600 rpm in [HNEt$_3$][HSO$_4$] is shown in Fig. 5.12B, and the calculated transferring electron number for ORR is 2.1~2.5, which means that 2e-ORR mainly occurs in [HNEt$_3$][HSO$_4$] and H_2O_2 is the main generated active oxygen species. This result is consistent with previous results in the [Bmim][BF_4]-H_2O system, which suggests that the presence of certain amounts of protons in the system promotes the occurrence of 2e-ORR.

5.4.2 Study on electrochemical lignin depolymerization using p-benzyloxyl phenol as model compounds

Since the molecular structure of an actual lignin is too complex, it is not conducive to the study of the depolymerization mechanism. A simple compound has a α-O-4 bond, *p*-benzyloxyl phenol (PBP), which was chosen to be as a lignin model compound to study before the actual lignin sample. The molecular structure of PBP is shown in Fig. 5.13.

5.4.2.1 Electrochemical behaviors of p-benzyloxyl phenol in [Bmim][BF₄]

The green line in Fig. 5.14A is the CV curve in [Bmim][BF₄] containing PBP under an Ar atmosphere. According to the Compton group [77], peak **1a** (at 0.51 V) can be attributed to the PCET process of PBP on the electrode surface, that is, the phenol group (Ph–OH)

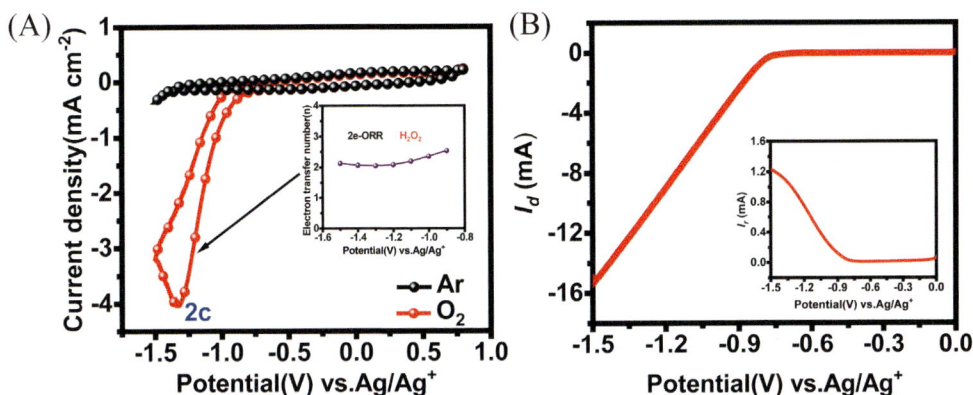

Figure 5.12 (A) CV curves of GCE in [HNEt₃][HSO₄] under Ar and O₂ atmospheres at 50 mV/s; (B) *I-V* curves of RRDE in [HNEt₃][HSO₄] under an O₂ atmosphere at 1600 rpm. *CV*, Cyclic voltammetry; *RRDE*, rotating ring-disk electrode.

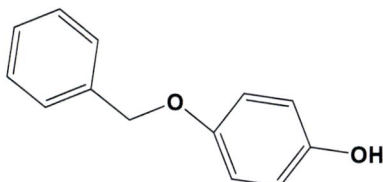

Figure 5.13 The molecular structure of PBP. *PBP*, *p*-Benzyloxyl phenol.

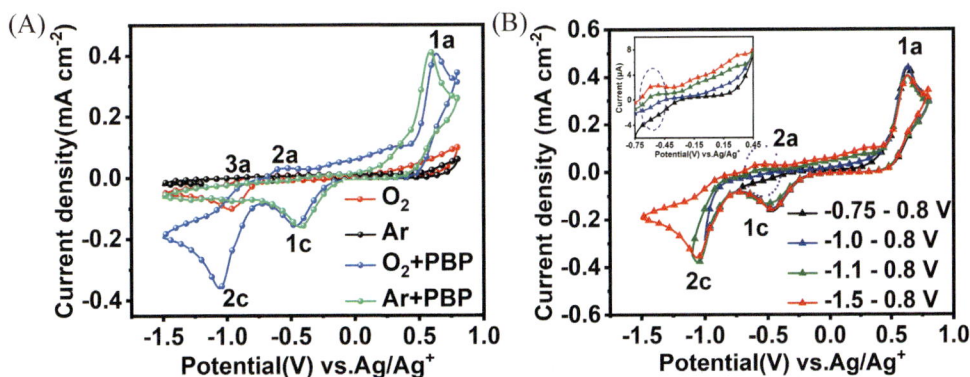

Figure 5.14 (A) CV curves of GCE in[Bmim][BF₄] in the absence and presence of PBP under O₂ and Ar atmospheres; (B) the CV curves of GCE in [Bmim][BF₄] in the presence of PBP under an O₂ atmosphere in different ranges of scanning potentials. *CV*, Cyclic voltammetry; *PBP*, *p*-benzyloxyl phenol.

in PBP releases a proton H^+ to be a phenolic anion (Ph-O$^-$) and then gives out an electron to form a phenoloxy radical (Ph-O$\cdot$). Peak **1c** (at –0.45 V) is related with the process that part of phenoloxy radicals (Ph-O$\cdot$) are reduced back to phenoloxy anions (Ph-O$^-$). The blue line in Fig. 5.14A is the CV curve in [Bmim][BF$_4$] containing PBP under an O$_2$ atmosphere; there are three anodic peaks **1a, 2a, and 3a** and two cathodic peaks **1c and 2c** appeared. Comparing that in [Bmim][BF$_4$] without PBP (red line), peak **2c** (at –0.98 V) and peak **3a** (at –0.87 V) can be respectively attributed to the 1e-ORR process of O$_2$ molecules dissolved in [Bmim][BF$_4$] and the oxidation process of some ORR products $\cdot$O$_2^-$ on the electrode surface. To clarify the origination of peak **2a** (at –0.50), the CV test was performed in different sweep potential ranges (–0.75 $\sim$ 0.8 V, –1.0 $\sim$ 0.8 V, –1.1 $\sim$ 0.8 V, and –1.5 $\sim$ 0.8 V), and the results are shown in Fig. 5.14B. It is obvious that peak **2a** could only be observed after the reaction corresponding peak **2c** (i.e., ORR) occurred. It is confirmed that peak **2a** is contributed by the oxidation process of the reaction product of PBP with reactive oxygen species. The increase of peak current in the potential range of 0 $\sim$ 0.5 V, indicating the involvement of active oxygen species in the oxidation reaction process of PBP.

The RRDE study result also confirmed that the presence of PBP in [Bmim][BF$_4$] effects on the sort of active oxygen species generated through ORR process. As shown in Fig. 5.15, the transferring electron number in ORR is 0.9 $\sim$ 1.2 in the range of –0.9 $\sim$ –1.5 V in [Bmim][BF$_4$] without PBP; while in the presence of PBP, the transferring electron number is 1.1 at –0.9 V, and it gradually increases to 2.2 as the potential negatively shifts to –1.2$\sim$1.5 V. This result shows that the presence of PBP as a weak proton donor promotes the reduction of O$_2$ on the electrode surface change to be a 2e-ORR process even in an aprotic IL, that is, the main ORR product becomes OOH$^-$. This phenomenon could also explain the negative shift of the peak **2c** in the presence of PBP and the decrease in the intensity of peak **3a** in Fig. 5.14A.

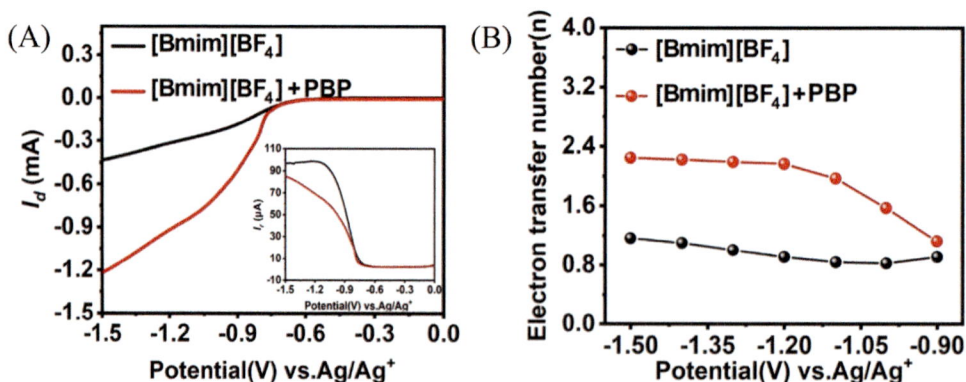

Figure 5.15 (A) The *I-V* curves of RRDE in [Bmim][BF$_4$] with or without PBP, the inset is the corresponding ring current recorded at potential of 1.2 V. (B) The calculated transferring electron numbers for ORR based on RRDE curves. *ORR*, Oxygen reduction reaction; *PBP*, p-benzyloxyl phenol; *RRDE*, rotating ring-disk electrode.

5.4.2.2 Degradation mechanism of p-benzyloxyl phenol in [Bmim][BF₄]

In a nondiaphragm electrolysis bath, 10 mL of [Bmim][BF$_4$] containing 1 mmol/L PBP was added as the electrolyte, a piece of graphite felt (20 mm × 20 mm) as the cathode, a piece of Ti/RuO$_2$-IrO$_2$ mesh (20 mm × 20 mm) as the anode, and a self-made Ag/Ag$^+$ electrode as the reference electrode. Under a O$_2$ atmosphere, electrolysis was operated at a constant cathode potential of –1.0 V for 18 h. The depolymerization products were extracted by ether from the electrolyte after electrolysis and analyzed by GC-Ms, and the results are shown in Table 5.1.

Table 5.1 The depolymerization products of PBP after electrochemical degradation in [Bmim][BF$_4$].

No.	Structure	Name	Relative content (%)
1		Benzaldehyde	73.8
2		Benzoquinone	7.3
3		Benzyl alcohol	15.3
4		p-Benzyloxyl phenol	3.6

Combined with the results of electrochemical study and the depolymerization product analysis, the degradation mechanism of PBP by electrolysis in [Bmim][BF$_4$] is deduced as follows (as shown in Fig. 5.16) [78]. Firstly, a PBP loses a proton to form a phenoloxy anion (PBP$^-$) in electrolyte, which diffuses to the anode surface and gives out an electron turning to be a phenoloxy group (PBP$_1 \cdot$); then PBP$_1 \cdot$ transfers to

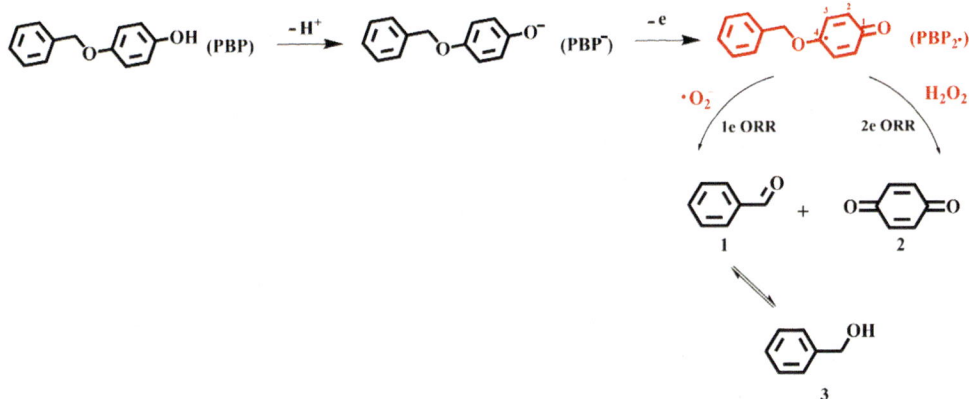

Figure 5.16 Scheme of the cleavage of ether bonds in PBP by active oxygen species in [Bmim][BF$_4$]. *PBP, p-Benzyloxyl phenol.*

$PBP_2 \cdot$ by quinone resonance, in which the 4-C atom in a phenolic hydroxyl group becomes an active site for nucleophilic attack. Next, $PBP_2 \cdot$ reacts homogeneously with the active oxygen species generated through the ORR process in the electrolyte. In the initial stage of the reaction, it is the main ORR product of $\cdot O_2^-$ as a nucleophile to attack the 4-C site in $PBP_2 \cdot$, resulting in the cleavage of ether bonds in PBP; Then, H_2O_2 becomes the main ORR product with the extension of electrolysis time, and the product overoxidation occurs at the same time as the ether bond breakage due to the strong oxidation capacity of H_2O_2, resulting in the formation of benzaldehyde, p-benzoquinone and benzyl alcohol and other products.

Fig. 5.17 shows the variation of conversion percent and Faraday efficiency in electrochemical degradation of PBP under different electrolysis conditions. The results showed that the degradation of PBP is related with electrolysis conditions such as current density, time, and temperature.

As shown in Fig. 5.17A, both of the conversion percent and Faraday efficiency of PBP degradation show a decreasing trend with the increase of current density. This may be due to the increase of bath voltage with the increase of current density, resulting in partial self-quenching of active oxygen in the electrolyte or diffusion to the anode to be oxidized. By monitoring the changes in cathodic potential under different current densities, it was observed that the cathodic potentials of the electrochemical system were -1.02 V, -1.16 V, -1.31 V, and -1.49 V at the current densities of 0.4, 0.7, 1.0, and $1.3\,mA/cm^2$, respectively. At the cathodic potential of -1.03 V and the current density of 0.4 mA/cm^2, the Faraday efficiency for ORR to provide active oxygen species is the highest, so is the conversion percent of PBP. The generated active oxygen species diffuses to the anode region and undergoes oxidation with the potential shifts negatively, resulting in a decrease in the conversion rate and Faraday efficiency of PBP.

As shown in Fig. 5.17B, the conversion percent of PBP increased with the prolongation of electrolysis time, keeping a constant current density of 0.4 mA/cm^2, indicating the active oxygen species ($\cdot O_2^- / \cdot OOH$) continuously generated in situ. However, the Faraday efficiency decreases with increasing time. The phenomenon can be explained by the fact that the valid reaction on the anode decreased because the concentration of PBP decreased as the reaction progressed, while the ORR process on the cathode remained unchanged, resulting in a decrease in total Faraday efficiency.

The efficiency of electrochemical degradation of PBP at different temperature was also studied. As shown in Fig. 5.17C, the conversion percent is 83.6% with Faraday efficiency of 6.4% at 20°C. When the reaction temperature rises to 80°C, the conversion rate of PBP is 66.4% with Faraday efficiency of 5.1%. According to the reported study [79], the $\cdot O_2^-$ produced by O_2 through 1e-ORR is an exothermic reaction, and the reaction is inhibited at elevated temperature. Moreover, at higher temperature, $\cdot O_2^-$ tends to decompose, which reduces the concentration of reactive oxygen species in the system and is detrimental to the degradation of PBP.

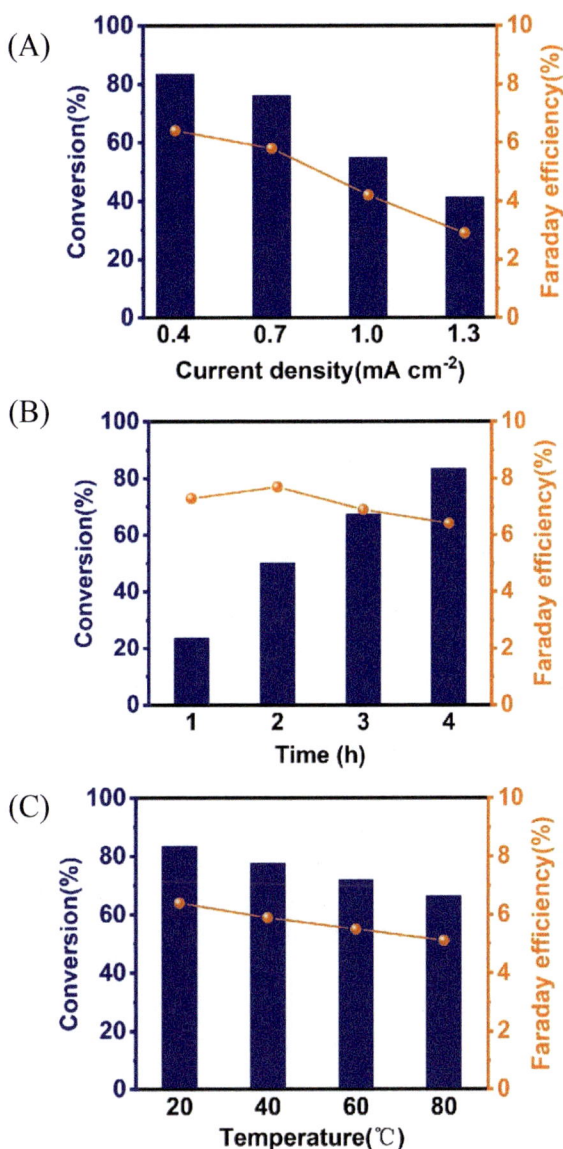

Figure 5.17 The conversion and the Faraday efficiency of electrolyzing PBP in [Bmim][BF$_4$]. *PBP, p-Benzyloxyl phenol.*

5.4.2.3 Electrochemical behaviors of p-benzyloxyl phenol in [HNEt$_3$][HSO$_4$]

The CV curves in [HNEt$_3$][HSO$_4$] containing PBP under an O$_2$ atmosphere (green line in Fig. 5.18A) appeared two cathodic peaks and one anodic peak. Comparing the CV curves without PBP (black and red lines), the cathodic peak **2c** at –1.28 V belongs to the ORR process. To further clarify the origination of the anodic peak **1a** at 0.51 V

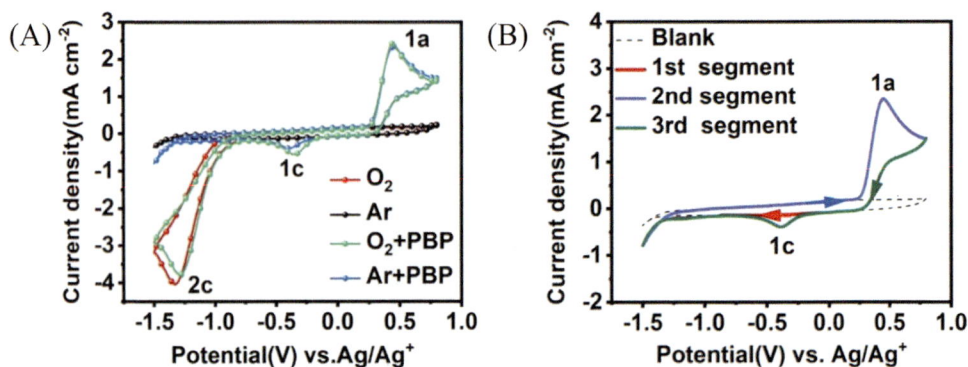

Figure 5.18 (A) CV curves of GCE in [HNEt₃][HSO₄] without or with PBP under Ar and O₂ atmospheres; (B) CV curves of GCE in [HNEt₃][HSO₄] with PBP under an Ar atmosphere; negative scanning started from the open circuit potential of -0.065 V to -1.50 V, then became positive (0.80 V), and finally became negative (-1.50 V). CV, cyclic voltammetry; PBP, p-benzyloxyl phenol. *Reprinted from Ghosh S, Singh T. Role of ionic liquids in organic-inorganic metal halide perovskite solar cells efficiency and stability. Nano Energy, 2019, 63: 103828. Open Access.*

and the cathodic peak **1c** at –0.45 V, CV tests were conducted in different regions. As shown in Fig. 5.18B, no peak appeared when the potential was negatively scanned starting from the open circuit potential of –0.065 V to –1.50 V (red line); however, peak **1a** at 0.51 V appeared in the following forward scanning (blue line), which can be attributed to the oxidation reaction of PBP on the anode; and peak **1c** at –0.45 V appeared during further negative scanning (green line). As peak **1c** only appeared after peak **1a**, it can be attributed to the reduction reaction of PBP oxidation products.

Similarly, the RRDE technology was used to study the transferring electron number of ORR in the [HNEt₃][HSO₄] system. As shown in Fig. 5.19, there is no significant change in the intensity of disk current and ring current with or without PBP in the electrolyte. The transferring electron numbers of ORR are calculated around $2.0 \sim 2.3$ in both cases, indicating that 2e-ORR is promoted and H_2O_2 is the main ORR products due to the presence of a large amount of free H^+ in the electrolyte system.

5.4.2.4 Degradation mechanism of p-benzyloxyl phenol in [HNEt₃][HSO₄]

The electrolysis was carried out using [HNEt₃][HSO₄] containing PBP as the supporting electrolyte, graphite felt as the cathode, and RuO_2/Ti electrode as the anode. The depolymerized products in the electrolyte after electrolysis with a constant cathode potential of –1.2 V in O_2 or Ar atmospheres for 24 hours were analyzed by GC-Ms after ether extraction. The results are shown in Fig. 5.20; benzyl alcohol, benzaldehyde, and p-benzoquinone were detected as degradation products of PBP when electrolysis under an O_2 atmosphere, while no products except for raw material PBP were

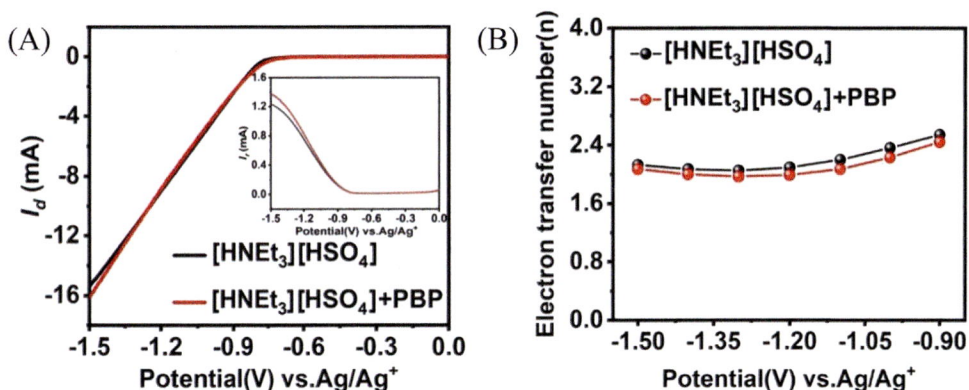

Figure 5.19 (A) The *I-V* curves of RRDE in [HNEt₃][HSO₄] with or without PBP; the inset is the corresponding ring current recorded at potential of 1.2 V; (B) the calculated transferring electron numbers for ORR based on RRDE curves. *ORR*, Oxygen reduction reaction; *RRDE*, rotating ring-disk electrode. *Reprinted from Jiang HM, Wang L, Qiao LL, et al. Improved oxidative cleavage of lignin model compound by ORR in protic ionic liquid. International Journal of Electrochemical Science, 2019, 14(3): 2645−2654. Open Access.*

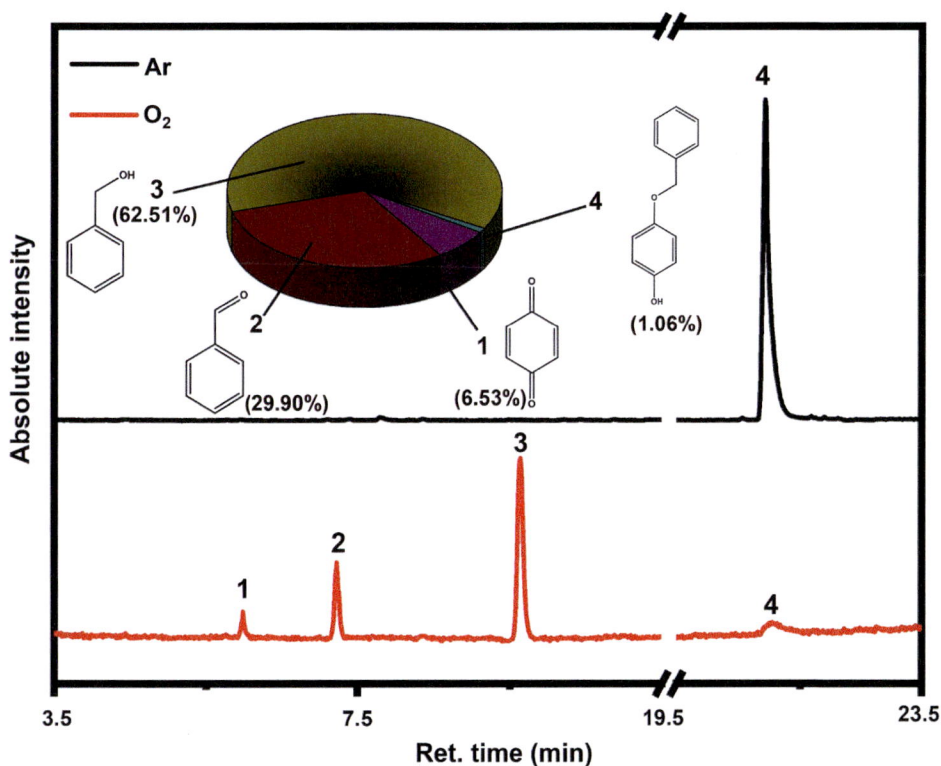

Figure 5.20 Gas chromatograms of the ether extractive of the reaction liquid after PBP electrochemical degradation under Ar or O₂ atmosphere. *PBP*, p-Benzyloxyl phenol. *Reprinted from Jiang HM, Wang L, Qiao LL, et al. Improved oxidative cleavage of lignin model compound by ORR in protic ionic liquid. International Journal of Electrochemical Science, 2019, 14(3): 2645−2654. Open Access.*

detected when electrolysis under an Ar atmosphere. This result indicates that H_2O_2 produced by ORR under an O_2 atmosphere is essential for PBP degradation.

Based on the above results of electrochemical research and product analysis, the deduced mechanism of electrochemical degradation of PBP in protonic IL [HNEt$_3$] [HSO$_4$] is shown in Fig. 5.21. At first, a PBP molecule in [HNEt$_3$][HSO$_4$] deprotonated to form a phenolic anions (PBP$^-$), which loses an electron at 0.51 V on the anode to form a phenolic radicals (PBP·) or its resonance quinone structure. Then, part of the PBP· radicals in the electrolyte diffuse to the cathode surface to be reduced to PBP$^-$; the others react with H_2O_2 generated through the ORR process homogeneously in the electrolyte, so that PBP transfers to benzyl alcohol and p-benzoquinone, in which benzyl alcohol may be further oxidized by H_2O_2 to produce benzaldehyde [80].

The conversion percent and Faraday efficiency of PBP degradation in [HNEt$_3$] [HSO$_4$] under different electrolysis conditions are shown in Fig. 5.22. Fig. 5.22A shows the results that electrolysis operated under different current densities but with applying

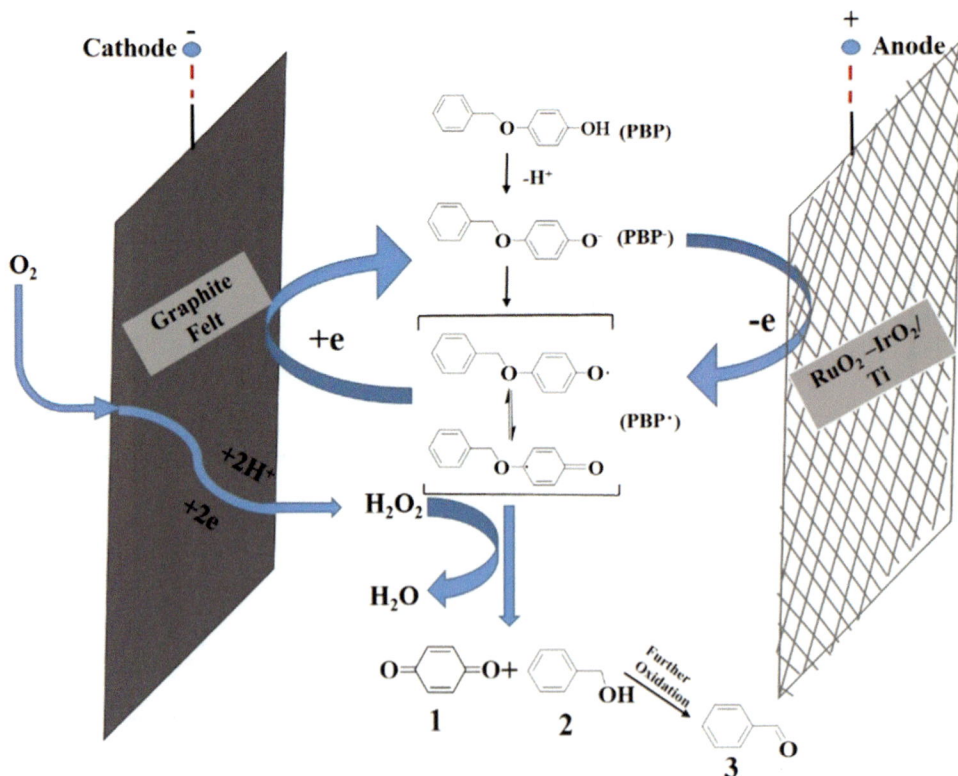

Figure 5.21 Scheme of the cleavage of ether bond in PBP by H_2O_2 in [HNEt$_3$][HSO$_4$]. *PBP, p*-Benzyloxyl phenol.

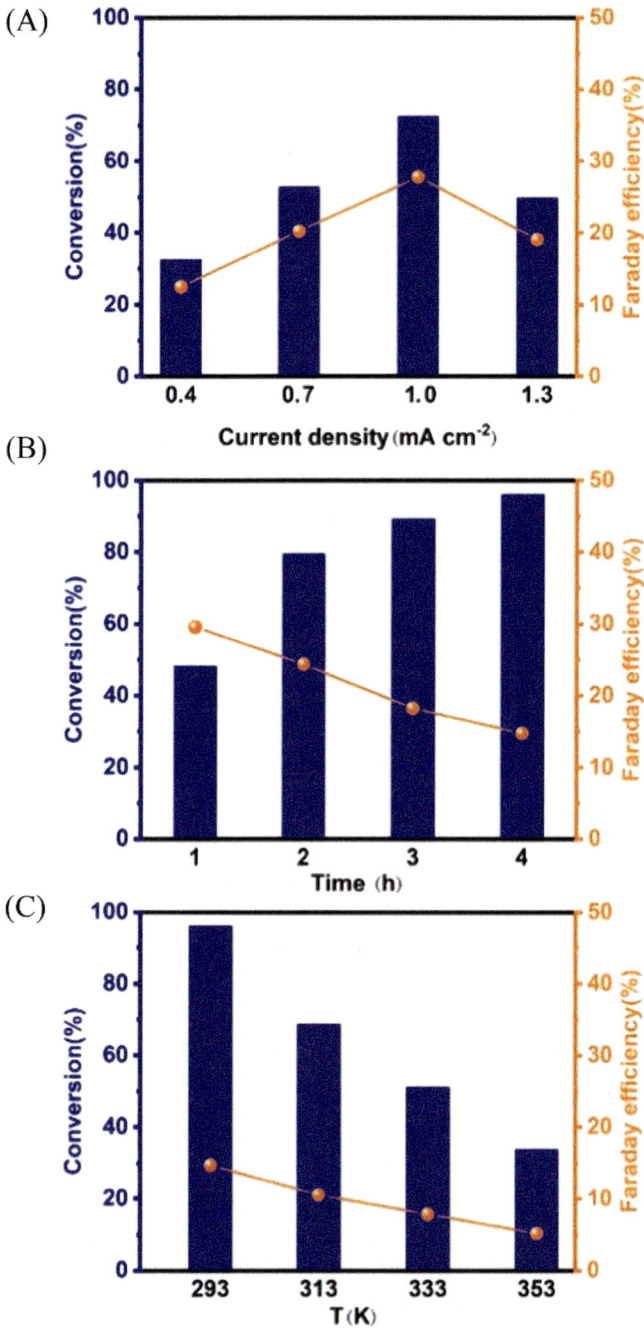

Figure 5.22 The conversion and the Faraday efficiency of electrolyzing PBP in [HNEt$_3$][HSO$_4$]. *PBP, p-Benzyloxyl phenol.*

the same amount of electricity (50.4 C). The highest conversion percent of 72.5% and Faraday efficiency of 27.8% were obtained at a current density of 1.0 mA/cm^2. The cathode potential during electrolysis at current densities of 0.4, 0.7, 1.0, and 1.3 mA/cm^2 was calculated as -0.54 V, -0.97 V, -1.25 V, and -1.62 V, respectively. According to the CV analysis results, the optimal potential for ORR in [HNEt$_3$][HSO$_4$] is around -1.20 V, resulting in more H$_2$O$_2$ generated at a current density of 1.0 mA/cm^2, thereby improving the Faraday efficiency.

As shown in Fig. 5.22B, in the first hour of PBP degradation at a constant current density of 1.0 mA/cm^2, the conversion percent of was 48.2% and the Faraday efficiency was 29.5%. As the electrolysis time extended to 4h, the conversion percent reached 96.2%, but the Faraday efficiency decreased to 14.7%. The decrease in Faraday efficiency may be due to the decreasing amount of PBP in the electrolyte as the reaction progresses, resulting in the extra H$_2$O$_2$ self-decomposed in the electrolyte or underwent oxidation near the anode area, resulting in a decrease in current efficiency.

The degradation percent and Faraday efficiency decreased when electrolyzed at the temperature changed from 20°C to 80°C, as shown in Fig. 5.22C. This phenomenon is consistent with the results observed in the [Bmim][BF$_4$] system, which is due to the shortened lifespan of active oxygen species at elevated temperature.

Under the similar electrolysis conditions (all with an applied charge of 50.4 C), the degradation percent of PBP in aprotic IL [Bmim][BF$_4$] was only 23.7% and the Faraday efficiency was only 7.3%, which is significantly lower than the degradation in protonic IL [HNEt$_3$][HSO$_4$]. In addition to the high viscosity and the slow diffusion rate of O$_2$ of [Bmim][BF$_4$] leading to a decrease in ORR reaction rate, another reason is deduced that the presence of an appropriate number of protons in [HNEt$_3$][HSO$_4$], which promotes the generation of H$_2$O$_2$ as the main active oxygen species in the system. H$_2$O$_2$ has stronger oxidation capability than O$_2{}^-$ and displays a better effect on bond breaking in PBP, thereby improving degradation percent and Faraday efficiency. However, it also led to overoxidation of degradation products.

5.4.2.5 Study on electrochemical lignin depolymerization using guaiacylglycol-β-guaiacyl ether as model compounds

There are both β-O-4 and C$_\alpha$-C$_\beta$ bonds in the molecular structure of guaiacylglycol-β-guaiacyl ether (GGE) (Fig. 5.23). Since the β-O-4 bond and C$_\alpha$-C$_\beta$ are the common linkage bonds in lignin molecules, GGE would be as an appropriate lignin model compound to study the bond-breaking mechanism by the active oxygen species generated through ORR process, which can help us further explore the factors for the product selectivity in lignin depolymerization.

The electrochemical behaviors of GGE in [Bmim][BF$_4$] was studied by the CV method. As shown in Fig. 5.24, the black line represents the CV curve of the blank electrolyte under an Ar atmosphere, no obvious redox peak appears in the range of

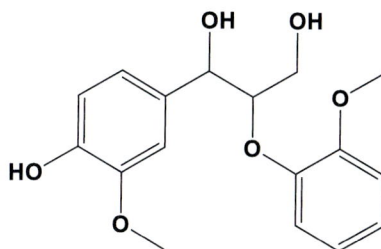

Figure 5.23 The molecular structure of GGE. *GGE*, Guaiacylglycol-β-guaiacyl ether.

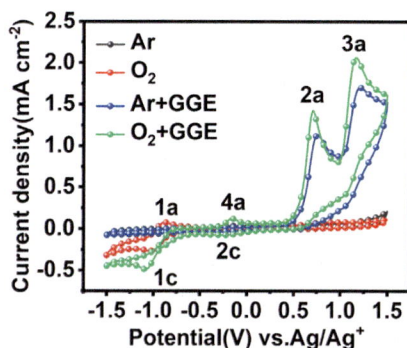

Figure 5.24 CV curves of GCE in [Bmim][BF$_4$] without or with GGE under Ar or O$_2$ atmosphere. *CV*, Cyclic voltammetry; *GGE*, guaiacylglycol-β-guaiacyl ether.

−1.5 to 1.5 V; while under an O$_2$ atmosphere a pair of redox peaks appear (red line), in which the cathodic peak **1c** near −0.97 V is attributed to the 1e-ORR process, and the anodic peak **1a** near −0.85 V is attributed to the oxidation process of oxygen reduction product of · O$_2$$^−$. The blue line in Fig. 5.24 shows the CV curve of GGE in [Bmim][BF$_4$] under an Ar atmosphere. There are two obvious anodic peaks (peak **2a** and **3a**) at 0.71 V and 1.21 V, which can be attributed to the direct oxidation process of GGE on the anode. Among them, the peak **2a** belongs to the PCET process of phenolic hydroxyl groups (Ph–OH) in GGE converting to phenolic oxygen radicals (Ph–O ·) on the anode; and the peak **3a** is originated from the PCET process of the hydroxyl groups on C$_\alpha$ (C$_\alpha$-OH) in GGE to C$_\alpha$-O· radicals. The "shoulder peak" appeared at 1.5 V is speculated to be related with the process of hydroxyl groups on C$_\gamma$ (C$_\gamma$- OH) to C$_\gamma$-O radicals [77]. It is worth noting that only a weak peak **2c** related to Ph–O · and no corresponding cathodic peak was observed in the CV curve, which indicates that the oxidation of GGE is an irreversible electrode process, indicating that the oxidized GGE is extremely unstable and prone to conversion. The green line in Fig. 5.24 is the CV curve of GGE in [Bmim][BF$_4$] under an O$_2$ atmosphere. The current intensity of peak **2a** and **3a** increases, accompanied by a decrease in intensity of peak **1a**, indicating that the active oxygen generated by the

ORR process is conducive to promoting the oxidation of GGE on the electrode surface. Similar to the CV curve of PBP in [Bmim][BF$_4$], peak **4a** was only observed under an O$_2$ atmosphere, suggesting that this is related to the oxidation process of a product produced by active oxygen species attacking GGE.

Using the RRDE technology, the number of transferring electrons during the ORR process in [Bmim][BF$_4$] in the absence of GGE is determined as 0.9−1.1 (black line in Fig. 5.25), indicating that the O$_2$ molecule mainly goes through the 1e-ORR pathway in the blank [Bmim][BF$_4$]; so $\cdot$O$_2^-$ are the main ORR products. When GGE exists in [Bmim][BF$_4$], the peak **1c** of ORR shifts negatively by about 100 mV to -1.09 V compared to the blank electrolyte, and the current intensity is significantly enhanced (as shown by the green line in Fig. 5.25). The RRDE results indicate that the number of transferred electrons in the ORR process rapidly changes from 1.2 to 1.8 within the range of -0.8 V to -1.0 V (red line in Fig. 5.25), while it remains unchanged at a more negative potential, with transfer electron numbers ranging from 1.8 to 2.1. This means that in the presence of GGE, the ORR process in [Bmim] BF$_4$ can transition to mainly 2e-ORR at a relatively negative potential, meaning that the main ORR product changes from $\cdot$O$_2^-$/$\cdot$OOH to HO$_2^-$/H$_2$O$_2$.

[Bmim][BF$_4$] containing GGE as the supporting electrolyte was electrolyzed by keeping the cathode potentials of -0.86 V or -1.12 V for 30 minutes; the electrolyte was extracted with butyl acetate; after centrifugation, the organic phase was detected by HPLC. As shown in Fig. 5.26, the substances with retention time of 9.3 min, 9.7 min, 11.3 min, and 13.4 min were determined by comparison of standards to be coniferyl alcohol, vanillin, GGE, and guaiacol, respectively, which were identified to be the electrochemical degradation products of GGE [81].

It is worth noting that only vanillin and guaiacol were detected when GGE was electrolyzed at -1.12 V; three degradation products such as coniferyl alcohol, vanillin,

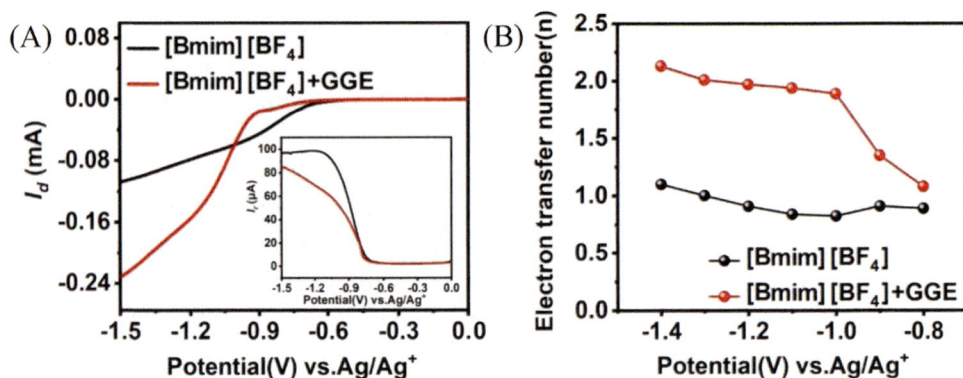

Figure 5.25 (A) The *I-V* curves of RRDE in [Bmim][BF$_4$] with or without GGE; the inset is the corresponding ring current recorded at potential of 1.2 V; (B) the calculated transferring electron numbers for ORR based on RRDE curves. *GGE*, Guaiacylglycol-β-guaiacyl ether; *ORR*, oxygen reduction reaction; *RRDE*, rotating ring-disk electrode.

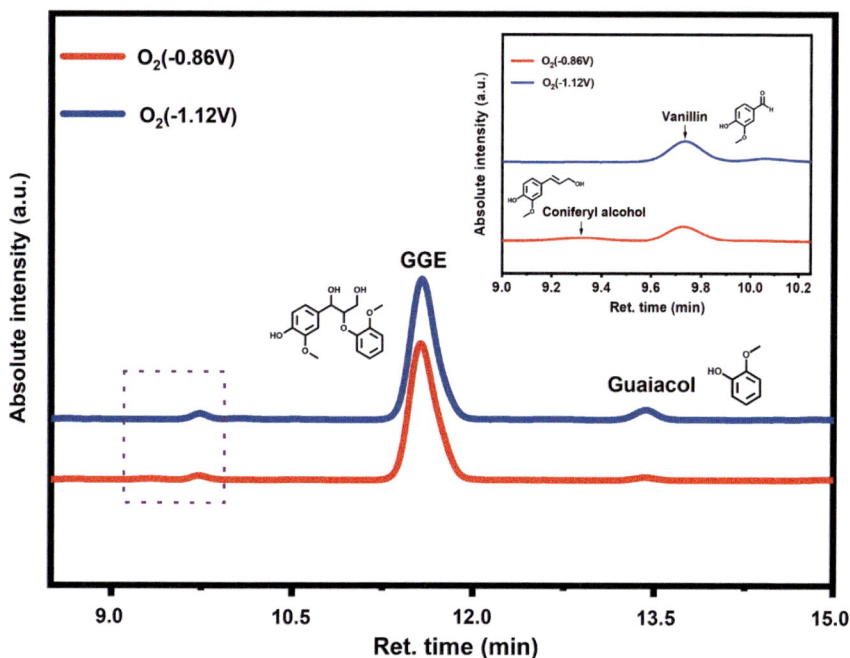

Figure 5.26 HPLC of the butyl acetate extractive after GGE electrolyzed in [Bmim][BF₄] at a cathodic potential of -0.86 V or -1.12 V (vs Ag/Ag$^+$) for 30 min. *GGE*, Guaiacylglycol-β-guaiacyl ether. *Reprinted from Ref. [120]. Open Access.*

and guaiacol were detected when electrolyzing at -0.86 V (as shown in Fig. 5.26). This result preliminarily confirms that the ORR process can be regulated by controlling the cathode potential, thereby in situ providing different types of active oxygen species ($\cdot O_2^-$ / $\cdot OOH$ or HO_2^- / H_2O_2). Due to their different oxidation abilities, GGE undergoes different pathways of reaction and obtains different degradation products.

Fig. 5.27 shows the conversion percent of GGE after electrolysis at different cathode potentials for 1 h under an O_2 atmosphere. As shown in Fig. 5.27A, the conversion percent of GGE of 25.6% was obtained at -1.12 V. This is because the conversion percent depends on the generation rate of active oxygen and the diffusion rate for GGE to the electrode, and the ORR process has the highest Faraday efficiency near -1.12 V according to the CV curve. It is worth noting that the Faraday efficiency decreased as the negative shift of the cathode potential, and the highest Faraday efficiency of 4.89% was obtained at -0.86 V. The decrease in Faraday efficiency as the negative shift of cathode potential is speculated to be due to the lower concentration of GGE (1 mmol/L) in [Bmim]BF₄ and the more difficulty for GGE to be degraded.

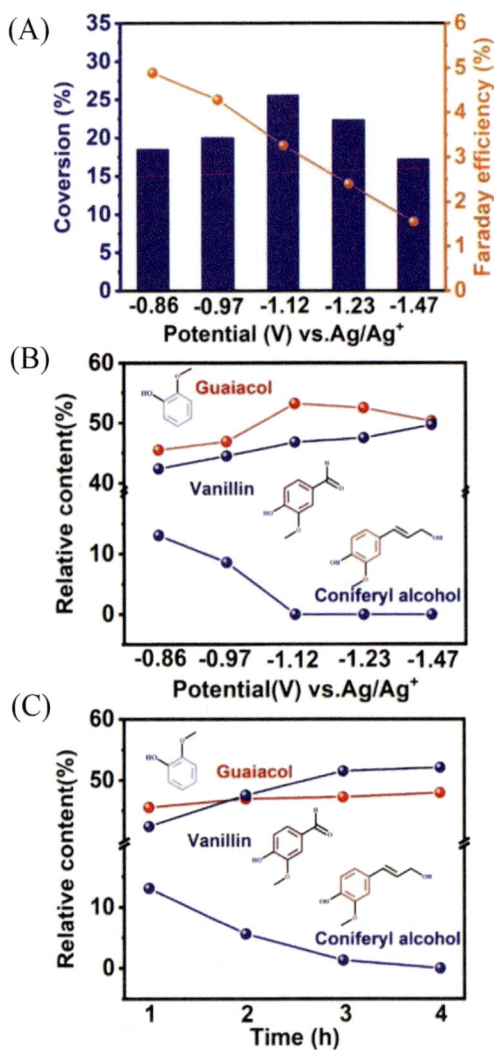

Figure 5.27 The electrolysis of GGE in [Bmim][BF$_4$] under an O$_2$ atmosphere. *GGE*, guaiacylglycol-β-guaiacyl ether.

An interesting phenomenon was found in analyzing the degradation products of GGE at different potentials by HPLC. As shown in Fig. 5.27B, the relative content of coniferyl alcohol decreased as the cathode potential shifted from -0.86 V to -0.97 V and disappeared when the cathode potential was -1.12 V. In addition, the relative content of coniferyl alcohol also decreased with the increase of electrolysis time, as shown in Fig. 5.27C. When the electrolysis time exceeds 3 h, the relative content of vanillin and guaiacol was about 50%, but coniferyl alcohol was not detected.

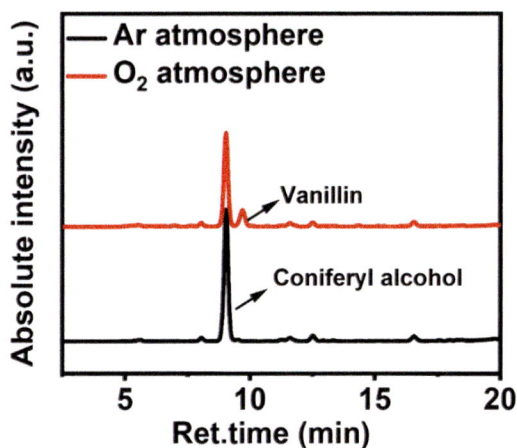

Figure 5.28 HPLC of the butyl acetate extractive of the reaction liquid after coniferyl alcohol electrochemical degradation under Ar or O_2 atmosphere at -0.86 V versus Ag/Ag$^+$ for 1 h.

In a special experiment, 1 mmol/L of coniferyl alcohol was degraded at a constant cathode potential of -0.86 V under an O_2 atmosphere for 1 h. The electrolyte after electrolysis was extracted by butyl acetate and analyzed by HPLC (as shown in Fig. 5.28), and a small amount of vanillin was detected. This result confirmed that vanillin was the degradation product of coniferyl alcohol. In summary, the 1e-ORR product ($\cdot$OOH) triggers the breakage of the β-O-4 bond to form coniferyl alcohol and guaiacol, while some coniferyl alcohol is further oxidized to form vanillin. This might be an indirect reason why three products were observed after the electrolysis of GGE at constant -0.86 V.

Based on the above product analysis and electrochemical research results, it is speculated that the selective degradation of GGE is triggered by $\cdot$OOH as a nucleophilic reagent attacking the GGE after anode activation. The electrochemical degradation process is speculated as follows (as shown in Fig. 5.29): Firstly, GGE is oxidized on the anode to form a C_α cationic intermediate (GGE$^+$), while O_2 undergoes the 1e-ORR process at -0.86 V on the cathode to generate $\cdot O_2^-/\cdot$OOH; Subsequently, $\cdot$OOH attacked the C_β site of GGE$^+$, leading to β-O-4 bond breaking to produce guaiacol and coniferyl alcohol; Furthermore, coniferyl alcohol is further oxidized by $\cdot$OOH to form vanillin. If the applied cathodic potential is -1.12 V, the $\cdot$OOH generated through the ORR process tends to accept another electron to form HO$_2^-$/H_2O_2 in the presence of free protons. Owing to H_2O_2 being a stronger oxidizing nucleophilic reagent, it directly attacks C_α cations, causing the bond of C_α-C_β in GGE$^+$ to be preferentially cleaved, leading to the production of vanillin and guaiacol.

5.4.3 Depolymerization of lignin by electrochemical method in ionic liquids

The actual lignin samples were electrolyzed in a bath using ILs as the supporting electrolyte. Due to the complexity of the sample itself, it is difficult to evaluate the

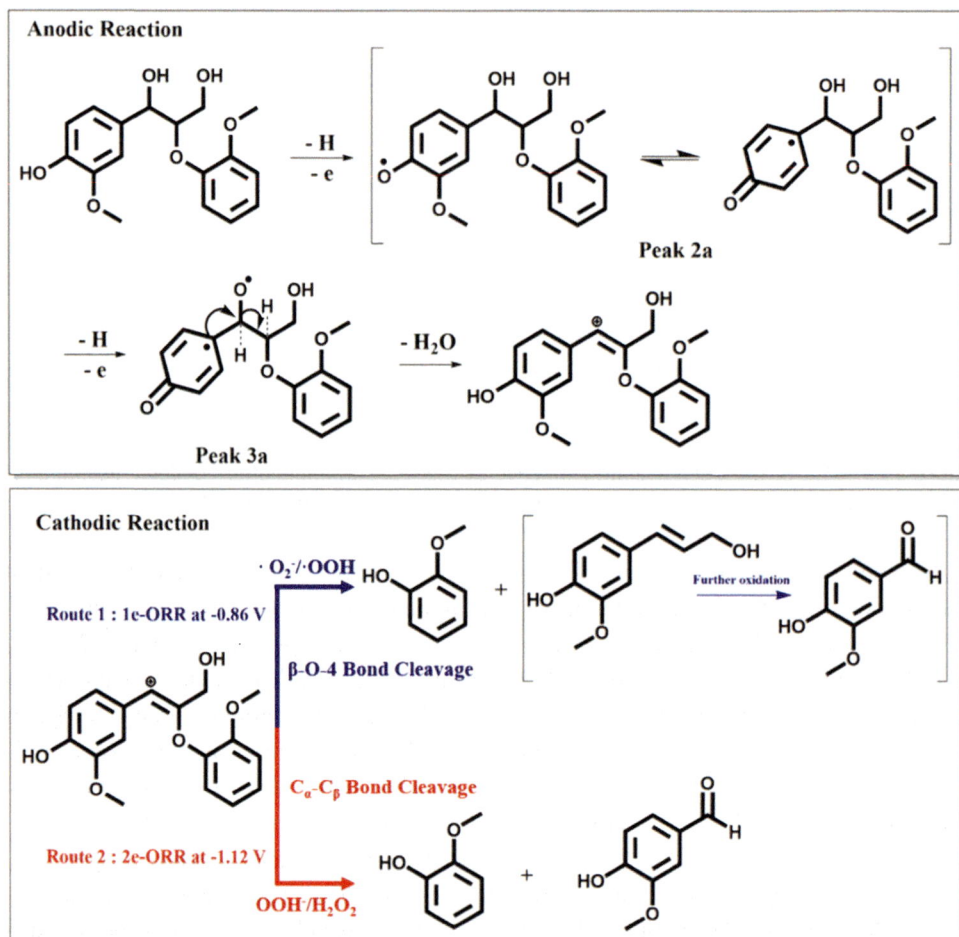

Figure 5.29 Reaction pathways of anodic oxidative reaction of GGE and the cleavage of β-O-4 and C_α-C_β bonds in GGE by $\cdot O_2^-/\cdot OOH$ or HO_2^-/H_2O_2 generated through the controlled ORR process. GGE, Guaiacylglycol-β-guaiacyl ether.

conversion percent by measuring the yield of all products. Therefore this study used the weighing method to calculate the conversion percent of lignin and using the relative content of the specified products determined by GC-Ms to evaluate the product selectivity.

5.4.3.1 Electrochemical depolymerization of lignin in [Bmim][BF₄]

An actual lignin sample was electrolyzed in [Bmim][BF₄] at a constant current density of 0.4 mA/cm², and the conversion percent was evaluated by the diffidence in weight of lignin before and after the electrolysis. As shown in Table 5.2, the conversion

Table 5.2 The results of the electrolysis depolymerization of lignin in [Bmim][BF$_4$].

No.	j/mA/cm^2	T/°C	t/h	Electrolyte	Q/C	Conversion/%
1	0.4	20	6	[Bmim][BF$_4$]	34.4	16.4
2	0.4	20	12	[Bmim][BF$_4$]	68.8	24.9
3	0.4	20	18	[Bmim][BF$_4$]	103.2	29.4
4	0.4	20	24	[Bmim][BF$_4$]	137.6	32.6
5	0.4	20	6	[Bmim][BF$_4$]/CH$_3$CN^a	34.4	22.3
6	0.4	20	12	[Bmim][BF$_4$]/CH$_3$CN^a	68.8	30.2
7	0.4	20	18	[Bmim][BF$_4$]/CH$_3$CN^a	103.2	41.9
8	0.4	20	24	[Bmim][BF$_4$]/CH$_3$CN^a	137.6	50.3

^{a}[Bmim][BF$_4$]/CH$_3$CN: v/v = 7:3.

percent of lignin increases with time, but the weight loss per unit time decreases with the prolongation of electrolysis time. That is, under the condition of applying the same amount of electricity, the Faraday efficiency significantly decreases with the prolongation of electrolysis time. This phenomenon is basically consistent with the other results of electrochemical lignin depolymerization reported in literature [82], and the reason is that monobenzene ring free radicals as an intermediate were formed during the lignin depolymerization process and they are readily to reaggregation with each other.

In addition, the comparison experiments in the [Bmim][BF$_4$]/CH$_3$CN system were performed, and the results show that the conversion percent and Faraday efficiency of lignin depolymerization improved due to the lower viscosity. After 24 hours of electrochemical depolymerization, a conversion percent of 50.3% was obtained, and the Faraday efficiency was about twice that of [Bmim][BF$_4$] under the same conditions.

5.4.3.2 Electrochemical depolymerization of lignin in [HNEt$_3$][HSO$_4$]

Similarly, the same actual lignin sample was electrolyzed in [HNEt$_3$][HSO$_4$] at a constant current density of 1.0 mA/cm^2. The results after 24 hours of electrochemical depolymerization are shown in Table 5.3. It is shown that the lignin depolymerization effect in [HNEt$_3$][HSO$_4$] is significantly improved compared to that in [Bmim][BF$_4$] system, and the conversion percent reaches up to 88.4%. The phenomenon that the amount of depolymerized lignin per unit time decreased with the prolongation of electrolysis time was also observed, but the Faraday efficiency in [HNEt$_3$][HSO$_4$] system is higher than that in the [Bmim][BF$_4$] system. This is because the presence of free protons in [HNEt$_3$][HSO$_4$] causing 2e-ORR occurred on the surface of the graphite felt cathode to generate a large amount of H$_2$O$_2$, which has stronger oxidation ability and promotes the depolymerization of lignin.

Table 5.3 The electrochemical depolymerization of lignin in [HNEt$_3$][HSO$_4$].

No.	$j/mA/cm^2$	$T/^{\circ}C$	t/h	Electrolyte	Q/C	Conversion/%
1	1.0	20	6	[HNEt$_3$][HSO$_4$]	86.4	46.3
2	1.0	20	12	[HNEt$_3$][HSO$_4$]	172.8	64.6
3	1.0	20	18	[HNEt$_3$][HSO$_4$]	259.2	79.7
4	1.0	20	24	[HNEt$_3$][HSO$_4$]	345.6	88.4

The comparison of the depolymerization products of the same lignin sample in the aprotic [Bmim][BF$_4$]/CH$_3$CN and the protonic [HNEt$_3$][HSO$_4$] are shown in Table 5.4. It is found that the depolymerization products in both systems are small-weighted aromatic compounds, but the types and the relative contents are different. Owing to the different sorts of active oxygen species generated through ORR in the two systems, the bond breaking mechanisms are different. H$_2$O$_2$ has stronger oxidation capacity, and there are more active sites in actual lignin molecules, resulting in more kinds of products. Comparably, $\cdot$O$_2^-$ has weaker oxidation ability so selective bond breaking is observed during the depolymerization process, resulting in only four kinds of products.

In fact, vanillin is the most efficient pure monoaromatic phenol currently produced from lignin on an industrial scale and plays an important role as a platform compound for the synthesis of other valuable products, mainly derived from the oxidative decarboxylation of petroleum–based guaiacol production. Taking vanillin as the representative depolymerization product, the quantitative relationship between the generated H$_2$O$_2$ using [HNEt$_3$][HSO$_4$] as the electrolyte and the produced vanillin was calculated in another typical study [83]. In the electrolysis of an actual lignin sample, it is evaluated that about 20 molecules of H$_2$O$_2$ are needed for producing one molecule of vanillin when taking vanillin as a representative depolymerization product of lignin.

Lignin depolymerization by active oxygen species generated in situ through ORR processes is intensified by using ILs as the support electrolyte. The electrochemical studies showed that the sorts of IL (protonic or aprotic), cathode potential, and other factors will affect the type and number of active oxygen species, and the main ORR product is superoxide radicals ($\cdot$O$_2^-$) in an aprotic IL, while the main ORR product is HOO$^-$ or H$_2$O$_2$ in a protonic IL or an aprotic IL containing trace water. The analysis of degradation products of lignin model compounds showed that the conversion percent and product selectivity were closely related to the sort and number of active oxygen species generated through the ORR process. Taking the same actual lignin as a sample, the 2e-ORR process was mainly produced in the IL [Bmim][BF$_4$] to produce H$_2$O$_2$, and the lignin conversion rate reached 88.4%, but a total of 8 kinds of small molecule products were detected due to the strong oxidation capacity of H$_2$O$_2$. In the aprotic IL [Bmim][BF$_4$], $\cdot$O$_2^-$ are mainly generated through the 1e-ORR

Table 5.4 Product distribution of electrochemical lignin depolymerization in [Bmim][BF$_4$]/CH$_3$CN and [HNEt$_3$][HSO$_4$].

No.	Structure	Depolymerization products		Relative content (%)	
		Name		[Bmim][BF$_4$]/CH$_3$CN	[HNEt$_3$]HSO$_4$
1		2,3-Dihydrobenzofuran		37.6	17.7
2		Vanillin		20.7	29.5
3		4-Hydroxy-3,5-dimethoxybenzaldehyde		24.3	10.4
4		1-(4-Hydroxy-3,5-dimethoxyphenyl)ethan-1-one		17.4	17.1
5		2-Methoxy-4-vinylphenol		—	6.3
6		1-(4-Hydroxyphenyl)ethan-1-one		—	5.2
7		1-(4-Hydroxy-3-methoxyphenyl)ethan-1-one		—	6.8
8		p-Hydroxyacetophenone		—	7.0

process, and the conversion rate was 50.3%, but only four kinds of products were detected. That is, the selectivity of products was improved. The difference between the melting boiling point and solubility of lignin depolymerization products and IL is conducive to the separation of lignin products and makes it easy to realize IL recovery. However, there is still a long way to realize commercially electrochemical depolymerization of lignin due to the complex mechanism. Nowadays, the development of advanced analytical instruments might help to reveal the complex mechanism of electrochemical depolymerization of lignin, which is of great significance to improve the selectivity and Faraday efficiency of value-added chemicals.

5.5 Application of ionic liquids in new energy batteries

IL is a kind of low-temperature molten salt, which has the advantages of low steam pressure, high conductivity, nonflammability, thermal stability, wide electrochemical window, and high electrochemical stability. It has become a hot spot in the research of new green electrolytes and has been widely used in the research of new electrolytes required by various electrochemical energy storage devices, such as electrodeposition, supercapacitors, lithium-ion battery, synthesis of new organic hydrolysates, and so forth.

5.5.1 Application of ionic liquids in lithium-ion batteries

Lithium-ion batteries, because of their high energy density, small size, no memory effect, and other advantages, are widely used in computers, mobile phones, cameras, and other portable electronic products [84]; however, the traditional organic electrolyte is prone to volatilization, leakage, flammability, and other safety issues, which would limit its application in power vehicles and large-scale energy storage [85].

5.5.1.1 Liquid electrolyte

At present, lithium-ion batteries mainly use carbonate as electrolyte solvents, and there are huge security risks in large-scale applications, especially in high-voltage and high-energy density battery systems. ILs have many advantages in lithium-ion battery electrolyte due to their high electrochemical stability, wide electrochemical window, high volatility, high safety, and easy recovery, so it can be used as a new solvent or additive to replace or partially replace the traditional carbonate electrolytes. The ILs applied in the electrolyte can be divided into imidazole ions, quaternary ammonium, and quaternary phosphine ILs according to the types of cations, as shown in Fig. 5.30.

Imidazole ILs have become the earliest ILs used in lithium-ion batteries because of their small viscosity, high conductivity, and difficulty in volatilization. However, the early imidazole ILs with $AICl_4^-$ as an anion have a narrow electrochemical stability window and cannot maintain the stability of lithium. Therefore the researchers

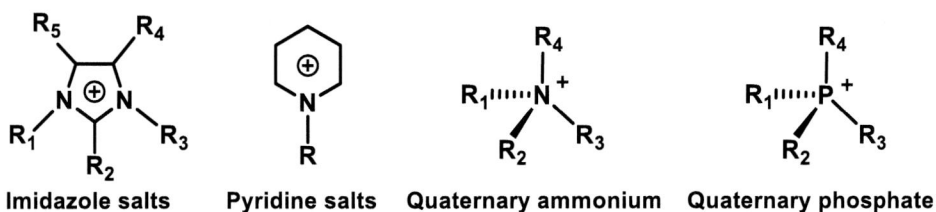

Imidazole salts   Pyridine salts   Quaternary ammonium   Quaternary phosphate

Figure 5.30 Structure diagram of common ionic liquid cations.

developed anionic groups such as BF_4^-, PF_6^-, and $TFSI^-$ to replace $AICI_4^-$ and applied them to imidazole electrolytes, which greatly improved the stability of imidazole ring.

Seki et al. [86,87] used $LiCoO_2$ as the cathode and Li as the anode and 3-propyl-1, 2-dimethylimidazole diimide salt ([DMPim][TFSI]) IL and 1-ethyl-3-methylimidazole diimsi ([Emim][TFSI]) IL as the electrolyte. They found that the reduction stability of imidazole IL cations could be improved by the introduction of methyl groups on the 2-carbon, the alkyl chain length of the 1-carbon can be increased, and then the cycle performance of the battery can be improved. In the $TFSI^-$ anion, the hydrogen bond between $TFSI^-$ and cation is weakened by the strong delocalization of fluorine substituents on negative charge, so $TFSI^-$ has the advantages of low viscosity, low melting point, high conductivity, and so forth. Therefore the IL formed by the combination of TFSI- and imidazole cation has the advantages of both. Sun Shanshan et al. [88] synthesized 1-methyl-3-ethylimidazoldiimide (EMI-TFSI) and 1-butyl-3-ethylimidazoldiimide (BMI-TFSI) IL as the electrolyte and then assembled half batteries using $LiCoO_2$ and $LiFePO_4$ as cathode materials. It was found that the two ILs had different electrochemical windows (4.8 V and 4.6 V respectively), and the compatibility between $LiCoO_2$ and the two electrolytes was poor, while the half battery assembled with $LiFePO_4$ had a higher specific capacity, and the specific discharge capacity remained above 120 mAh/g after 20 charge/discharge cycles (0.1 C). It indicates that different IL electrolytes need to be prepared for different battery systems. Wang et al. [89] designed and synthesized vinyl-functionalized imidazole IL. The kind of IL was applied as an electrolyte additive in high-voltage lithium-ion battery system. They found that after introducing vinyl or allyl group into the imidazole cation, the IL can stabilize the carbonate electrolyte on the surface of the cathode material $LiNi_{0.5}Mn_{1.5}O_4$ without being decomposed at a high voltage of 5.0 V. The battery performance test found that the cycle life and rate performance of the battery were greatly improved, and the discharge capacity of the battery reached above 130 mAh/g when adding 1-allyl-3-vinylimidazole bis (trifluoromethane sulfonyl) imide ([AVIm][TFSI]) in the electrolyte.

Quaternary ammonium IL has good electrochemical stability, and its electrochemical window is usually $0 \sim 5$ V (vs Li/Li^+), which can withstand the electrochemical deposition and dissolution of lithium without its own reduction decomposition, and it is not easy to occur oxidation decomposition on the cathode surface. However, compared with imidazole ILs, quaternary ammonium ILs have higher viscosity and lower conductivity. Their electrochemical properties can be improved by introducing some specific groups into the IL.

Sakaebe et al. [90] compared quaternary ammonium IL with traditional imidazole IL using $Li/LiCoO_2$ half battery and found that quaternary ammonium IL n-methyl-n-propyl piperidine bis (trifluoromethyl sulfonyl) imide had better electrochemical properties than 1-ethyl-3-methylimidazole IL. The discharge capacity of the electrolyte system reaches 130 mAh/g, and the charge/discharge efficiency is greater than 97% at 0.1 C and remains at about 85% after 50 cycles at 0.5 C rate. Fang et al. [91] synthesized quaternary ammonium cations and TFSI⁻ anions with 3—4 ether functional groups. They evaluated the viscosity, thermal stability, electrical conductivity, and electrochemical stability of the electrolyte. They found that the IL was used as electrolyte without adding other components in $Li/LiFePO_4$ battery, and the battery had better capacity and cycle performance at 0.1 C.

When the central atom N of quaternary ammonium IL is replaced by P, it also has good electrochemical stability, but its viscosity is larger. With the emergence of low viscosity quaternary phosphorus IL containing an ether group and functional group at room temperature, it has also been applied to the electrolytic liquid system of lithium-ion battery. Katsuhiko Tsunashima et al. [92] designed and synthesized a quaternary phosphate-based IL P222(2ol)-TFSI. Compared with quaternary ammonium IL DEME-TFSI, this IL has lower viscosity and higher ionic conductivity, showing better electrochemical performance. It was found that the initial capacity of the battery was 141 mAh/g (0.05 C), and the capacity retention rate was 85% after 50 cycles at the $Li/LiCoO_2$ battery system.

Pyrrole IL has good electrochemical stability, low viscosity, and ionic conductivity. Wongittharom et al. [93] synthesized PYR14-TFSI IL and used LiTFSI as lithium salt to prepare electrolyte for half battery. They found that the cycle performance and capacity retention rate of the battery were lower at 20°C due to the larger viscosity of PYR14-TFSI. However, the viscosity of the IL decreases and its conductivity increases as the temperature increases. The specific capacity of the battery for the first discharge reaches 140mAh/g at 50°C, and the rate is 0.1 C. The capacity retention rate of the battery can still reach 45% at 5 C. Pyrrole also has the problem of cationic coembedding behavior. The coembedding phenomenon can be reduced by adding film-forming additives to form stable SEI film on the anode surface of graphite. LiFSI/PP_{13}TFSI electrolyte was synthesized by Reiter et al. [94]. A stable SEI film was formed on the surface of the graphite anode by the FSI anion, which inhibited the occurrence of cationic codeposition. It was found that the specific discharge capacity of the graphite anode could reach 340—345mAh/g at the 0.1 C rate in the graphite/

$LiCo_{1/3}Mn_{1/3}Ni_{1/3}O_2$ battery, and the specific discharge capacity of NMC(111) could also reach 160mAh/g; the specific capacity of the battery did not decrease significantly after 25 cycles.

Electrolytes with excellent performance are the key materials for the preparation of lithium-ion batteries with high energy density, long cycle life, and high safety. Currently, lithium-ion batteries with traditional organic solvents as electrolytes have been commercialized, but carbonate-based organic solvents still have safety problems such as volatilization, flammability, and poor resistance to high pressure. However, IL shows great potential in the application of electrolyte because of its high electrochemical stability, wide electrochemical window, low volatilization, high safety, and easy recovery. It is mixed with organic solvents in a certain proportion and used as an electrolyte additive or a complex distribution solution, which not only solves the problem of large viscosity and high cost of IL, but also improves the problem of poor safety of pure organic solvent. However, IL is still in the research stage in lithium-ion battery electrolyte, and some of its properties are still unable to meet the requirements of commercialization, so there is still a long way to go to develop a new IL electrolyte with a better comprehensive performance.

5.5.1.2 Solid electrolyte

Compared with liquid electrolyte, polymer electrolyte realizes the functions of traditional separator and electrolyte and then saves the cost of consumables. It can also avoid electrolyte leakage problems and improve the safety of the battery. The size and shape of the polymer electrolyte are controllable, which is convenient for battery diversification and lightweight design. It can effectively inhibit the growth of lithium dendrites and greatly reduce the probability of short circuit in the battery. However, due to the low vapor pressure, thermal stability, and designability of traditional polymer electrolytes, ILs can be successfully transferred to polymer chains to form polymeric ILs, which effectively improves the conductivity of polymer electrolytes. The main factor of IL to improve the conductivity of polymer electrolyte is the positive and negative ion properties of IL monomers in polyelectrolyte. The influence of anions on the conductivity of polyionic liquid depends on the delocalization ability of anions, ionic scale, and the ability to form hydrogen bond with polyelectrolyte. The anions with the characteristics of small volume, strong mobility, delocalized charge, and weak interaction with skeleton can provide high conductivity of the polyionic. The introduction of substituents with very large steric resistance in cations can reduce the bulk density after IL polymerization and thus improve the conductivity of the polymer electrolyte. In addition, the aprotonated cations have higher conductivity.

For example, Vygodskii et al. [95] reported that the conductivity of the original polymerized IL was increased by four orders of magnitude after independent insertion of the propyl phosphonate group. The protonation characteristics of ions also affect the conductivity of their polymers. Although the mechanism is not clear, it can still be found that aprotonated cations have higher conductivity without considering the purity and molecular weight through a lot of literature comparison. When comparing the anions of pyrrole IL, it was found that the two factors that determine the conductivity of polymerized IL are delocalization ability and ionic scale. The strong protonation brought about by sulfonic acid can increase the conductivity of the corresponding polymer to 1.1×10^{-4} S/cm, and the modified sulfonic acid anion has a significantly higher effect on the conductivity than the common sulfonic acid group [96].

In recent years, the research of polymeric IL is increasing, and the polymeric IL has been widely used in lithium battery. Li et al. [97] synthesized guanidinium IL monomer and methyl acrylate to form polymeric IL. The gel electrolyte formed by the combination of the polymeric IL with LiTFSI and nano-SiO_2 still has good compatibility with lithium metal at 80°C, high ionic conductivity, wide electrochemical window, and good lithium dissolution/deposition performance.

Similar to the polymer solid electrolyte, the gel electrolyte uses the matrix structure of organic matter to fix the organic solvent, which can effectively improve the conductivity of the organic solid electrolyte. However, many organic solvents will reduce the mechanical properties of the solid electrolyte, resulting in the collapse of the structure of the organic polymer and some safety problems such as electrolyte leakage. The combination of IL and organic solid electrolyte can effectively improve the conductivity of electrolyte and reduce safety risks. IL plays the role of a plasticizer in the gel electrolyte, and it can reduce the glass transition temperature of the polymer electrolyte and improve the ionic conductivity of the electrolyte. The ionic conduction mechanism of the gel electrolyte is similar to that of the liquid electrolyte, but the transport mechanism is different from that of the liquid electrolyte. The gel electrolyte absorbs the liquid electrolyte through the swelling of the organic solid electrolyte, forming a number of solution cavities inside the electrolyte film, and the movement of the end of the organic chain on the edge of the solution cavity will also be accelerated because of the presence of the liquid electrolyte [98], so the conductivity of the gel electrolyte can reach to 10^{-3} S/cm, which is generally higher than that of the organic solid electrolyte. For example, K. Karuppasamy et al. [99] used a new type of perfluoroanion IL BmimNfO mixed with LiNfO and PVDF-HFP to prepare a gel electrolyte by solution casting. This method not only broadened the electrochemical window but also significantly improved the conductivity. The conductivity can reach 10^{-2} S/cm at 100°C, and the discharge capacity of the Li/LiCO$_2$ battery can reach up to 138 mAh/g at low current density. The application of IL in the gel electrolyte can not only solve the safety problem of traditional organic solvents, but also improve the ionic conductivity. ILs have a wide electrochemical window and good thermal stability, so the application of ILs in lithium batteries will be more and more extensive, and they play a great role in flexible energy storage equipment. Therefore solid electrolytes based on ILs are a big development prospect for the future energy storage industry.

5.5.2 Application of ionic liquids in supercapacitors

This section mainly introduces the application of IL in the synthesis of electrode materials and electrolyte in supercapacitors and its influence on the electrochemical performance of supercapacitors.

5.5.2.1 Ionic liquids as a supercapacitor support electrolyte

Supercapacitors can be divided into two categories, namely, double layer capacitors and Faraday capacitors, which work by storing charge through a high specific surface area, electrode/interface double layer, and rapid charge transfer reactions between the electrode and the electrolyte, such as electrochemical adsorption/desorption and the Faraday process of redox reaction. A supercapacitor is a new type of energy storage component between traditional capacitors and batteries, which has the characteristics of fast charging and discharging, higher energy density than physical capacitors, higher power density, and longer cycle life than batteries [100]. Fig. 5.31 shows the mechanism and working principle of the double electric layer capacitor.

The electrodes connected with the positive and negative electrodes carry the corresponding electrode charge in the charging state, and the cation and anion in the electrolyte move to the electrode with opposite charge under the action of the electric field. During the charging process, the electrolyte ions are oriented at the interface of the electrode material with opposite charge, forming a double electric layer to store charge. After the end of charging, electrostatic force maintains the double electric layer

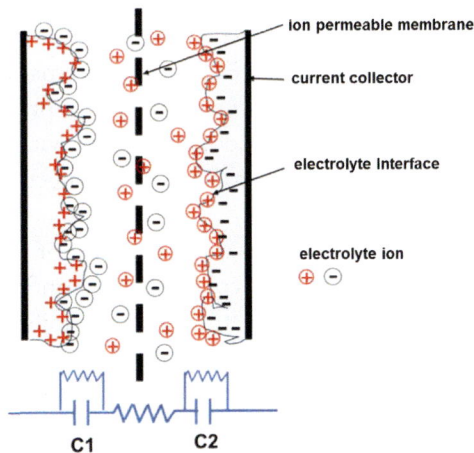

Figure 5.31 Schematic diagram of a double electric layer capacitor.

structure of the electrode material and the electrolyte. When discharging, the electrostatic force between the electrode material and the electrolyte disappears, the charged ions return to the electrolyte solution, and the electrons move from the anode to the cathode. Eq. (5.1) is the capacitance equation of the capacitor: $C = Q/U = \varepsilon A/d$, where C is the capacitance (F), Q is the electric quantity (C), U is the applied voltage (V), ε is the dielectric constant, A is the specific surface area of the electrode, and d is the thickness of the electrode sheet. Supercapacitors use porous materials with a high specific surface area or materials with redox reaction ability as electrode materials, so they have high specific capacitance. This working principle of supercapacitors makes them have the advantages of specific power greater than the battery, specific energy better than the traditional capacitor, and long cycle life.

The electrolyte of traditional supercapacitors is composed of quaternary ammonium salt dissolved in acrylic carbonate (PC) or acetonitrile (ACN), and the operating voltage range of this traditional supercapacitor is 2.3−2.7 V. With the increasing demand for power density and energy density of electrochemical capacitors, traditional electrolytes such as aqueous electrolyte and organic electrolyte are more and more difficult to meet the demand. Therefore the research and development of high-performance electrolytes is a main direction of supercapacitor research, and the research goal is to obtain superior electrochemical properties, such as high energy and high power, high safety, and environmentally friendly electrolytes.

The formulas for calculating the energy (E) and power (P) of supercapacitors are Eqs. (5.1) and (5.2) respectively:

$$E = 1/2CV \tag{5.1}$$

$$P = 4V^2/\text{ESR} \tag{5.2}$$

Where C is the capacitance, V is the effective voltage, and ESR is the equivalent resistance of the supercapacitor. Among them, ESR is the internal resistance of the supercapacitor, which is closely related to its power density and practical application. From the above two formulas, it can be concluded that improving the operating voltage window and conductivity of the supercapacitor is an effective way to improve its energy density and power density.

ILs are composed of anions and ions. As the electrolyte of the supercapacitor, it has the advantages of low steam pressure, high conductivity, nonflammability, thermal stability, wide chemical window, and high electrochemical stability, which can effectively improve the energy density and safety of the supercapacitor. The main reason why ILs can increase the energy density of a capacitor is that the addition of ILs can effectively improve the electrochemical window of the electrolyte, and the voltage window size mainly depends on the electrochemical stability of cations and anions of ILs, in which cations mainly affect the reduction potential, and anions determine the oxidation potential. The cation stability order of common ILs is: pyridine $<$ pyrazole $<$ imidazole $<$ sulfur salt $<$ quaternary ammonium salt, and the stability order of anions is: halogen ion $<$ chloroaluminate ion $<$ fluorinated ion $<$ sulfonamide. Most ILs have an electrochemical window of 4−6 V.

In addition, ILs are composed entirely of positive and negative ions with high conductivity. Their conductivity varies widely at room temperature, generally in the range of 0.1−18 mS/cm. The conductivity of IL is comparable to that of organic electrolytic liquid, such as LiPF6 in ethyl carbonate and dimethyl carbonate electrolytes, which have a conductivity of 16.6 mS/cm. However, IL itself has a high viscosity, and its conductivity is greatly affected by the viscosity of the system, so pure IL is rarely used as the supercapacitor electrolyte. To reduce the viscosity of the electrolyte and improve the conductivity of the electrolyte, researchers mainly adopted the method of adding organic solvents in the system, and the cation and anion of the IL as a conductive salt are separated under the action of organic solvents. The concentration is smaller, the degree of ionization is higher, the free mobility of ions is higher, and the conductivity is increased. At the same time, some researchers use the sol-gel method to polymerize IL with polymer material to prepare a solid electrolyte, which can further improve the conductivity of electrolyte and the safety of supercapacitor. Different ILs play different roles in improving the electrochemical performance of supercapacitors. For example, pyrrole ILs have a higher voltage window, while imidazolyl ILs show higher conductivity and lower viscosity. To achieve a collaborative output of energy density and power density, the respective advantages of different ILs are effectively combined. The electrochemical performance of supercapacitors can be further improved by preparing mixed electrolytes composed of different ILs. Introducing new ILs with high conductivity, low viscosity, and high voltage windows at room temperature is an ideal solution for achieving high energy density and power density cooutput and high safety of supercapacitors.

Supercapacitors mainly store charge at the interface between electrode material and electrolyte, so the capacitance value of supercapacitors is related to the aperture distribution and structure of electrodes and the size of electrolyte ions. Electrolyte ions present different electrochemical properties in holes of different sizes, and the matching degree of electrolyte ion size and electrode material hole size directly affects capacitance value and other electrochemical properties. Therefore further exploring the influence of electrolyte ion size on the capacity of supercapacitors is an effective way to improve the energy density of supercapacitors. The IL usually has a large ionic radius, so to improve the contact area between the electrode material and the IL, it is necessary to consider the matching of the ion size in the IL and the electrode material.

5.5.2.2 Ionic liquids is used to prepare electrode materials for supercapacitors

The capacitance value of the supercapacitor is not only related to the size of the electrolyte ion/molecule and the aperture of the electrode material, but also closely related to the surface properties of the electrode material, and the hydrophilic and hydrophobic properties of the electrode material directly affect the interface resistance of the electrolyte and the electrode material. The IL not only can be used as the electrolyte of a supercapacitor, but also can be used to prepare the electrode material of a supercapacitor. For example, C. Arbizzani and his collaborators used [Emim]TFSI as a solution to prepare polymeric IL as the electrode and obtained a high capacitance value [101]. ILs can also be used as carbon precursors to directly prepare carbon materials with excellent properties, and the preparation process is simple and environmentally friendly. The synthesis of multifunctional carbon materials by direct carbonization of IL has the following characteristics: (1) The low vapor pressure and high thermal stability of IL determine the high yield of carbon materials, (2) the designability and regulation of ILs guide the preparation of carbon materials with different structural properties, (3) the hydrophilic groups of ILs enable them to dissolve in different reagents, (4) the heteroatoms carried by the IL itself can introduce phosphorus, sulfur, nitrogen, and other heteroatoms into the prepared carbon material skeleton, and heteroatom doping can increase the infiltration of electrolytes to carbon material, while increasing the defects and active sites in the structure of carbon materials. Therefore IL as a precursor can be used to prepare carbon materials with high conductivity, adjustable pore size, and surface structure.

In addition to direct carbonization, IL can also be used as a solvent in ionic thermal synthesis to prepare nanocarbon materials. The ionic thermal method using the IL as a solvent to prepare carbon material is efficient, green, and simple, avoiding the high temperature and high-pressure conditions of traditional hydrothermal or solvothermal. IL is used not only as a solvent, but also as a soft template and stabilizer to regulate the structure of carbon nanomaterials. The morphology of the prepared materials can be regulated by designing different types of ILs. At the same time, in the process of preparing graphene-based polyporous carbon materials, the IL can also be used as a stabilizer to prevent the restack/superposition of the graphene layer and prepare high specific surface area graphene-based carbon materials [102,103].

5.5.3 Application of ionic liquids in flow batteries

Among the various electrochemical energy storage batteries, the flow battery is a preferred energy storage system for harnessing renewable energy due to its advantages such as the separation of active materials from the cell stack, scalability, long cycle life, low operational cost, and geographical flexibility [104,105]. Currently, research on the flow batteries is focusing on zinc-bromine, sodium polysulfide-bromine, all-vanadium, lithium-ion, and lithium-sulfur systems. The main systems of flow batteries consist of reaction stack, electrolyte solution, electrolyte storage tanks, pumps, and charge—discharge system. The stacks are assembled by multiple cells according to series and parallel in which the conversion of chemical and electrical energy is taking place, forming the crucial component. Flow batteries can be categorized as all-liquid, mixed-liquid, single-liquid, and semisolid flow batteries based on the flow mode and structure as shown in Fig. 5.32. In all-liquid flow batteries, both the positive and negative electrodes are flowable liquid electrolytes (Fig. 5.32A), and the all–vanadium system is the typical example. In mixed-liquid flow batteries, the positive electrode is typically a fluid electrode, while the negative electrode is either a metal or air electrode (Fig. 5.32B), such as the zinc-bromine and hydrogen-bromine systems. Single-liquid flow batteries have only one flowable electrode (Fig. 5.32C), such as the zinc-nickel hydroxide and lithium-sulfur systems. Semisolid flow batteries have active material suspensions as the positive and negative electrodes (Fig. 5.32D), with lithium-ion flow batteries being the main example. In oxidation—reduction flow batteries, the active materials are separated from the cell stack, and the battery capacity is determined by the size of the storage tanks and the concentration of active materials. The system power, on the other hand, is determined by the size of the reaction stack. This unique characteristic allows for separate optimization of system energy and power, thereby expanding the application versatility of flow batteries.

Currently, traditional flow batteries are generally utilizing water-based electrolytes due to their high safety, high conductivity, and low cost, such as the commercially available all-vanadium flow battery. However, the electrochemical window of water-

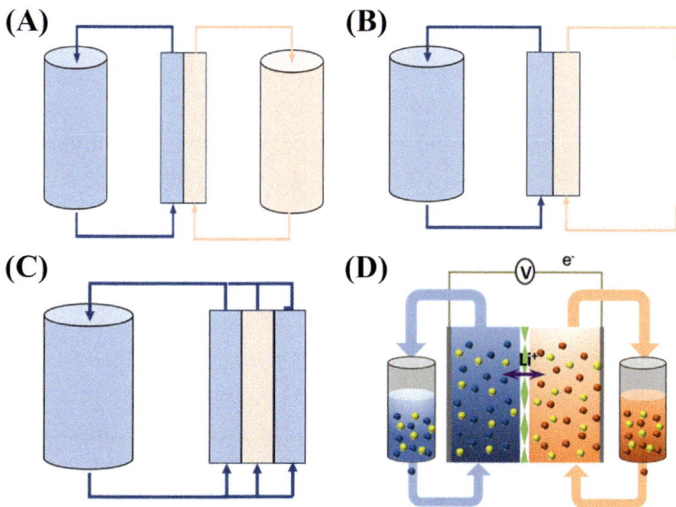

Figure 5.32 Flow batteries with different structures.

based systems is generally below 2 V, severely limiting the energy density of flow batteries. For example, the energy density of the all-vanadium flow battery is generally below 50 Wh/kg [106]. To increase the operating voltage and energy density of flow batteries, some organic solvents, such as acetonitrile (CH_3CN), have been applied. Thus the electrochemical window can reached 5.0 V, significantly increasing the battery's energy density [107]. However, the volatility, flammability, and toxicity of organic solvents hinder their practical applications. In comparison, ILs possess excellent properties such as nonvolatility, nonflammability, and a wide electrochemical window. In comparison to molten salts, ILs have a lower liquid state temperature. These characteristics make them useful as electrolyte solvents or additives in various battery systems, improving the working voltage and safety of traditional lithium-ion batteries, inhibiting lithium dendrite growth and enhancing battery cycle performance and safety in lithium-sulfur batteries [108−110]. The unique properties of ILs also offer new insights and a wide range of materials for flow battery research.

5.5.3.1 *Water-based flow battery*

Developing a stable vanadium-water solution system that operates at high potentials is crucial for improving the energy density of vanadium redox flow batteries. The electrochemical window of water (~ 1.5 V) severely limits the energy density of vanadium redox flow batteries in which the positive and negative electrodes are full of different vanadium ionic solutions. Adjusting the pH of the electrolyte solution can enhance the electrode potentials for H_2 and O_2 evolution, but this affects the solubility, stability, and electrochemical performance of vanadium ions in the electrolyte [111]. Adding an IL to the aqueous vanadium redox flow battery can effectively increase the operating voltage and maintain a high concentration of active species.

This "water-in-IL" concept was proposed according to the number of ions presented in the given solvent. The broad electrochemical stability window (ESW) of "water-in-IL" is suitable for most reported organic redox species [112]. For instance, Chen reported a mixed system of the nonelectroactive IL 1-butyl-3-methylimidazolium chloride (BmimCl) and water, which exhibited higher water decomposition voltage [113]. The Cl^- ions in the IL facilitate the migration of active species ions by reducing the concentration of free water in the electrolyte, leading to the significant suppression of H_2 and O_2 evolution, thereby widening the electrochemical window. By dissolving highly hydrophilic ILs with high melting points in water, an IL-based aqueous electrolyte with high ESW from 3 to 4.4 V can be obtained [114]. Moreover, the addition of ILs in the aqueous system can efficiently expand the temperature stability window, and the solubility of the organic molecules in the proposed supporting electrolyte will be enhanced [115]. The primary role of the IL in the vanadium redox flow battery is to improve the stability of vanadium ions in the electrolyte, inhibit cross-contamination, and enhance charge−discharge efficiency. Furthermore, due to the various kinds of ILs, by functional design and selection, appropriate ILs can be utilized as solvents or electrolyte additives in flow battery to address the aforementioned issues and promote the development of vanadium redox flow batteries. Additionally, a more important aspect of "water-in-ILs" application is developing membraneless redox flow batteries, in which the membranes are removed but not replaced, resulting in a direct fluid—fluid interface membranes [116]. Though there are many advantages of "water-in-ILs" electrolyte, the issues of self-discharge arising from the direct contact of redox pairs at the liquid—liquid interface and cross-mixing of active species through the interface that remain need to be solved in the membrane-free redox flow batteries in the future, and the studies in this aspect are still at an early stage.

Zinc-bromine flow battery is another type of aqueous flow battery. It uses zinc bromide solution as the electrolyte. During the charging process, zinc is deposited on the surface of the carbon plastic electrode, while bromine forms complex compounds and is stored at the bottom of the positive electrode. The theoretical energy density of zinc-bromine flow battery can reach 430 Wh/kg, and the open circuit voltage of the battery can reach 1.82 V. The battery also has a long cycle life, lasting for several thousand cycles. Zinc-bromine batteries can operate near room temperature without the need for complex thermal control systems. Both the anode and cathode materials, as well as the electrolyte, are inexpensive, resulting in low manufacturing costs. Therefore zinc-bromine flow batteries are competitive in terms of both energy density and manufacturing costs, making them a highly competitive energy storage technology. However, zinc dendrites are produced during the charging and discharging process of the zinc electrode, and the bromine electrode is unstable, easily migrating to the negative electrode and reacting with zinc, leading to significant self-discharge. These issues have hindered the development of zinc-bromine batteries. Gobinath P. Rajarathnam et al. [117] evaluated the role of six ILs as stabilizers for the bromine electrode, among which $[C_2Py]Br$, $[C_2mim]Br$, and $[C_2OHPy]Br$ significantly improved the electrochemical performance of the zinc electrode and alleviated the growth of zinc dendrites. Currently, there is limited research on the application of ILs in the zinc-bromine flow battery system, focusing mainly on inhibiting zinc dendrite growth and stabilizing the bromine electrode. There is a huge research space in this area, and it is believed that more scholars will pay attention to the application of ILs in zinc-bromine flow batteries and zinc hybrid flow batteries in the future.

5.5.3.2 Organic flow battery

Adding ILs to the organic-based lithium-sulfur flow battery and lithium-ion flow battery can enhance the high/low-temperature performance and safety characteristics of the electrolyte. The lithium-sulfur flow battery and lithium-ion flow battery combine the features of lithium-sulfur and lithium-ion batteries with flow batteries, resulting in a promising energy storage technology known as semisolid flow batteries. These batteries possess high energy density, high power density, as well as the flexibility, controllability, and scalability of flow batteries. However, certain challenges can be seen in lithium-sulfur batteries, such as the shuttle effect of polysulfides and lithium dendrite growth, which persist in lithium-sulfur flow batteries. Zhang et al. [118] reported the use of pyridine-based IL nanoparticles as additives for the positive electrode fluid in lithium-sulfur flow batteries (Fig. 5.33). The IL nanoparticles act as a bridge between sulfur/polysulfides and carbon-loaded materials, effectively immobilizing the polysulfides in the positive electrode region and significantly controlling the shuttle effect. Additionally, the IL enhances the migration of lithium ions in the flow system and suppresses lithium dendrite growth. With the introduction of these IL nanoparticles, the cycle life of the lithium-sulfur flow battery reaches more than 1000 cycles. ILs have been extensively studied as solvents or electrolyte additives in lithium-ion and lithium-sulfur batteries. Combining the favorable properties of ILs with functionalized designs, it is believed that ILs will play a significant role in the development of lithium-ion flow batteries and lithium-sulfur flow batteries, driving their progress forward.

When conducting electrochemical studies using ILs as supporting electrolytes, it is necessary to understand their electrochemical stability in advance. This can be estimated by determining the maximum potential range in which no electrochemical reactions occur, known as the "electrochemical window" of the electrolyte. The expansion of the electrochemical window is also an important symbol in the development of ILs, despite of the continuous development of ILs with better electrochemical

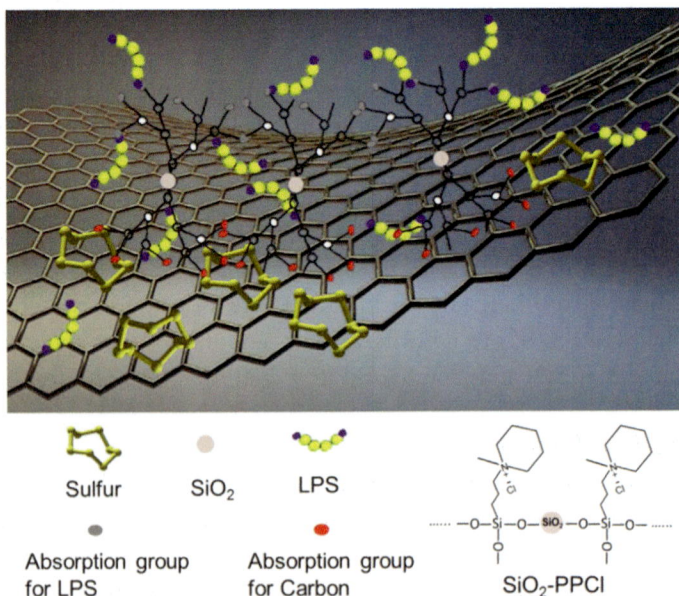

Figure 5.33 The function of ionic liquid nanoparticles in lithium-sulfur flow battery [118].

stability. Similar to traditional supporting electrolytes, the electrochemical window of ILs is typically determined using CV and linear sweep voltammetry. However, even for the same IL, the voltammograms can vary in different research reports. This variation can be attributed to several factors, including inherent factors such as voltage drops caused by solution resistance, as well as different measurement conditions such as working and/or reference electrodes, scan rates, and cut-off current densities for determining cathodic and anodic limits. Additionally, the purity of the IL can also influence the voltammograms. For example, even the presence of a small amount of water in the IL, such as 100 ppm, should not be overlooked, as electrochemical measurements are often highly sensitive to impurities. Despite these challenges, the determination of the electrochemical window still provides important information for finding electrochemically stable ILs.

References

[1] Lv RB, Zhang M, Huang H, et al. Self-doped TiO_2 nanotubes with surface modification by ils for enhanced photoreduction of CO_2 to acetic acid. Applied Surface Science 2023;621:156897.
[2] Lin JL, Ding ZX, Hou YD, et al. Ionic liquid co-catalyzed artificial photosynthesis of CO. Scientific Reports 2013;3:1056.
[3] Liu Y, Sun JH, Huang HH, et al. Improving CO_2 photoconversion with ionic liquid and co single atoms. Nature Communications 2023;14:1457.
[4] Sun L, Zhang ZQ, Bian J, et al. A Z-scheme heterojunctional photocatalyst engineered with spatially separated dual redox sites for selective CO_2 reduction with water: insight by in situ µs-transient absorption spectra. Advanced Materials 2023;35:2300064.
[5] Can E, Uralcan B, Yildirim R. Enhancing charge transfer in photocatalytic hydrogen production over dye-sensitized Pt/TiO_2 by ionic liquid coating. ACS Applied Energy Materials 2021;4:10931−9.
[6] Naikwade AG, Jagadale MB, Kale DP, et al. Photocatalytic degradation of methyl orange by magnetically retrievable supported ionic liquid phase photocatalyst. ACS Omega 2020;5:131−44.
[7] Gao Y, Meng QB, Wang BX, et al. Polyacrylonitrile derived robust and flexible poly(ionic liquid)s nanofiber membrane as catalyst supporter. Catalysts 2022;12:266.
[8] Wang YN, Deng KJ, Zhang LZ. Visible light photocatalysis of BiOI and its photocatalytic activity enhancement by in situ ionic liquid modification. Journal Of Physical Chemistry C 2011;115:14300−8.
[9] Valverde D, Porcar R, Izquierdo D, et al. Rose bengal immobilized on supported ionic-liquid-like phases: an efficient photocatalyst for batch and flow processes. ChemSusChem 2019;12:3996−4004.
[10] Fabry DC, Ronge MA, Rueping M. Immobilization and continuous recycling of photoredox catalysts in ionic liquids for applications in batch reactions and flow systems: catalytic alkene isomerization by using visible light. Chemistry A European Journal 2015;21:5350−4.
[11] Barange SH, Raut SU, Bhansali KJ, et al. Biodiesel production via esterification of oleic acid catalyzed by brønsted acid-functionalized porphyrin grafted with benzimidazolium-based ionic liquid as an efficient photocatalyst. Biomass Conversion and Biorefinery 2020;13(3):1873−88.
[12] Li DM, Hua T, Li XM, et al. In-situ fabrication of ionic liquids/MIL-68(In)−NH2 photocatalyst for improving visible-light photocatalytic degradation of doxycycline hydrochloride. Chemosphere 2022;292:133461.
[13] Zeng GT, Qiu J, Hou BY, et al. Enhanced photocatalytic reduction of CO_2 to CO through TiO_2 passivation of inp in ionic liquids. Chemistry A European Journal 2015;21:13502−7.
[14] Miró R, Guzmán H, Godard C, et al. Solar-driven CO_2 reduction catalysed by hybrid supramolecular photocathodes and enhanced by ionic liquids. Catalysis Science & Technology 2023;13(6):1708−17.

[15] Lu WW, Jia B, Cui BL, et al. Efficient photoelectrochemical reduction of carbon dioxide to formic acid: a functionalized ionic liquid as an absorbent and electrolyte. Angewandte Chemie—International Edition 2017;56:11851−4.

[16] Zhou FL, McDonnell-Worth C, Li HT, et al. Enhanced photo-electrochemical water oxidation on mnox in buffered organic/inorganic electrolytes. Journal of Materials Chemistry A 2015;3:16642−52.

[17] Shen XJ, Chen L, Li JN, et al. Silicon microhole arrays architecture for stable and efficient photo-electrochemical cells using ionic liquids electrolytes. Journal of Power Sources 2016;318:146−53.

[18] Park SM, Wei M, Xu J, et al. Engineering ligand reactivity enables high-temperature operation of stable perovskite solar cells. Science (New York, N.Y.) 2023;381(6654):209−15.

[19] Park J, Kim J, Yun HS, et al. Controlled growth of perovskite layers with volatile alkylammonium chlorides. Nature 2023;616(7958):724−30.

[20] Lin YH, Sakai N, Da P, et al. A piperidinium salt stabilizes efficient metal-halide perovskite solar cells. Science (New York, N.Y.) 2020;369(6499):96−102.

[21] Wang F, Duan D, Singh M, et al. Ionic liquid engineering in perovskite photovoltaics. Energy & Environmental Materials 2022;6(5):e12435.

[22] Xia R, Gao XX, Zhang Y, et al. An efficient approach to fabricate air-stable perovskite solar cells via addition of a self-polymerizing ionic liquid. Advanced Materials 2020;32(40):2003801.

[23] Bai S, Da P, Li C, et al. Planar perovskite solar cells with long-term stability using ionic liquid additives. Nature 2019;571(7764):245−50.

[24] Ran J, Wang H, Deng W, et al. Ionic liquid-tuned crystallization for stable and efficient perovskite solar cells. Solar RRL 2022;6(7):2200176.

[25] Zhang F, Zhu K. Additive engineering for efficient and stable perovskite solar cells. Advanced Energy Materials 2020;10(13):1902579.

[26] Zhu X, Du M, Feng J, et al. High-efficiency perovskite solar cells with imidazolium-based ionic liquid for surface passivation and charge transport. Angewandte Chemie—International Edition 2021;60(8):4238−44.

[27] Wan Y, Dong S, Wang Y, et al. Ionic liquid-assisted perovskite crystal film growth for high performance planar heterojunction perovskite solar cells. RSC Advances 2016;6(100):97848−52.

[28] Liu D, Shao Z, Gui J, et al. A polar-hydrophobic ionic liquid induces grain growth and stabilization in halide perovskites. Chemical Communications 2019;55(74):11059−62.

[29] He D, Zhou T, Liu B, et al. Interfacial defect passivation by novel phosphonium salts yields 22% efficiency perovskite solar cells: experimental and theoretical evidence. EcoMat 2022;4(1):e12158.

[30] De Marco N, Zhou H, Chen Q, et al. Guanidinium: a route to enhanced carrier lifetime and open-circuit voltage in hybrid perovskite solar cells. Nano Letters 2016;16(2):1009−16.

[31] Hu P, Huang S, Guo M, et al. Ionic liquid-assisted crystallization and defect passivation for efficient perovskite solar cells with enhanced open-circuit voltage. ChemSusChem 2022;15(15):e202200819.

[32] Ghosh S, Singh T. Role of ionic liquids in organic-inorganic metal halide perovskite solar cells efficiency and stability. Nano Energy 2019;63:103828.

[33] Liu Y, Gao Y, Lu M, et al. Ionic additive engineering for stable planar perovskite solar cells with efficiency >22%. Chemical Engineering Journal 2021;426:130841.

[34] Chen Z, Wang L, Fang J, et al. Effective defect passivation via molecular interactions based on amine-functionalized ionic liquid for perovskite solar cells. ACS Applied Energy Materials 2023.

[35] Imran T, Raza H, Aziz L, et al. High performance inverted RbCsFAPbI$_3$ perovskite solar cells based on interface engineering and defects passivation. Small (Weinheim an der Bergstrasse, Germany) 2023;19(25):2207950.

[36] Seo JY, Matsui T, Luo J, et al. Ionic liquid control crystal growth to enhance planar perovskite solar cells efficiency. Advanced Energy Materials 2016;6(20):1600767.

[37] Zhou X, Wang Y, Li C, et al. Doping amino-functionalized ionic liquid in perovskite crystal for enhancing performances of hole-conductor free solar cells with carbon electrode. Chemical Engineering Journal 2019;372:46−52.

[38] Guo Y, Shoyama K, Sato W, et al. Chemical pathways connecting lead(II) iodide and perovskite via polymeric plumbate(II) fiber. Journal of the American Chemical Society 2015;137(50):15907−14.

[39] Öz S, Burschka J, Jung E, et al. Protic ionic liquid assisted solution processing of lead halide perovs-kites with water, alcohols and acetonitrile. Nano Energy 2018;51:632−8.

[40] Gu L, Ran C, Chao L, et al. Designing ionic liquids as the solvent for efficient and stable perovskite solar cells. ACS Applied Materials & Interfaces 2022;14(20):22870−8.

[41] Huang C, Lin P, Fu N, et al. Ionic liquid modified SnO_2 nanocrystals as a robust electron transport-ing layer for efficient planar perovskite solar cells. Journal of Materials Chemistry A 2018;6 (44):22086−95.

[42] Zhuang Q, Zhang C, Gong C, et al. Tailoring multifunctional anion modifiers to modulate interfacial chemical interactions for efficient and stable perovskite solar cells. Nano Energy 2022;102:107747.

[43] Cheng H, Li Y, Zhang M, et al. Self-assembled ionic liquid for highly efficient electron transport layer-free perovskite solar cells. ChemSusChem 2020;13(10):2779−85.

[44] Cao F, Zhu Z, Zhang C, et al. Synergistic ionic liquid in hole transport layers for highly stable and efficient perovskite solar cells. Small (Weinheim an der Bergstrasse, Germany) 2023;19(27):2207784.

[45] Zhang W, Liu X, He B, et al. Interface engineering of imidazolium ionic liquids toward efficient and stable $CsPbBr_3$ perovskite solar cells. ACS Applied Materials & Interfaces 2020;12(4):4540−8.

[46] Zhu X, Yang S, Cao Y, et al. Ionic-liquid-perovskite capping layer for stable 24.33%-efficient solar cell. Advanced Energy Materials 2022;12(6):2103491.

[47] Huang X, Guo H, Wang K, et al. Ionic liquid induced surface trap-state passivation for efficient perovskite hybrid solar cells. Organic Electronics 2017;41:42−8.

[48] Xiong Z, Chen X, Zhang B, et al. Simultaneous interfacial modification and crystallization control by biguanide hydrochloride for stable perovskite solar cells with PCE of 24.4%. Advanced Materials 2022;34(8):2106118.

[49] Kulkarni A, Siahrostami S, Patel A, et al. Understanding catalytic activity trends in the oxygen reduction reaction. Chemical Reviews 2018;118(5):2302−12.

[50] Guo SJ, Zhang S, Sun SH. Tuning nanoparticle catalysis for the oxygen reduction reaction. Angewandte Chemie—International Edition 2013;52(33):8526−44.

[51] Yang YL, Tang Y, Jiang HM, et al. 2020 Roadmap on gas-involved photo- and electro-catalysis. Chinese Chemical Letters 2019;30(12):2089−109.

[52] Andrieux CP, Hapiot P, Saveant JM. Mechanism of superoxide ion disproportionation in aprotic-solvents. Journal of the American Chemical Society 1987;109(12):3768−75.

[53] Khan A, Lu XY, Aldous L, et al. Oxygen reduction reaction in room temperature protic ionic liquids. Journal of Physical Chemistry C 2013;117(36):18334−42.

[54] Nie Y, Li L, Wei ZD. Recent advancements in Pt and Pt-free catalysts for oxygen reduction reac-tion. Chemical Society Reviews 2015;44(8):2168−201.

[55] Siahrostami S, Li GL, Viswanathan V, et al. One- or two-electron water oxidation, hydroxyl radi-cal, or H_2O_2 evolution. Journal of Physical Chemistry Letters 2017;8(6):1157−60.

[56] Guo YJ, Yang MJ, Xie RC, et al. The oxygen reduction reaction at silver electrodes in high chloride media and the implications for silver nanoparticle toxicity. Chemical Science 2021;12:397−406.

[57] Townshend A. Standard potentials in aqueous solutions. Analytica Chimica Acta 1987;198:333−4.

[58] Sawyer DT, Chiericato G, Angelis CT, et al. Effects of media and electrode materials on the elec-trochemical reduction of dioxygen. Analytical Chemistry 1982;54(11):1720−4.

[59] Chin DH, Chiericato G, Nanni EJ, et al. Proton-induced disproportionation of superoxide ion in aprotic media. Journal of the American Chemical Society 1982;104(5):1296−9.

[60] Katsounaros I, Cherevko S, Zeradjanin AR, et al. Oxygen electrochemistry as a cornerstone for sus-tainable energy conversion. Angewandte Chemie—International Edition 2014;53(1):102−21.

[61] Khan A, Gunawan CA, Zhao C. Oxygen reduction reaction in ionic liquids: fundamentals and applications in energy and sensors. ACS Sustainable Chemistry & Engineering 2017;5:3698−715.

[62] Allen CJ, Hwang J, Kautz R, et al. Oxygen reduction reactions in ionic liquids and the formulation of a general ORR mechanism for Li-air batteries. Journal of Physical Chemistry C 2012;116 (39):20755−64.

[63] Sekhon S, Krishnan P, Singh B, et al. Proton conducting membrane containing room temperature ionic liquid. Electrochimica Acta 2006;52(4):1639−44.

[64] Wang Z, Lin P, Baker GA, et al. Ionic liquids as electrolytes for the development of a robust amperometric oxygen sensor. Analytical Chemistry 2011;83(18):7066−73.

[65] Jiang HM, Xue AG, Wang ZH, et al. Electrochemical degradation of lignin by ROS. Sustainable Chemistry 2020;1(3):345−60.

[66] Wang L, Chen YM, Liu SY, et al. Study on the cleavage of alkyl-O-aryl bonds by in situ generated hydroxyl radicals on an ORR cathode. RSC Advances 2017;7:51419−25.

[67] Zhu HB, Chen YM, Qin TF, et al. Lignin depolymerization via an integrated approach of anode oxidation and electro-generated H_2O_2 oxidation. RSC Advances 2014;4:6232−8.

[68] Jiang HM, Zhang L, Wang Z, et al. Electrocatalytic methane direct conversion to methanol in electrolyte of ionic liquid. Electrochimica Acta 2023;445:142065.

[69] Li M, Pu Y, Ragauskas AJ. Current understanding of the correlation of lignin structure with biomass recalcitrance. Frontiers in Chemistry 2016;4:45.

[70] Song JW, Chen CJ, Zhu SZ, et al. Processing bulk natural wood into a high-performance structural material. Nature 2018;554(7691):224−8.

[71] Ragauskas AJ, Beckham GT, Biddy MJ, et al. Lignin valorization: improving lignin processing in the biorefinery. Science (New York, N.Y.) 2014;344(6185):709.

[72] Li CZ, Zhao XC, Wang AQ, et al. Catalytic transformation of lignin for the production of chemicals and fuels. Chemical Reviews 2015;115(21):11559−624.

[73] Jiang HM, Li A, Sun YZ, et al. Enhanced ORR performance to electrochemical lignin valorization in a mixture of ionic liquid/organic solvent binary electrolytes. New Journal Of Chemistry 2023;47(40):18682−9.

[74] Zigah D, Wang A, Lagrost C, et al. Diffusion of molecules in ionic liquids/organic solvent mixtures. Example of the reversible reduction of O_2 to superoxide. Journal of Physical Chemistry B 2009;113(7):2019−23.

[75] Watanabe M, Thomas ML, Zhang SG, et al. Application of ionic liquids to energy storage and conversion materials and devices. Chemical Reviews 2017;117(10):7190−239.

[76] MacFarlane DR, Tachikawa N, Forsyth M, et al. Energy applications of ionic liquids. Energy & Environmental Science 2014;7(1):232−50.

[77] Chen A, Rogers EI, Compton RG. Abrasive stripping voltammetric studies of lignin and lignin model compounds. Electroanalysis 2010;22(10):1037−44.

[78] Wang L, Liu S, Jiang HM, et al. Electrochemical generation of ROS in ionic liquid for the degradation of lignin model compound. Journal of the Electrochemical Society 2018;165(11):H705−10.

[79] Wu Z, Li M, Howe J, et al. Probing defect sites on CeO_2 nanocrystals with well-defined surface planes by raman spectroscopy and O_2 adsorption. Langmuir: the ACS Journal of Surfaces and Colloids 2010;26(21):16595−606.

[80] Jiang HM, Wang L, Qiao LL, et al. Improved oxidative cleavage of lignin model compound by ORR in protic ionic liquid. International Journal of Electrochemical Science 2019;14(3):2645−54.

[81] Jiang HM, Cheng YJ, Wang ZH, et al. Degradation of a lignin model compound by ROS generated in situ through controlled ORR in ionic liquid. Journal of the Electrochemical Society 2021;168(1):016504.

[82] Tolba R, Tian M, Wen J, et al. Electrochemical oxidation of lignin at IrO_2-based oxide electrodes. Journal of Electroanalytical Chemistry 2010;649(1):9−15.

[83] Han ZW, Jiang HM, Xue AG, et al. H_2O_2 generated through ORR on cathode in a protic ionic liquid and its utilization in lignin valorization. Journal of Electroanalytical Chemistry 2022;923:116814.

[84] Armand M, Endres F, MacFarlane DR, et al. Ionic-liquid materials for the electrochemical challenges of the future. Nature Materials 2009;8(8):621−9.

[85] Finegan DP, Scheel M, Robinson JB, et al. In-operando high-speed tomography of lithium-ion batteries during thermal runaway. Nature Communications 2015;6:6924.

[86] Seki S, Ohno Y, Kobayashi Y, et al. Imidazolium-based room-temperature ionic liquid for lithium secondary batteries. Journal of the Electrochemical Society 2007;154(3):173−7.

[87] Seki S. lithium secondary batteries using modified-imidazolium room-temperature ionic liquid. The Journal of Physical Chemistry. B 2006;110(21):10228−30.

[88] Sun S. Study on ionic liquid electrolytes for lithium ion batteries. Power Technology 2010;34 (1):55−8.

[89] Wang Z, Cai Y, Wang Z, et al. Vinyl-functionalized imidazolium ionic liquids as new electrolyte additives for high-voltage Li-ion batteries. Journal of Solid State Electrochemistry 2013;17(11):2839−48.

[90] Sakaebe H, Matsumoto H. N-Methyl-N-propylpiperidinium bis(trifluoromethanesulfonyl)imide (PP13-TFSI) novel electrolyte base for Li battery. Electrochemistry Communications 2003;5 (7):594−8.

[91] Fang S, Jin Y, Yang L, et al. Functionalized ionic liquids based on quaternary ammonium cations with three or four ether groups as new electrolytes for lithium battery. Electrochimica Acta 2011;56(12):4663−71.

[92] Tsunashima K, Yonekawa F, Sugiya M. A lithium battery electrolyte based on a room-temperature phosphonium ionic liquid. Chemistry Letters 2008;37(3):314−15.

[93] Wongittharom N, Lee TC, Hsu CH, et al. Electrochemical performance of rechargeable Li/LiFePO$_4$ cells with ionic liquid electrolyte: effects of Li salt at 25°C and 50°C. Journal of Power Sources 2013;240:676−82.

[94] Reiter J, Nádherná M, Dominko R. Graphite and LiCo$_{1/3}$Mn$_{1/3}$Ni$_{1/3}$O$_2$ electrodes with piperidinium ionic liquid and lithium bis(fluorosulfonyl)imide for Li-ion batteries. Journal of Power Sources 2012;205:402−7.

[95] Vygodskii YS, Mel'nik OA, Lozinskaya EI, et al. The influence of ionic liquid's nature on free radical polymerization of vinyl monomers and ionic conductivity of the obtained polymeric materials. Polymers for Advanced Technologies 2007;18(1):50−63.

[96] Ohno H, Yoshizawa M, Ogihara W. Development of new class of ion conductive polymers based on ionic liquids. Electrochimica Acta 2004;50(2−3):255−61.

[97] Li M, Yang L, Fang S, et al. Polymerized ionic liquids with guanidinium cations as host for gel polymer electrolytes in lithium metal batteries. Polymer International 2012;61(2):259−64.

[98] Saito Y, Kataoka H, Stephan AM. Investigation of the conduction mechanisms of lithium gel polymer electrolytes based on electrical conductivity and diffusion coefficient using NMR. Macromolecules 2001;34(20):6955−8.

[99] Karuppasamy K, Reddy PA, Sriniyas G, et al. Electrochemical and cycling performances of novel nonafluorobutanesulfonate (nonaflate) ionic liquid based ternary gel polymer electrolyte membranes for rechargeable lithium ion batteries. Journal of Membrane Science 2016;514:350−7.

[100] Pandolfo AG, Hollenkamp AF. Carbon properties and their role in supercapacitors. Journal of Power Sources 2006;157:11−27.

[101] Arbizzani C, Soavi F, Mastragostino M. A novel galvanostatic polymerization for high specific capacitance poly(3-methylthiophene) in ionic liquid. Journal of Power Sources 2006;162:735−7.

[102] Bag S, Samanta A, Bhunia P, et al. Rational functionalization of reduced graphene oxide with imidazolium-based ionic liquid for supercapacitor application. International Journal of Hydrogen Energy 2016;41(47):22134−43.

[103] Kim TY, Lee HW, Stoller M, et al. High-performance supercapacitors based on poly(ionic liquid)-modified graphene electrodes. ACS Nano 2011;5(1):436−42.

[104] Soloveichik GL. Flow batteries: current status and trends. Chemical Reviews 2015;115 (20):11533−58.

[105] Duduta M, Ho B, Wood V, et al. Sem-solid lithium rechargeable flow battery. Advanced Energy Materials 2011;1(4):511−16.

[106] Skyllas-Kazacos MH, Chakrabarti M, Hajimolana YS, et al. Progress in flow battery research and development. Journal of the Electrochemical Society 2011;158:55−79.

[107] Sleightholme AES, Shinkle AA, Liu Q, et al. Non-aqueous manganese acetylacetonate electrolyte for redox flow batteries. Journal of Power Sources 2011;196(13):5742−5.

[108] Huie MM, DiLeo RA, Marschilok AC, et al. Ionic liquid hybrid electrolytes for lithium-ion batteries: a key role of the separator-electrolyte interface in battery electrochemistry. ACS Applied Materials & Interfaces 2015;7(22):11724−31.

[109] Lu Y, Korf K, Kambe Y, et al. Ionic-liquid-nanoparticle hybrid electrolytes: applications in lithium metal batteries. Angewandte Chemie—International Edition 2013;53(2):488−92.

[110] Wang J, Chew SY, Zhao ZW, et al. Sulfur-mesoporous carbon composites in conjunction with a novel ionic liquid electrolyte for lithium rechargeable batteries. Carbon 2008;46(2):229–35.

[111] Gu S, Gong K, Yan EZ, et al. A multiple ion-exchange membrane design for redox flow batteries. Energy & Environmental Science 2014;7(9):2986–98.

[112] Huang Z, Zhang P, Gao X, et al. Unlocking simultaneously the temperature and electrochemical windows of aqueous phthalocyanine electrolytes. ACS Applied Energy Materials 2019;2:3773–9.

[113] Chen R, Hempelmann R. Ionic liquid-mediated aqueous redox flow batteries for high voltage applications. Electrochemistry Communications 2016;70:56–9.

[114] Zhang Y, Ye, Henkensmeier RD, et al. "Water-in-ionic liquid" solutions towards wide electro-chemical stability windows for aqueous rechargeable batteries. Electrochimica Acta 2018;263:47–52.

[115] Xue B, Wu X, Guo Y, et al. Review-ionic liquids applications in flow batteries. Journal of the Electrochemical Society 2022;169:080501.

[116] Bamgbopa MO, Almheiri S, Sun H. Renew prospects of recently developed membraneless cell designs for redox flow batteries. Renewable and Sustainable Energy Reviews 2017;70:506–18.

[117] Rajarathnam GP, Easton ME, Schneider M, et al. The influence of ionic liquid additives on zinc half-cell electrochemical performance in zinc/bromine flow batteries. RSC Advances 2016;6 (33):27788–97.

[118] Xu S, Cheng Y, Zhang L, et al. An effective polysulfides bridge builder to enable long-life lith-ium-sulfur flow batteries. Nano Energy 2018;51:113–21.

CHAPTER 6

Synthesis of ionic liquid-based materials

Contents

6.1 Overview   203
6.2 Preparation of nanomaterials with ionic liquids   204
6.3 Ionic liquids intensify the synthesis of molecular sieve materials   225
6.4 Synthesis of ionic liquid-based metal organic complexes   229
    6.4.1 Ionic liquids as charge compensation and structure guiding agent action   230
    6.4.2 Mixed ionic liquids as solvents for synthesis of porous metal organic complexes   231
    6.4.3 The role of ionic liquid solvents in the synthesis of rare earth complexes   231
    6.4.4 Synthesis of metal organic complexes using carboxyl-functionalized ionic liquids   232
6.5 Synthesis and application of polymerized ionic liquids   236
    6.5.1 Synthesis of poly(ionic liquid)s   237
    6.5.2 Applications of poly(ionic liquid)s   239
References   245

6.1 Overview

Due to the designable structure and excellent physicochemical properties, ionic liquids have significant advantages in the synthesis of functional materials. In this chapter, the highly promising application of ionic liquids in the preparation of nanoparticles and nanotubes, ionic liquids-based nanoparticle hybrids, molecular sieves, poly(ionic liquid)s, and ionic liquids-based metal complexes are summarized. In general, unique cationic and anionic structures, as well as their structure dependent physicochemical properties, facilitate the interactions between ionic liquids and reactants, including covalent interaction, electrostatic interactions, van der Waals, steric hindrance, hydrogen bonding, and solvation, etc. In the presence of ionic liquids, different self-organizing or post-modified structures are generated due to the mutual influence of these specific and local interactions, including ionic liquid short-range ordered colloidal glass and gel, long-range ordered liquid crystals, ionic liquids-based emulsion, as well as poly(ionic liquid). Therefore, ionic liquids can serve as solvents, surfactants, a charge compensator and a structure directing agent, or a reactant to participate the formation of functional materials. In particularly, the designability of ionic liquids is also conducive to the synthesis of metal organic complexes, molecular sieves, and poly(ionic liquid) with excellent performances. It is evident that ionic liquids play a crucial role in the synthesis and application of functional materials.

Here, we reviewed various synthesis methods of functional materials based on ionic liquids, and then focused on the role of ionic liquids in the synthesis of metal organic complexes, molecular sieves and poly(ionic liquid). Interpreting the interaction between ionic liquids and substances in reaction systems from a microscopic perspective, with the aim of obtaining a systematic understanding of the synthesis and application of ionic liquid enhanced functional materials, and providing theoretical support for the synthesis of high-performance ionic liquid involved functional materials. Furthermore, their application in sensing, catalysis, biomedicine, and energy sources etc. are discussed. This chapter presents the current research status of ionic liquid enhanced functional material synthesis, and introduces the latest developments and promising directions in the utilization of ionic liquids.

6.2 Preparation of nanomaterials with ionic liquids

Nanomaterials are an important research trend in nano science and technology, which are composed of nanostructures with sizes between atoms, molecules, and macroscopic systems. Nanomaterials have a distinctive size effect, which can change their physico-chemical properties. With the smaller size, nanomaterials usually exhibit higher surface area, higher molecular diffusivity, higher electron mobility, and stronger light absorption rate. It makes nanomaterials show unique thermal, optical, electrical, and magnetic properties, which have been widely used in various scientific and technological fields. For example, nanomaterials play an important role in catalytic reaction, which can be used as the carrier of catalyst to improve the dispersion and stability of catalysts, thus improving the efficiency of catalytic reaction. Moreover, they also can be directly used as catalysts for catalytic oxidation, hydrogenation, decomposition, and other reactions.

ILs are a new type of environmentally friendly and green solvent. Based on their excellent physiochemical properties, they have obvious advantages over the conventional solvents in the synthesis of nanomaterials. According to their structure and structure-dependent properties, ILs are used as significantly promising group of compounds in synthesis of nanomaterials, for example, nanoparticles and nanotubes, and nanoparticle hybrids. The distinct cation and anion structures and structure-dependent physicochemical properties of ILs are beneficial for interactions between ILs and nanostructures. Just based on these interactions including electrostatic, van der Waals, steric hindrance, hydrogen bonding, and solvation, ILs play a pivotal role in the synthesis and applications of nanomaterials. As a consequence of mutual effect of these specific and local interactions among numerous constituents of nanomaterials in the presence of ILs, different self-organized structures have been generated involving nanofluids [1], short range ordered colloidal glasses and gels [2], IL surface-functionalized nanoparticles (NPs) [1,2], IL-based emulsions [1,3], long range ordered liquid crystals [2,4], and stable colloidal nanodispersions [5−22].

In the design and synthesis process of nanomaterials, ILs have excellent regulatory effects on the morphology, structure, and properties of nanomaterials. The designability of ILs and combining different anions and cations, a vast variety of ionic liquid species can be obtained, offering more challenges for their preparation and application in nanomaterials. Here, we have focused on the synthesis and applications of nanomaterials according to the working of ILs.

ILs can serve as solvent media to balance of intra- and intermolecular interactions between the constituents of ILs and nanostructures, facilitating the dispersion of nanomaterials, such as metal or metal oxide nanoparticles, carbon nanotubes, graphene, and nanocomposites [2]. The colloidal dispersion of the nanostructured materials profit from ILs without the need to add classical stabilizers (surfactants and/or polymers) because of electrostatic and steric forces. Around the nanostructures, a large number of amphiphilic structures and inherent charges of ILs could avoid agglomeration of nanostructures via the formation of a protective layer [1]. In addition, alkyl side chains of cations further impede the nanostructures from approaching each other. Hydrogen bonds between ILs and nanostructures can also stabilize nanostructures and obviously keep nanostructures from aggregating. Based on these intra- and interactions, many functional ILs and nanomaterials which are dispersed in ILs are designed and fabricated [1,2]. A novel IL−nanomaterial hybrid consisting of 1-dodecyl-3-methylimidazolium bromide-functionalized nanosilica was synthesized and used as a highly efficient solid phase extraction material [5]. Choosing water-soluble ILs and nanosilica with a wide surface area as supporting materials promotes the separation process and shows excellent extraction performance of inorganic arsenic in complex water samples. Using [Omim]Br as a solvent and template, ZnO nanostructures with nanoparticle morphology have been synthesized at ambient pressure and different temperatures. The experiment results show that higher annealing temperature in the ILs produced ZnO nanoparticles with an increased crystalline structure and enlarged band gap [2]. Through the additive imidazolium-based ILs with different cation and anion sizes, a proper amount for the additive ILs was found to promote polyvinylidene fluoride (PVDF) crystallization [3]. By comparison, it is found that the presence of ILs reduced the size of the NPs and thus IL can be very useful to control the size of ZnO NPs and the appearance of novel properties [4]. Until now, imidazolium-based ILs have been widely utilized as versatile solvents for metal nanoparticles preparation. 1-butyl-3-methylimidazolium bromide ([Bmim]Br)-based IL was used to prepared the silver-nanoparticle-complexed *G. applanatum*. Due to improved solubility of [Bmim]Br, the skin permeability of the silver NP-complexed *G. applanatum* was improved by about 1.7 times by the IL [5]. To obtain special size and morphology nanomaterials, the effects of the molecular geometry and composition of the ILs are investigated. Therefore, it can be concluded that ILs can modulate the sizes and shapes, as well as the thickness of nanoplates [10,11]. Many nanomaterials obtained in ILs systems are reported to act as catalysts in styrene hydrogenation reaction [13], alcohol oxidation [16], CO_2 conversion [15,17], isomerization [18], and other rodex reaction [19,20]. Furthermore, nanomaterials could be also used as sensors [12,14,22] to detect the harmful product and an adsorbent for adsorbing a variety of mycotoxins in corn and wheat [21]. These well-dispersed, narrow-sized nanomaterials with different morphologies exhibit significantly superior performances in contrast with the nanomaterials obtained via a traditional approach. Many reports presented regarding these nanomaterials are listed in Table 6.1.

In addition to being solvent media-dispersing nanostructures through physical processes or PILs, nanomaterials could be prepared by the chemical grafted method. ILs are chemically bonded to the nanostructures via chemical interactions including covalent and coordination interaction. Due to high stability of the grafted layers, it is most prominent to use the grafting method for surface modification of nanostructures [23,24]. In 2023, Ji et al. reported that $[C_{12}mim][NTf_2]$-β-CD/ATP nanocomposites were prepared by the grafting method by immobilizing IL on the surface of β-CD/ATP, exhibiting efficient Thiamethoxam adsorption, satisfactory reusability, and stability [23]. The grafted technology can be used to form different types of hybrid nanostructures (short range ordered colloidal glasses and gels etc.). IL-functionalized nanostructures can have tunable properties, such as better dispersibility, thermostability, good solubility, lower melting points, good flowability, abrasion resistance, high humidity tolerance, gas-like diffusivity, chemical and solvent resistance, and other desirable physicochemical properties. A versatile method to disperse cellulose nanofibrils (CNFs) via one-step grafting of PILs onto CNFs (PIL@CNF) was proposed by Yuan [25]. The PIL@CNF hybrids display excellent dispersibility in water and various organic solvents. Then they are applied to reinforce the mechanical properties of porous PIL membranes. It shows that using ILs to functionalize the surface of high-performance nanocomposites is an effective strategy in biomedicine [26], adsorption and separation [24,25,27−34], catalysis [32−38], electronics [39], and other application fields. M. Valcarcel et al. modified the surface of magnetic nanoparticles with ILs via covalent interaction, offering $Fe_3O_4@SiO_2@mim$-Cl composites. These methylimidazolium chloride-functionalized magnetic nanoparticles were practically used for the magnetic solid phase extraction of polycyclic aromatic hydrocarbons (PAHs) from water samples. The covalent immobilization provides a high stability for the magnetic NPs avoiding the loss in the process of the extraction and elution [40]. Many IL nanostructure hybrids have been generated by the grafting of ILs on nanostructures by means of available functional groups from ILs and nanostructures, which minimizes the use of organic solvents and offers several advantages like those mentioned above, ultimately reducing reaction time and steps involved in conventional solvent-based grafting techniques. The reported synthesis process and application are shown in Table 6.2.

Table 6.1 Nanomaterial dispersions in ionic liquids.

Material	Synthetic method	Application	References
IL–SiO$_2$	Add [C$_{12}$mim]Br to the nanosilica suspension aqueous solution and stir by vortex for 10 s, followed by centrifugation to remove the supernatant.	Separation and determination of inorganic selenium in different water samples.	[5]
ZnO	Use [Omim]Br as a solvent and template, Zn (OAc)$_2 \cdot 2H_2O$ was dispersed into IL at a certain temperature, NaOH was added, and the reaction was stirred for 4 h.	Effect of IL on structure and optical properties of ZnO NPs at different temperatures.	[6]
PVDF	PVDF was dissolved into DMF, and different ILs (such as [DecMI$^+$][TFSI$^-$]) were added to the solution and magnetically stirred at 70°C for 2 h, and the mixture was then spun.	The additive IL has a promoting effect on the crystallization of PVDF.	[4]
ZnO–IL	[M(CH$_2$)$_3$IM^{2+}][2Br$^-$] was dissolved into methanol, and then zinc acetate and sodium hydroxide were added to obtain ZnO-IL nanoparticles after 1 h of ultrasound.	Effect of IL on the structure, optics and morphology of ZnO nanoparticles.	[8]
Ag nanoparticles	The flat extract and 1-butyl-3-methylimidazole bromide were mixed with silver nitrate solutions with different mass ratios, stabilized with soluble starch, stirred by magnetic force at 80°C for 2 h, and cooled to room temperature to prepare nanosilver complex flat.	Films for the treatment of diseases.	[9]
TiS$_2$ NSs	TiS$_2$ powder was put into liquid phase stripping medium IL (such as [C$_{14}$mim][Ben]) solution, sonicated for 7 h, and then separated and centrifuged to obtain TiS$_2$ NSs.	As catalysts in the electrocatalytic CO$_2$ to CO conversion.	[10]
AuPt FNBs	HAuCl$_4 \cdot 4H_2O$ and H$_2$PtCl$_6 \cdot 6H_2O$ were added to a 5.0 mL aqueous solution containing [C$_4$mim]$_n$[PSS] and magnetically stirred for 1 h.	With component-dependent SERS sensitivity and catalytic activity.	[11]

(Continued)

Table 6.1 (Continued)

Material	Synthetic method	Application	References
[Pt(mesBIAN)(tda)]	Place [Pt(mesBIAN)]Cl₂ and CuI in a sealed reaction vessel and then add the solution of degassed CH₂Cl₂ and diisopropylamine (volume ratio 5:1) and excess (OBET)H₂ and stir for 24 h.	Solvent chromic properties in conventional molecular solvents and ionic liquids.	[12]
RuNPs	[Ru(COD)(COT)] and IL([MEmim][NTf₂] or [MMEIm][NTf₂]) were introduced into the Fisher–Porter reactor under argon and stirred for 24 h at 40°C.	As a catalyst in styrene hydrogenation reaction.	[13]
Ni/Cu-MOF@Au/ [Mbim][PF₆]/ GCE	The Ni/Cu-MOF@Au prepared by hydroxylamine hydrochloride reduction method was dispersed in ultrapure water, and [Mbim][PF₆] solution was added to the ultrasonic dispersion uniformly and coated on the surface of GCE and dried at room temperature to obtain Ni/Cu-MOF@Au/ [Mbim][PF₆]/GCE.	As a novel enzyme sensor to detect organophosphorus pesticides such as dichlorvos.	[14]
Ru/[Bmim][BF₄]/ SiO₂	The Ru/SiO₂ solution was dispersed in [Bmim] [BF₄] solution and aged at 60°C for 3 h.	As a catalyst for CO₂ methanation.	[15]
Pd(C₆H₅NH₂)₂Cl₂	Aniline was added to the hydrochloric acid solution of palladium chloride and heated for 30 min to obtain trans-Pd(C₆H₅NH₂)₂Cl₂ and [Bmim][BF₄] as an effective dispersant for palladium nanoparticles.	Highly active and recyclable catalyst for alcohol oxidation.	[16]
[Bmim]NO₃/Cu	The Cu sheet was dispersed into [Bmim] [NO₃] solution to prepare [Bmim] [NO₃]/Cu nanoparticles, and the dispersion was coated on a polysulfone carrier to obtain a thin film.	Films for CO₂ transport.	[17]

ILs@silica	The active silica gel and C_2H_5OH were added to the round bottom flask, and the IL ([Hmim][HSO_4]/[Hmbim][HSO_4]) solution was added dropwise to the flask at N_2, followed by stirring for 3 h at 25°C, and the solvent was evaporated at low pressure to obtain IL@silica.	Efficient heterogeneous catalyst for n-heptane and n-octane isomerization. [18]
Au$_{75}$Pt$_{25}$ NDs	HAuCl$_4$ and H$_2$PtCl$_6$ were dissolved in ViEtImBr aqueous solution, and then ascorbic acid solution was added dropwise to the above mixed solution and stirred for 1 h at 25°C.	Catalyst for ethylene glycol redox reaction. [19]
Au@Pt NPs/rGO	Using IL as a shape-directing agent, add HAuCl$_4$ and H$_2$PtCl$_6$ solutions to [C$_4$(mim)$_2$]$_2$Br aqueous solution with constant stirring, then add hydrazine solution and stir for 30 min, and hold the reaction for 1 h.	It exhibits excellent electrocatalytic performance in both hydrogen evolution reaction and redox reaction. [20]
PDA–IL–NFsM	PS NFsM was prepared by electrospinning as a template, and then PDA NFsM was prepared by in situ oxidation and self-polymerization of dopamine. Finally, [Hmim][PF$_6$] was selected to physically dip PDA NFsM to prepare PDA-IL–NFsm.	As an adsorbent for adsorbing a variety of mycotoxins in corn and wheat. [21]
Co–CeO$_2$/C	Using Ce(NO$_3$)$_3 \cdot$ 6H$_2$O and Co(NO$_3$)$_2 \cdot$ 6H$_2$O as precursors, Co–CeO$_2$/C nanocomposites were synthesized with the coprecipitator KOH, and they were added to 1-butyl-3-methylimidazolium tetrafluoroborate solution and physically impregnated with mortar and pestle for 40 min.	As a biosensing probe for colorimetric analysis of hydrogen peroxide in urine samples from cancer patients. [22]

Table 6.2 Ionic liquid-grafted nanomaterials.

Material	Synthetic method	Application	References
$[C_{12}mim][NTf_2]$-β-CD/ATP	The precursor β-CD/ATP was directly immersed in an ethanol solution containing $[C_{12}mim][NTf_2]$ by grafting, immobilized, reacted at room temperature for 12 h, and dried.	Adsorbs thiamethoxam in water samples.	[23]
IL–SiO_2	The methoxysilane group of ILs was covalently bonded with the hydroxyl group on mesoporous SiO_2 by grafting, that is, the mixture was heated at 90°C for 24 h.	As an adsorbent for industrial engineering dehydration.	[24]
PIL@CNF	CNF is ultrasonically polymerized in water and 2-ethyl-2-vinylimidazolium bromide (IL-Br) is added in situ, the gel state is reached during polymerization, and water is used as eluate for ultrafiltration concentration after polymerization.	Used to enhance porous PIL membranes.	[25]
[i-C_5TPim][NTf_2]-MS	IL ([BTPM][Cl] or [i-C_5TPim][Cl]) was dissolved in toluene and reacted in a glass reactor filled with N_2 at 95°C for 48 h, and [Cl$^-$] was exchanged to [NTf_2^-], [PF_6^-], and [DCA^-] on a [i-C_5TPim][Cl]IL-grafted carrier.	Adsorbent for CO_2	[27]
SiO_2–MPTS-[C_nVim]Br	SiO_2 MPTS was vacuum-sealed in a polyethylene bag, then a deoxidized [C_nVim]Br (n = 2,4,6,8,10) aqueous solution was injected into the polyethylene bag, and the sample was irradiated by an electron beam accelerator and washed and dried to obtain the final product.	Adsorption and separation of TcO_4^-/ReO_4^- in radioactive waste liquid.	[28]
Fe_3O_4@SiO_2@PIL	Fe_3O_4@SiO_2@VTES by reaction of Fe_3O_4@SiO_2 with vinyltriethoxysilane; using AIBN as the initiator, PIL was fixed on the surface of magnetic nanoparticles by copolymerization of 1-allyl-3-methylimidazolium chloride salt and Fe_3O_4@SiO_2@VTES radicals.	As an adsorbent to remove Congo red.	[29]

CMPS-triethylammonium	CMPS and triethylamine were refluxed in acetonitrile for 48 h, filtered, and washed with CH_3CN to obtain pale yellow globules CMPS-triethylammonium.	Used for $AuCl_4^-$ adsorbent.	[30]
$Fe_3O_4@SiO_2@mim–PF_6$ MNPs	Fe_3O_4 was prepared by coprecipitation, MNP was covered with silica, the IL was covalently fixed on the surface of silica-coated Fe_3O_4 to obtain $Fe_3O_4@SiO_2@mim–Cl$, and $Fe_3O_4@SiO_2@mim–PF_6$ MNPs were obtained by simple metathesis reaction using KPF_6 as reagent.	Used as an adsorbent for the extraction of PAHs in water samples.	[40]
UiO-66-IL– PF_6	UiO-66-IL-Br was synthesized by covalent polymerization of imidazolyl salt IL-functionalized dicarboxylic acid ligands with $ZrCl_4$, and Br- in ILs was further replaced with different anion X^- to obtain a series of UiO-66-IL-X membranes ($X = PF_6^-$, $SO_3CF_3^-$, ClO_4^-).	Bifunctional membranes for CO_2 adsorption and catalytic conversion.	[35]
ILs-grafted SiO_2	SiO_2 nanoparticles were added to the solution of 3-chloropropyltriethyloxysilane and triethylamine and stirred and refluxed for 6 h, then dispersed in anhydrous 1-methylimidazolium acetonitrile solution, refluxed for 8 h, and ILs-grafted SiO_2 was obtained by the substitution reaction of 1-methylimidazole and chlorine functional groups.	Adsorbs phenol as a solid adsorbent.	[41]
SBA-CIL-CS-Lac	The carboxy-functionalized IL was used as a bridging agent to combine SBA-15 with chitosan to obtain SBA-CIL-CS and then dispersed in citric acid-disodium hydrogen phosphate buffer, and laccase solution was added dropwise at 45°C for 3 h to obtain immobilized laccase SBA-CIL-CS-Lac.	As a vehicle for physical adsorption of immobilized enzymes and removal of 2, 4-dichlorophenol.	[31]

(Continued)

Table 6.2 (Continued)

Material	Synthetic method	Application	References
[Pmim]Cl-FeCl$_3$-MCM-41	MCM-41 and [Pmim]Cl-FeCl$_3$ were dispersed in acetonitrile solution and then refluxed under N$_2$ for 24 h, and after acetonitrile removal, the resulting mixture was extracted by Soxhlet with boiling dichloromethane for 48 h.	Catalyst for synthesis of diphenylmethane and its derivatives.	[32]
Ti−Zr-SBA−15-IL	The chloropropyl chemically bonded on the surface of silica reacts with N- methylimidazole, that is, Ti-Zr-SBA−15-Cl and N-methylimidazole are dissolved in 60 mL of toluene and stirred for 12 hours at reflux temperature in a nitrogen environment.	Catalytic cycloaddition reaction of carbon dioxide with epoxide.	[33]
PIL-POM	1,3-propane-sultone was slowly added to a dispersion of poly(4-vinylpyridine) cross-linked with DVB in toluene and stirred for 2 h, and the temperature was raised to 60°C and stirred for 24 h to obtain VPyPS. The PMo$_{10}$V$_2$ solution was dropped into VPyPS methanol solution and refluxed for 24 h to obtain VPyPS-PMo$_{10}$V$_2$.	As a catalyst for sulfide oxidation.	[34]
DFNT@IL/Mo$_{132}$	IL and dendritic fiber nanotitanium DFNT were stirred and refluxed in methanol solution for 18 h, and DFNT@IL were obtained due to covalent interaction. [N$_2$H$_4$ · H$_2$SO$_4$] was added to the mixture of NH$_4$CH$_3$CO$_2$ and (NH$_4$)$_6$Mo$_7$O$_{24}$ · 4H$_2$O, acidified with DFNT@IL in distilled water, sonicated for 38 min and stirred for 9 h to obtain DFNT@IL/Mo$_{132}$.	As a catalyst for the synthesis of dimethyl carbonate from methanol and carbon dioxide.	[36]

NGO/IL–Fe	NGO/IL was obtained by grafting IL(1-methyl-3 (3-trimethoxysilylpropyl)–imidazolium chloride) onto nano graphene oxide NGO by ultrasound for 24 h, and then NGO/ IL–Fe was obtained by stirring the mixed solution of NGO/IL, dimethyl sulfoxide and $FeCl_3 \cdot 6H_2O$ at room temperature for 24 h.	As a stable and efficient nanocatalyst for the synthesis of tetrahydrobenzo pyrans. [37]
Pd(0)/GO-ILCS	At 75°C, under a N_2 atmosphere, chitosan HCS was dispersed into a mixed solvent of water and isopropanol, and then BHMIB was added to magnetic stirring for 40 h, and ILCS was obtained by grafting. Pd(0)/GO-ILCS is obtained by in-situ reduction of $PdCl_2$ with $NaBH_4$ layer by layer assembly.	Catalyst for hydrogen production by hydrolysis of ammoniaborane. [38]
Thin film transistor	AB blocks are prepared by random copolymerization of highly conductive anionic (ILMA) or cationic (ILMC) monomers and PEGM by RAFT. Using PhEtM as raw material, the C block with mechanical properties of a copolymer was synthesized after polymerization. The A block copolymer with an AB − C structure is obtained.	As a self-assembled gating material for single–walled carbon nanotube thin–film transistors. [39]

The sol-gel method is a very common and promising technology in the preparation of nanomaterials [42,43]. Alkyl chain and nonpolar/polar behavior of ILs offer good affinity and selectivity toward polymeric materials, producing new nanomaterials. In the sol-gel system, ILs act as solvents, a stabilizer, or structuring agent and/or additives, which improve some physicochemical properties of nanomaterials (shape-memory performance, toughness, thermal stability, and acid and alkali resistance etc.) [43,48]. Polymerizing ILs, polymer or nanomaterials via strong interactions, such as hydrogen bonds, covalent and coordination bonds, ILs and polymer enables the preparation of ionogels, and them fabricate the IL-functionalized nanomaterials for biomedicine [44], environmental protection [45], catalysis [46,47]. Multifunctional calcium phosphate-IL bionanocomposites were obtained based on the loading of 1-alkyl-3-alkylimidazolium chloride ILs ([$C_n$mim]Cl (n = 4, 10, 16) and [(C_{10})$_2$mim]Cl) during the *in situ* sol-gel synthesis of calcium phosphates. The study of CaP crystallization and biological properties shows that the N-alkyl chain length of ILs influenced the crystallization of CaP and, consequently, the biological properties, affording these multifunctional materials able to simultaneously inhibit microbial infections and induce osteogenic activity [44]. Obviously, IL provides a shortcut for the synthesis of nanomaterials with the sol-gel method. The size of the nanomaterials can be changed by controlling physical conditions such as stirring speed and temperature, or the types of cations and anions of ILs. Furthermore, the new functionalized nanomaterials present improved functional manifestations, which combine the excellent properties of both ILs and polymers, such as magnetic properties [49–51], electrolyte [52], and conductivity [53]. The relative reprorts prepared via the sol-gel method are listed in Table 6.3.

In the synthesis of nanomaterials, the hydro- and solvothermal method is widely used to control the morphology and structure through tuning of temperature and pressure. Based on the electrostatic attraction, $\pi-\pi^*$ stacking interactions, hydrogen bonding, self-assembled mechanism, as well as the efficient template effect between ILs and nanostructures, the structure and component of ILs significantly affect the size and shape of nanomaterials [54]. During the preparation of ZnO nanoparticles, the size and morphology of the nanoparticles varied interestingly with different structural ILs used during preparation, and they also exhibited a obvious change with varying concentrations of the ILs. Due to the concentration change of ILs, the band gap energy varied with the change in size of the ZnO nanostructures [55]. Via the hydrothermal method using ILs as the structure directing agents, IL-Eu-MOF nano-/micromaterials were obtained and successfully applied to the highly effective removal of CR from aqueous solutions [56]. Similarly, by an environmentally friendly and facile hydrothermal carbonization procedure, new nanomaterials were synthesized, as it involves the use of ILs at low temperatures ($130°C-250°C$) in aqueous medium under self-generated pressure [57]. In the facile yet efficient process, PILs are used as

Material	Synthetic method	Application	References
Dye-sensitized solar cells (DSC)	Gel electrolyte is composed of IL and gel, and DSC are prepared by surface-modified TiO_2 nanoparticles and IL (methylimidazole iodide) gel electrolyte in a ratio of 1:1.	Liquid electrolyte curing of dye-sensitized solar cells.	[42]
SiO_2	C_{16}mimCl was dissolved in EtOH, mixed with TMOS, and then stirred with HCl aqueous solution dropwise for 30 min, and the sample was placed in an open flask at room temperature to complete gelation.	An easy way to prepare porous silica glass using IL as a template.	[43]
CaP–IL	Bio-nanocomposite CaP-IL is obtained by incorporating multiple ILs ([$C_n$mim]Cl (n = 4, 10, 16) and [(C_{10})$_2$mim]Cl) on CaP in situ by sol–gel technology at room temperature.	The material has antibacterial, anti-inflammatory, and regenerative properties and can be used in minimally invasive surgery for bone and maxillofacial defects.	[44]
Fe_3O_4–SiO_2	[Omim][BF_4] was used as a solvent and template to synthesize nano-Fe_3O_4, and directly in the system, with TMOS as the silicon source, using the hydrogen bonding between BF_4^- and Si–OH and interionic Coulomb force, nano-SiO_2 was orderly recombined on the surface of Fe_3O_4 by sol–gel technology, forming a porous composite material.	Adsorption of dyes such as methylene blue.	[45]
CeO_2–TiO_2–graphene	CeO_2–TiO_2 nanoparticles were prepared by the sol–gel method using ammonium 2-hydroxyethylcarboxylate as a sol and then assembled on graphene nanosheets by the hydrothermal method to form a composite material.	As a photocatalyst to catalyze the degradation of reactive dyes RR195 and 2,4-dichlorophenoxyacetic acid.	[46]
Co_3O_4	Add $CoCl_2$ to [Emim][TfO] solution and stir for about 6 h, and add distilled water dropwise and stir for hydrolysis for 3 h. Adjust the pH to neutral using NH_4OH solution, followed by aging for 6 h to produce a gel, centrifugation and filtration, followed by calcined crystallization at different temperatures.	Used in catalysis, battery, and sensor fields.	[47]

(Continued)

Table 6.3 (Continued)

Material	Synthetic method	Application	References
PMMA/TiO$_2$/TBPC	PMMA/TiO$_2$/TBPC was synthesized by the sol–gel method, that is, MMA-co-MSi was dissolved in THF and added dropwise to the mixture of titanium tetraisopropanol (TiOPr) and TBPC, and HCl was added dropwise and stirred at 20°C for 15 min. The solution is cast in a Petri dish and allowed to stand for 2 days, then heat-treated at 100°C for 90 min under an argon atmosphere to obtain a hybrid film.	The addition of IL improves the transparency and toughness of the hybrid film.	[48]
Ti$_{15}$TB$_{40}$Fe$_5$	Dissolve MMA-co-MSi in THF and add dropwise to TiOPr, TBPC, and FeCl$_3$ mixed solutions, and add HCl and stir at 20°C for 15 min. The solution is cast in Petri dishes, allowed to stand at 25°C for 2 days, and heat-treated at 100°C for 90 min in an Ar atmosphere to form a Ti$_{15}$TB$_{40}$Fe$_5$ hybrid film.	As a magnetic material.	[49]
Polycarbonate/titania hybrid films	The mixture of PHMCD, 3-PTES, and toluene was reacted under an argon atmosphere at 100°C for 6 hours to obtain ET-PHMCD, and ET-PHMCD was dissolved in THF. TiOPr and HCl are added dropwise to this solution and stirred at 20°C for 15 min, cast in Petri dishes to dry for 3 days and in an argon atmosphere for 24 min at 80°C to obtain hybrid films.	As a photochromic magnetic material.	[50]
Ti$^{IV/III}$NSs	An aqueous dispersion of Ti^{IV}NSs was transferred into a quartz cuvette with a screw cap and deoxygenize with a constant stream of nitrogen gas for 30 min. Then, the cuvette was irradiated at 25°C with a UV light for 3 h to afford an aqueous dispersion of Ti$^{IV/III}$NSs.	As an optical switch that can be operated remotely by magnets and light.	[51]

solvents to assist the preparation of nanostructured porous carbon materials Au−Pd@N-Carbon under template-free conditions. PILs are believed to behave as a stabilizer, pore-generating agent, and nitrogen source in the hydrothermal carbonization process to improve the formation of nanomaterials. Uniform flower-like hollow microsphere BiOBr with porous nanosphere structures was synthesized through the solvothermal process in the presence of ionic liquid 1–hexadecyl-3-methylimidazolium bromide ([C_{16}mim]Br). The BiOBr porous nanospheres and hollow microsphere samples showed much higher photocatalytic activity than the BiOBr synthesized by conventional method. During the reactive process, IL [C_{16}mim]Br played the role of solvent, reactant, and template at the same time [58] (Table 6.4).

Microwave heating is essentially a process of absorbing microwave electromagnetic fields by the reaction materials, which is quite different from the traditional heating method. Due to the fact that the entire process does not require any heat conduction and is completed in the dipole molecules, both the internal and external parts of the materials are heated simultaneously during the whole microwave-heating process. Compared with traditional heating, microwave heating not only has high energy efficiency, but also is a low-temperature and environmentally friendly synthesis method. ILs have a strong absorption for microwaves [59], and the microwave itself can accelerate the reaction and nucleation process. Therefore, combining ILs with microwave radiation heating has become a green and efficient synthesis method in recent years. Ni-MOF nanorods in an NMP-based coordination IL were synthesized, and then the Ni-MOF nanorods were anchored on rGO, producing Ni−MOF@rGO composites with mesoporous 3D nanostructures. Ni−MOF@rGO composites were applied as electrocatalysts in the urea oxidation reaction, and they exhibited a lower onset potential, increased current responses, faster kinetics of urea oxidation, and lower charge transfer resistance [60]. In this example, NMP-$NiCl_2$ (N-methyl-2-pyrrolidone-nickel chloride) coordination IL is a good alternative to imidazolium-based ILs and very useful in many applications such as drug delivery, desulfurization, extraction, catalysts in organic reactions, and so forth. In the microwave-assisted IL synthesis technique, ILs act as solvents and templates. It not only retains the low melting point and nonvolatile characteristics of ILs, but also provides a new development perspective for material synthesis with the high-efficiency and energy-saving characteristics of microwave, which is in line with the carbon neutrality development goal. Here, the application and synthesis of nanomaterials through the microwave-assisted IL heating method is discussed, for example, hybrid nanomaterials [60,61], metal oxides [62−67], nanoparticles [68,69], and metal fluorides and sulfides [70−73]. In some synthesis system, ILs could be a source of F^- to participate in the construction of target nanomaterials. It is pointed out that microwave heating synthesis has good development prospects.

The degradation activity and selectivity of P25@ IL were investigated by the photocatalytic degradation experiment and selectivity experiment, and its selective photodegradation mechanism was clarified [62]. The results showed that the type II

Table 6.4 Nanomaterials synthesized by hydro- and solvothermal method.

Material	Synthetic method	Application	References
ZnO	Three different hydrophilic ionic liquids ([Emim][CH₃SO₄], [Bmim][CH₃SO₄], and [Emim][C₂H₅SO₄]) were used as soft templates, and nano-zinc oxide was prepared by the hydrothermal method, such as Zn (OAc)₂ · 2H₂O, [Emim][CH₃SO₄], and NaOH mixtures were put into an autoclave and reacted at 121 °C for 2 h.	Hydrophilic IL acts as a directing agent and soft template to regulate the morphology and optical properties of ZnO nanostructures.	[55]
IL-Eu-MOF	A mixture of EuCl₃ · 6H₂O, 1,10-phenanthroline, isophthalic acid, IL ([Cₙmim]Cl, n = 4, 8, and 12), NaOH, and deionized water was placed in a 50 mL iron-fluorite stainless steel autoclave and heated at 150 °C for 2 h.	For efficient removal of Congo red.	[56]
Au-Pd@N-Carbon	PdCl₂, HAuCl₄ · 3H₂O, PIL-b, and CH₃CN were added to the reactor and stirred in an oil bath at 60 °C for 12 h; after the solvent was removed, the obtained product was loaded with a mixture of PIL-b, fructose, and deionized water into a PTFE-lined autoclave for 23 h, hydrothermal reaction at 160 °C, and finally freeze–dried Au-Pd@N-Carbon.	Used as an effective catalyst for selective hydrogenation of phenylacetylene.	[57]
BiOBr	Bi(NO₃)₃ · 5H₂O was dissolved in the ethanol solution of [C₁₆mim]Br, stirred for 30 min, and then transferred to a PTFE-lined autoclave and heated at 140 °C for 24 h.	Catalytic degradation of rhodamine B as a photocatalyst.	[58]

heterostructure formed between IL and P25 improved the separation performance of electron—hole pairs and then improved the photocatalytic activity of the materials. The spherical-shaped Ag nanoparticles with improved morphology have been effectively synthesized via the microwave-assisted method using IL 1-dodecyl-3-methylimidazolium chloride [68]. Using ILs as a solvent via the microwave-assisted method, the surface morphology of synthetic Ag NPs has been improved. Ag NPs have shown advantages for cell activity; they also improve the material performances with VEGF for the regeneration of femoral fractures. In a water and IL (1-butyl-3-methylimidazolium tetrafluoroborate, [Bmim][BF$_4$]) system, a graphene-TiO$_2$ hybrid was synthesized by a solvothermal microwave-assisted method, which had a high surface area and exhibited high photocatalytic degradation of methylene blue [63]. Carbon NPs (CNPs) were prepared from ILs using a green and rapid microwave-assisted synthetic approach for the first time [69]. The CNPs being a fluorescence probe could act as a promising candidate to detect the trace levels of quercetin due to their advantages in low-cost production, low cytotoxicity, strong fluorescence, and excellent biocompatibility. The one-step green preparation process is simple and effective; neither a strong acid solvent nor surface modification reagent is needed, which makes this approach very suitable for large-scale production. During the formation of CNPs, 1-butyl-3-methylimidazolium tetra-fluoroborate ([Bmim][BF$_4$]) was used in the process as a solvent and reactant. Similarly, ILs were involved in the preparation of Nano-EuF$_3$ as solvent and a source of F$^-$. With this fast and facile synthesis route, pure nanofluoride materials were obtained without the use of dangerous HF, which makes this method superior to other reported synthesis routes [70].

Under microwave reaction conditions, different morphology nanomaterials can be synthesized using IL media. Metal oxides, such as ZnO [64], ZrO$_2$ [65], and TiO$_2$ [66] can be controlled by changing the structure of IL anions and cations and the reaction conditions, including the microwave absorption rate of the reaction system to change the oxide morphology. ILs themselves are excellent microwave absorbers, while also playing the role of surfactants and coating agents. During the formation of crystal morphology, specific crystal surfaces are selected for adsorption to induce the formation of crystal morphology, preventing grain growth and obtaining nanocrystals simultaneously [71]. MnF$_2$ nanoparticles were synthesized through the facile and rapid microwave-assisted method in binary mixtures of ethylene glycol and imidazolium-based IL. The experimental results show that the length of the alkyl chain of imidazolium could affect the morphologies and crystalline structures of MnF$_2$, which facilitate the synthesis of high-quality MnF$_2$ nanoparticles with controllable crystalline phase and morphology. A facile approach toward spherical SnSe$_2$ nanodots/graphene nanocomposites via an ionic liquid media assembly process was reported [61]. The nanocomposites are superior to the capacities of the previously reported SnSe$_2$ nanoplate/graphene composite and many other tin selenide electrodes. Different ILs anions and cations can change the microwave heat absorption rate, and with the prolongation of heating time, the reactants tend to develop into being sheet-, rod-, and plate-like [67,74]. Thus, microwave-assisted IL approach represents a promising, simple, and scalable synthetic protocol for the fabrication of nanocomposites. Some nanomaterials produced by the microwave-assisted IL approach are listed in Table 6.5.

Microemulsion is an effective method for preparing nanoparticles, nanowires, and nanorods. Furthermore, it has been demonstrated that ILs could substitute water or conventional organic solvents to form novel microemulsion systems in the presence of a surfactant, and these novel microemulsions incorporate the advantages of ILs and microemulsion. In an IL microemulsion system, ILs can be used as the oil phase, "amphi-solvent," and/or water phase, promoting their wider substitution of traditional microemulsions in material synthesis, separation engineering, and pharmaceutical industry [75] (Table 6.6).

In IL/water microemulsion, the polyaniline core decorated with TiO_2 (PANI-TiO_2) nanocomposite particles was successfully synthesized. PANI-TiO_2 had a novel structure with nanocrystalline TiO_2 deposited onto the surface of PANI, which is different from the reported structure of TiO_2-PANI nanocomposites [76]. The microemulsions act as soft templates for nanoparticle growth, which allows reliable control over sizes and compositions. $Ag_{1-x}Pd_x$ nanoparticles with a broad size range and a tunable composition by controlling the proportion and composition of the aqueous component are achieved by this method. The obtained $Ag_{40}Pd_{51}$ NPs exhibit extraordinary mass activities toward ethanol oxidation for alcohol fuel cells [77]. In situ in a H_2O/TX-100/[Bmim][PF_6] microemulsion, Pd nanoparticles were successfully prepared. Therein, IL [Bmim][PF_6] stabilized the Pd nanoparticles together with surfactant TX-100. Pd nanoparticles were found to be an excellent catalyst system for the ligand-free Heck reaction [78]. Using IL [C_{12}mim]Cl as the surfactant and MMA-AM mixture as the oil phase, AgCl nanoparticles were synthesized in a W/O microemulsion, and the size and amount of the AgCl nanoparticles could be adjusted via the parameters of the microemulsion. The AgCl/poly(MMA-co-AM) hybrid membranes exhibited potential applications in the separation of aromatic/aliphatic hydrocarbons [79]. Luo et al. prepared [C_{16}mim]Br/butan-1-ol/cyclohexane/[C_8mim]Ac IL microemulsions using 1-hexadecyl-3-methylimidazolium bromide ([C_{16}mim]Br) and 1-octyl-3-methylimidazolium acetate ([C_8mim]Ac) as substitutes for surfactants and the polar phase. Starch nanoparticles with small size and homogeneous distribution were prepared in an IL/O microemulsion system as the reaction system. This work might provide an efficient method to synthesis starch nanoparticles [80]. In a reverse IL-SFME of [Bmim][PF_6]-PAF-DEAF, La-based complex nanorods with large mesopores were synthesized. La(COOH)$_3$ nanorods have the maximum adsorption capacity of 381.6 mg PO_4^{3-}/g, which is obviously higher than those of previously reported adsorbents and foresees possible application in the

Material	Synthetic method	Application	References
Ni-MOF@rGO	C_5H_9NO, $NiCl_2$, and waste PET were added to the graphene oxide solution dispersed in NMP and microwaved for 180 s. Subsequently, the reaction mixture was stored in an oven at 100°C for 8 h.	Application as an electrocatalyst in urea–enhanced water oxidation reaction.	[60]
P25@IL	IL was prepared from 1,4-butanesulfonic lactone and N-allylimidazole; LEV, IL, and P25 were stirred and dissolved in an aqueous solution for 4 h, and under N_2, the crosslinker MBA and initiator ABVN were added and stirred for 30 min, and then the solution was put into a microwave reactor and reacted at 600 W, 60°C for 120 min.	As a photocatalyst to degrade levofloxacin hydrochloride residues in an aqueous environment.	[62]
rGO-TiO$_2$	Disperse GO in a mixture of water and [Bmim][BF$_4$] by ultrasound, add TTIP and hydrolyze for 2 h, and then irradiate the mixture in a microwave oven at 130 W for 10 min.	As a photocatalyst to catalyze the degradation of methylene blue.	[63]
Ag NPs	Add AgNO$_3$ solution dropwise to [C$_{12}$mim]Cl solution with continuous stirring, add NaBH aqueous solution and continue stirring for 30 min, and then move into an autoclave in an oven at 120°C for 12 h.	It is used to improve wound healing and reduce aseptic necrosis treatment of fracture healing.	[68]
C NPs	[Bmim][BF$_4$] aqueous solution was heated in a microwave (700 W) for different periods of time, and the color-changing solution was centrifuged for 30 min to remove deposits with less fluorescence, and finally a clear yellow-brown aqueous solution containing CNPs was obtained.	As a fluorescent probe for the detection of trace quercetin.	[69]
Nano-EuF$_3$	The reaction mixture of europium acetate hydrate, ethylene glycol, and IL ([Bmim]Cl/[Bmim][BF$_4$]/ [P$_{44414}$][BF$_4$]) was placed in a glass container containing a PTFE diaphragm, heated with single-mode microwave for 5 min to 120°C, and stirred continuously at 120°C for 10 min, and the reaction mixture was cooled to room temperature by pressurized air.	As inorganic luminescent materials.	[70]

(Continued)

Table 6.5 (Continued)

Material	Synthetic method	Application	References
ZnO	Using [Bmim]Cl and [Bmim][BF$_4$] as a solvent and template agent, Zn(OH)$_4$$^{2-}$ solution was dispersed into IL and microwaved for 7 min at 700 W, 120°C.	Used as a semiconductor material in photoelectric devices.	[64]
ZrO$_2$	Ammonia was added dropwise to ZrOCl$_2$ · 8H$_2$O and vacuum-dried at 60°C, and the dried product was added to [Bmim][BF$_4$] to assist microwave heating for 3 min.	It can be used as ceramic materials or in the fields of catalysis and sensing.	[65]
TiO$_2$	Dispersed titanium tetraisopropoxide into [Bmim][BF$_4$] and heated at 800 W microwave for 30 min.		[66]
MnF$_2$	Mn(CH$_3$COO)$_2$ · 4H$_2$O was added to the mixture of ethylene glycol and fluoroimidazolyl IL and stirred for 2 h, and microwave irradiation was performed at 500 W and 180°C.		[71]
SnSe$_2$/graphene	After mixing tin powder and selenium powder with [Bmim]Cl, they were heated in a microwave oven at 300 W, 120°C for 5 min, and then heated to 180°C for 55 min.	As a nanomaterial for lithium–ion batteries.	[61]
Mn$_5$O$_8$	MnCl$_2$ and [Bmim]Cl were dissolved in deionized water; NaOH solution was slowly added dropwise and stirred; and the mixture was heated at microwave 120 W, 75°C for 5 min.	Catalyst for 5-hydroxymethylfurfural oxidation.	[67]
LiMnPO$_4$\C	Combined LiOH · H$_2$O, H$_3$PO$_4$, and MnSO$_4$ · H$_2$O was synthesized by atmospheric pressure microwave heating with choline chloride/ethylene glycol low eutectic solvent as the reaction medium and sucrose as the carbon source.	As nanomaterials for lithium–ion batteries.	[74]
ZnF(OH)	Dissolve zinc acetate and thiourea in water, add NaBF$_4$ and mix well, stir for 2 h, and then heat in a microwave for a certain period of time.		[72]
M$_2$S$_3$ (M = Bi, Sb)	Sb$_2$O$_3$, ethanolamine, HCl, and Na$_2$S$_2$O$_3$ were mixed, dispersed into [Bmim][BF$_4$], and		[73]

Table 6.6 Nanomaterials prepared via the microemulsion method.

Material	Synthetic method	Application	References
PANI-TiO$_2$	[Bmim][PF$_6$]/aniline was added to OP-10/N-butanol/ HNO$_3$ solution and stirred for 1 h, and then TiO$_2$NPs were added and stirred for another 1 h to obtain oil-in-water microemulsion. Stir continuously to add APS solution to polymerize to form PANI-TiO$_2$.	Used in photocatalysis and electrochemistry.	[76]
AgPd NPs	IL microemulsion (ILM) was prepared by adding n-butanol, CTAB, [Bmim]Cl, and aqueous solutions of AgNO$_3$ and K$_2$PdCl$_4$ with different weight ratios, and AgPd NPs were prepared by electrochemical deposition with ILM as a soft template.	Electrocatalyst for ethanol oxidation reactions.	[77]
Pd NPs	H$_2$O/TX-100/[Bmim][PF$_6$] microemulsion was heated to 50°C, PdCl$_2$ aqueous solution was added, and PdCl$_2$ was rapidly reduced by surfactant TX-100 to obtain Pd NPs.	As a catalyst to catalyze the Heck reaction between butyl acrylate and iodobenzene.	[78]
AgCl/poly(MMA-co-AM)	The microemulsion was prepared by placing deionized water, [C$_{12}$mim]Cl, methacrylate (MMA), and acrylamide (AM) in an ultrasonic bath at 40°C, and AgNO$_3$ aqueous solution was added to sonicate for 30 min at 40°C to obtain AgCl nanoparticles. AIBN/ (MMA + AM) was added to the microemulsion and stirred at 60°C and coated on glass and polysulfone film and placed in a 60°C oven to obtain AgCl/poly (MMA-co-AM) hybrid film.	Adsorption and separation of benzene/cyclohexane mixtures.	[79]
Starch nanoparticles	The OSA starch was dissolved in [Omim]Ac and heated in an oil bath at 135°C for 2.5 h, followed by the addition of cyclohexane and surfactants [C$_{16}$mim]Br, butan-1-ol to form IL/O microemulsion, and after adding the cross-linking agent epichlorohydrin, the mixture was stirred at 50°C for 3 h and precipitated and centrifuged.	It has certain drug-loading properties for mitoxantrone hydrochloride.	[80]

(Continued)

Table 6.6 (Continued)

Material	Synthetic method	Application	References
La(COOH)$_3$	The reversed-phase microemulsion SFME was composed of ammonium propylformate (PAF), 1-butyl-3-methylimidazolium hexafluorophosphate ([Bmim] [PF$_6$]), and ammonium diethylformate (DEAF); H$_3$BTC and La(NO$_3$)$_3$ were added and stirred at 35°C for 24 h, and the precipitate was collected by centrifugation.	Used to remove phosphate from water samples.	[81]
PMMA/TiO$_2$/IL	TiO$_2$ nanoparticles were added to [Bmim][BF$_4$] solution and dispersed ultrasonically; methyl methacrylate (MMA) monomer solution and 1-butanol of TX-100 were added and stirred for 20 min to obtain a microemulsion; and the initiator benzoyl peroxide (BPO) was added under vigorous stirring for 8 h at 60°C.	Photocatalytic degradation of methylene blue dyes.	[82]
Bovine serum albumin (BSA) NPs	BSA aqueous solution was added dropwise to water/Tween 20/[Bmim][PF$_6$] (10:50:40 wt.%) microemulsion and stirred for 90 min; glutaraldehyde solution was added and stirred overnight to obtain BSA NPs with methanol precipitation.	For drug delivery.	[83]
Mg$_2$Al–Cl LDH nanosheets	The aqueous solution of NH$_3$ was added dropwise to the microemulsion containing MgCl$_2$ · 6H$_2$O and AlCl$_3$ · 9H$_2$O, and a certain proportion of [Bmim] [PF$_6$], DMF, and water were added. Stir the reaction mixture for 12 h at room temperature and then age at 25°C or 75°C for 12 or 24 h.	It is used as an adsorbent to remove phosphate compounds in water.	[84]

phosphate adsorption from effluent [81]. Using an IL-based microemulsion system, a new visible light-responsive PMMA/TiO$_2$/IL photocatalyst was synthesized. In this microemulsion system, 1-butyl-3-methylimidazolium tetrafluoroborate ([Bmim][BF$_4$]) was used as the dispersant phase. The results showed that PMMA/TiO$_2$/IL exhibits excellent photocatalytic activity under visible light irradiation and considerably higher photocurrents than bare TiO$_2$ [82]. In the same way, BSA NPs widely used in drug delivery studies were prepared in nanosized water droplets of water-in-IL (W/IL) microemulsion systems. Therein, a hydrophobic IL of 1-butyl-3-methylimidazolium hexafluorophosphate ([Bmim][PF$_6$]) was used as an oil component in the emulsification system and BSA NPs can be used in drug delivery studies [83]. By a double-microemulsion technique, the small sized and ultrathin Mg$_2$Al-Cl LDH nanosheets are prepared in a SFME of [Bmim][PF$_6$]−dimethylformamide−water [84]. By comparison with the large sized LDH particles basing on the traditional coprecipitation method, these nanosized LDH sheets exhibit the superior phosphate removing ability toward low concentration phosphate solution.

6.3 Ionic liquids intensify the synthesis of molecular sieve materials

Molecular sieves are one of the most important classes of inorganic materials, exerting a significant influence on industrial applications due to their roles as adsorbents, ion exchangers, and catalysts [85]. Aluminosilicate molecular sieves, also known as zeolite, possess pores or channels created by alumina and silica tetrahedra connected via oxygen bridges [86]. The replacement of Al and/or Si with other elements (e.g., B, Fe, Ti, Ga, or Ge) in the molecular sieve framework can result in the creation of a wide range of novel materials. In 1982, Wilson and coresearchers documented the synthesis of microporous aluminophosphate molecular sieves (AlPO$_4$-n) [87]. The AlPO$_4$ frameworks are built from an alternating arrangement of AlO$_2^-$ and PO$_2^+$ tetrahedra, rendering the resulting frameworks neutral.

Typically, molecular sieves are created through hydro-/solvothermal techniques that involve significant quantities of water or alcohol-based solvents within a sealed autoclave. The elevated autogenous pressure and the production of raffinate waste during the synthesis may pose safety and environmental concerns [88]. In comparison, ILs offer several advantageous physicochemical characteristics, such as their low volatility, high thermal and chemical stability, and adjustable cation-anion combinations for the synthesis of molecular sieves [89,90].

In 2004, Professor R. E. Morris and his colleagues from the University of St Andrews, UK, published groundbreaking research on utilizing IL/DES as both a solvent and a structure-directing agent (SDA) for the synthesis of AlPOs including several frameworks with unknown structures, coining the term "ionothermal synthesis" for this innovative approach [91]. In comparison to hydro/solvothermal methods, the key advantage of ionothermal synthesis lies in its ability to be conducted at ambient pressure, significantly reducing associated safety risks. In addition, a key advantage of ionothermal synthesis is the elimination of the competition between the solvent and the SDA [92], thus potentially resulting in more instances of "true" templating [93]. More interestingly, the recovery of ILs can be achieved [91]. In this case, ionothermal synthesis has garnered substantial attention from researchers in recent years owing to its convenience and versatility. To date, numerous publications, encompassing numerous outstanding reviews [92,94−97], have been issued in this field. This section aims to offer a practical guide on ionothermal synthesis, focusing specifically on aluminophosphates-based molecular sieves. It is noted that water is essential for the ionothermal synthesis, as water may influence the phase selectivity of the reaction [94]. Nevertheless, adding too much water into ILs compromises the unique properties associated with ionothermal synthesis to some extent, and in this case, it is not a true ionothermal preparation but a hydrothermal process. In accordance with this objective, we will summarize the advancements made in the aforementioned areas, presenting detailed accounts and discussions on the methodologies employed in relevant experiments.

To date, a majority of molecular sieves synthesized via the ionothermal method are metal phosphates, encompassing aluminophosphates and phosphates incorporating various heteroatoms such as Co, Zn, Cu, Si, Mg, Fe, Mn, and Ti [98−101]. The integration of these heteroatoms into the molecular sieve framework has emerged as a vital approach to enhance their efficacy in catalytic and separation processes [102]. The synthesis methods for these metal phosphates share similarities. Consequently, this section documents only a few standard frameworks.

$AlPO_4−11$ (AEL-type structure) and $AlPO_4−5$ (AFI-type structure) molecular sieves have been successfully produced through ionothermal synthesis without the presence of water, utilizing [Emim]Br or [Bmim]Br as both the reaction medium and template [103]. Investigations indicate that water profoundly influences the synthesis process, and the addition of specific amounts of water ($H_2O/Al = 1$, molar ratio) significantly accelerates the crystallization kinetics. Additionally, other organic species added as additives can markedly impact the ionothermal synthesis of molecular sieves. Tian and coworkers delved into the influence of seven diverse amines on the crystallization dynamics and resulting product in the ionothermal synthesis of aluminophosphate molecular sieves in [Bmim]Br [104]. The findings show that the addition of amines can enhance the selectivity of the crystallization reaction, resulting in the formation of pure AFI and ATV structures. The amine could potentially act as an SDA in synergy with the IL and create hydrogen-bonds with the IL's imidazolium ring.

Bats and colleagues introduced a novel ionothermal approach for the crystallization of LTA-type zeolite based on phosphates, utilizing 1-benzyl-3-methylimidazolium chloride ([Benzmim]Cl) [105]. Evidence indicates that Benzmim$^+$ ions are present in pairs within the α cage of $AlPO_4$-LTA. Further investigations revealed that co-SDAs in the form of tetramethylammonium and fluoride ions occupy the sodalite cages and

D4R units, respectively. This underscores the significance of a deliberate choice of templates. The targeted elimination of Benzmim$^+$ ions yields a stable AlPO$_4$-LTA molecular sieve. The α cages within the AlPO$_4$-LTA framework can be emptied while preserving structural integrity.

Utilizing deep eutectic solvents (DESs) as solvent and template is another viable option for ionothermal synthesis of phosphate molecular sieves. Morris and colleagues conducted the synthesis of various cobalt aluminophosphates in a DES comprising choline chloride and one of several carboxylic acids [106]. Among these obtained molecular sieves, distinctive ring-opened double-four-ring units are observed in a novel framework (SIZ-13), where a cobalt atom establishes a terminal Co−Cl bond. Unlike many other molecular sieves, the cobalt ions in the SIZ-13 structure are arranged at a single crystallographic site at the building unit's corner. They are coordinated to three oxygen atoms and one chlorine atom, differing from the usual coordination to four oxygen atoms. Typically, in traditional synthetic approaches in the presence of water, Co−Cl bonds, similar to those found in SIZ-13, are highly susceptible to hydrolysis. Consequently, such metal−chlorine bonds have not been observed in those compounds when created hydrothermally. The distinctive cobalt arrangement and the Co−Cl bond in SIZ-13 can be attributed to the deactivation of water and the milder hydrolysis conditions prevalent in DESs. Harrison also reported the presence of such metal−chlorine bonds in zinc phosphates, which were prepared via an ionothermal method based on choline chloride−urea DES [107].

ILs have demonstrated effective microwave absorption properties, thus opening up an opportunity to combine the environmentally friendly solvent from ILs with the eco-friendly chemistry aspect of microwave irradiation [108]. Xu and coworkers presented a microwave-enhanced ionothermal synthesis of AEL-type (AlPO$_4$-11 and SAPO-11) aluminophosphate molecular sieves [108]. This approach displays the advantages of accelerated crystal growth, excellent product selectivity, and reduced pressure. During a traditional heating ionothermal synthesis process, there are byproducts with a chabazite structure aside from generating AEL-type molecular sieves. In contrast, the utilization of microwave heating led to a product primarily consisting of AEL-type molecular sieve. The significant structural selectivity observed in microwave synthesis could be attributed to the local superheating from the microwaves. Such conditions likely facilitate the accelerated growth of AEL-type molecular sieves, thus suppressing the growth of other byproducts. The report also suggested that the existence of fluoride played a crucial role in the generation of molecular sieves, as the formation of AEL-type aluminophosphate molecular sieves did not occur without HF addition. Nonetheless, an excess of fluoride reduces the yield of crystalline products. The reported microwave-assisted ionothermal synthesis will hold tremendous promise for the future of molecular sieve preparation.

Yan and colleagues utilized a microwave-assisted ionothermal process at ambient pressure to fabricate an oriented SAPO-11 film coating on an aluminum alloy [109].

This coating exhibits outstanding resistance to corrosion. In a recent report, Zhao and coworkers detailed the successful synthesis of iron-containing aluminophosphate molecular sieves presenting AFI and LEV topologies [110,111]. They achieved this through a microwave-assisted ionothermal approach using a small quantity of eutectic mixture solvent composed of succinic acid, choline chloride, and tetraethylammonium bromide. Both choline and tetraethylammonium cations served as the SDAs. The resulting molecular sieve demonstrates elevated iron content and possesses a distinctive hierarchical structure and acidity. This leads to excellent catalytic performance in several industrially significant applications such as phenol hydroxylation reactions.

In addition to the well-known molecular sieve frameworks, the ionothermal method can yield various metal phosphates with innovative compositions and structures. SIZ-7 $(Co_{12.8}Al_{19.2}(PO_4)_{32})$ stands as a representation of an ionothermally synthesized CoAlPO featuring a unique zeolite framework denoted as SIV [112]. In the asymmetric unit, the structure of SIZ-7 is composed of eight crystallographically independent tetrahedral atoms, and it holds potential interest for applications in gas separation and storage.

Molecular sieves with large or extra-large pores remain highly important in both industrial and academic fields for their potential applications in processing large molecule [113]. Interestingly, the ionothermal synthesis can lead to large or extra-large pore molecular sieves with novel structures. Tian and coworkers reported the synthesis of aluminophosphate (DNL-1) with 20-ring pore openings in [Emim]Br with 1,6-hexanediamine (HDA) as the co-SDA [114]. DNL-1 is the first AlPO with 20-membered pore openings. Its framework exhibits several characteristics of the –CLO structure, including two separate three-dimensional channel systems with 20-ring and 8-ring windows, four terminal hydroxy groups protruding into the 20-ring opening, super cages located at the crossroads of the 20-ring channels, and partial connection of D4R units and leaving them incompletely connected. Furthermore, DNL-1 displays remarkable stability and boasts a substantial BET surface area and micropore volume. These features strongly imply the considerable potential of DNL-1 in applications like separation, catalysis, and gas storage.

Despite the successful application of the ionothermal method in the formation of aluminophosphates, their utilization in the synthesis of silicon-containing molecular sieves is challenging [115,116]. The major hurdle lies in the remarkably slow hydrolysis of silicon sources within the particular ionothermal environment. Successful synthesis of some silica-based zeolites has been achieved by performing hydrolysis of raw materials and subsequent crystallization under diverse conditions, including ionothermal treatment of prehydrolyzed silicon-containing reaction mixtures.

Morris and coworkers successfully synthesized pure silica zeolites with MFI and TON topologies by utilizing the basic IL [Bmim]OH$_{0.65}$Br$_{0.35}$, serving as a solvent, SDA, and mineralizer [115]. The task-specific IL [Bmim]OH$_{0.65}$Br$_{0.35}$ was obtained through anion exchange of [Bmim]Br and showed quite a crucial role for the successful ionothermal preparation of the zeolites. In addition, a small amount of HF was also added, aiding the dissolution of a silicate precursor. As the IL is notably abundant in the reaction, it is the genuine ionothermal synthesis siliceous zeolites. In the case of slicalite-1, the [Bmim] cation is incorporated within the structure's pores, which was confirmed by ^{13}C MAS NMR. The fluoride anions in the structure, confirmed by ^{19}F MAS NMR and ^{29}Si MAS NMR, severed as counter ions to balance the charge of the framework. Finally, it is important to highlight that this anion exchange process resulted in the degradation of Bmim cations. The inadequate basic stability of ILs has considerably limited their application for the genuine ionothermal synthesis of siliceous zeolites.

Recently, Xiao and coworkers reported a general ionothermal approach for the production of silica-based zeolites including MTT, TON, ITW, and MFI structures, employing NH$_4$F as a mineralizing agent and IL as a solvent and/or structure directing agent [116]. The authors investigated the relationship between silica solubility and fluoride quantity in the presence of ILs, which was examined through ^{29}Si MAS NMR spectra of silica when utilizing N,N'-diisopropylimidazolium iodide (DIPI) with differing mass ratios of NH$_4$F/SiO$_2$. The ^{29}Si MAS NMR spectrum displays a singular signal ranging from -120 to -90 ppm, corresponding to solid silica. The samples manifest a signal at -187.4 ppm upon NH$_4$F addition, ascribed to SiF$_6$$^{2-}$ ions dissolving in the DIPI. The intensity of this signal markedly augments with increasing NH$_4$F/SiO$_2$ ratios, demonstrating an improved solubility of silica species in ILs.

Based on this, zeolite with a typical MTT structure was obtained. Note that only a trace amount of water is added into the synthetic system. Thus, it can be regarded as a ionothermal process. It should also be pointed out that the mixtures consistently maintain a gel-like state during the crystallization process, suggesting the adoption of ionothermal synthesis rather than solvent-free synthesis. This ionothermal route can further be extended to the synthesis of other zeolites such as TON, ITW, and MFI.

6.4 Synthesis of ionic liquid-based metal organic complexes

In the synthesis of functional materials, ILs serve not only as solvents, SDAs, and modifiers, but also as reactants to participate in the formation of materials, forming an important component of the material skeleton. Now, based on the role of ILs in the synthesis of metal organic complexes, the synthesis and application of metal organic complexes strengthened by ionic liquids will be discussed.

In 2002, KunJin [117] first reported the synthesis of the first metal organic frameworks (MOFs) [Cu(bpp)][BF$_4$] (bpp = 1,3-di(4-pyridine)propane) using ILs [Bmim][BF$_4$] (Bmim = 1-butyl-3-methylimidazole) as a solvent. In this synthesis process, [Bmim][BF$_4$] serves not only as a solvent, but also as a charge compensating agent in the skeleton to balance the charge. In 2004, Kimoon et al. [118] synthesized the first three-dimensional metal organic complexes using ILs. In the past two decades, many research groups have synthesized various metal organic coordination polymers using ILs [119].

According to the role of ILs in the synthesis of complexes, four parts will be discussed:

6.4.1 Ionic liquids as charge compensation and structure guiding agent action

In Ling Xu's [120] report, three-dimensional metal manganese complexes were synthesized using ILs, with [rmi][Mn(btc)], where [rmi]X(rmi = 1-alkyl-3-methylimidazole; r = ethyl or propyl, X = Cl or Br) as a solvent and charge compensating agent, as well as an SDA. In [rmi][Mn(btc)], [emi]$^+$ or [pmi]$^+$ act as guests and stably exist in the pores of the complex through hydrogen bonding with the skeleton, playing a charge compensation role in the negatively charged three-dimensional skeleton and enhancing its thermal stability. Due to the different sizes of the substituents on [emi]$^+$ or [pmi]$^+$, the spatial orientations present differently in the pores, resulting in different pore sizes in [rmi][Mn(btc)]. Therefore, in the process of synthesis, [emi]$^+$ or [pmi]$^+$ simultaneously play a template role. Liu Qing Yan [121] used 1-ethyl-3-methylimidazolium bromide as a solvent to synthesize a three-dimensional complex $\{(Emim)[Dy_3(BDC)_5]\}_n$ through ligand terephthalic acid. This complex has a significant difference in the structure from the metal terephthalic acid complex synthesized by hydrothermal or solvothermal methods previously reported. [Emim]$^+$ enters the holes of the skeleton as a charge compensating agent and SDA through electrostatic interactions. In Ling Xu's [122] report, a novel tetranuclear zinc pyromellitic acid complex was first synthesized using 1-butyl-3-methylimidazole iodide [Bmim]I as a solvent. For the first time, the tetranuclear zinc unit "$Zn_4(\mu_3\text{-}O)_2\text{-}(\mu_1\text{-}O)_{12}\text{-core}$" was discovered in the complexes, and [Bmim]$^+$ was used as a charge compensating agent existing in the negatively charged Zn-O coordination skeleton through the electrostatic interaction. At the same time, the regulatory effects of the halogen ion size and the length of the imidazole substituent on the pore size were also discussed. ILs served as charge compensation and SDAs also appear in the synthesis of $(Bmim)_2[Cd_3(BDC)_3Br_2]$ (Bmim = 1-butyl-3-methylimidazole, BDC = 1,4-terephthalic acid) [123], wherein, [Bmim]$^+$ mainly exists in the holes of the two-dimension skeleton through the electrostatic interaction.

In the synthesis of complexes, the anions of ILs serve as charge compensators occupying the gaps of two-dimensional layered structures through electrostatic interactions [117]. For example, in the three-dimensional $[Cu_3(tpt)_4]$ $(BF_4)_3(tpt)_{2/3} \cdot 5H_2O$ [118], [Bmim][BF$_4$] serves as the solvent and [BF$_4$]$^-$ is the charge compensating agent. It exists in pores with a diameter of approximately 5 Å in the three-dimensional cationic framework. In the $(C_4C_1py)[Cu(SCN)_2]$, $(C_4C_1py$ = 1-butyl-4-methylpyridine) [124], C_4C_1py was used as the solvent, and its cations play a role of balancing charges. As a ligand, the $\mu_{1,3}$-SCN coordination mode was first discovered, which has never been seen in other synthesis methods.

6.4.2 Mixed ionic liquids as solvents for synthesis of porous metal organic complexes

Generally, when synthesizing metal organic complexes, ILs are usually used as templates or SDAs with cations to participate in the final product structure. Anions of ILs rarely play a structural guiding role in the preparation of coordination polymer materials. However, it can still produce interesting structures. In 2007, Morris et al. [125] synthesized anionic controlled coordination polymers, $[(Emim)_2][Co_3(TMA)_2(OAc)_2]$, $[(Emim)_2][Co(TMA)]$, $[Co_5(OH)_2(OAc)_8](H_2O)_x$, and $[(Emim)_2][Co_2(TMA)_4H_7(2,2\text{-}bpy)_2]$, (TMA = pyromellitic acid, 2,2-bpy = 2,2-bipyridine). When the solvent is highly polar [Emim]Br, the reaction product is $[(Emim)_2][Co_3(TMA)_2(OAc)_2]$, where Co ions form an octahedral coordination environment. It is found that the polarity of ILs solvents significantly impact the coordination behavior of metal ions: because of the different polarity of ionic liquids, cobalt ions present different coordination environments and different products. The poor reactivity between metal ions and organic ligands with poor polarity makes it difficult to form metal organic complexes. Moreover, in the process of forming complexes using ILs as solvents, cations serve as SDAs, and anions play a crucial role in the formation of structures. In addition, when two ILs [Emim]Br and [Emim]NTf$_2$ are mixed in a ratio of n([Emim]Br):n([Emim] NTf$_2$) = 1:1 as solvents, a third structure is generated that is different from two structures obtained before when the previous two ILs were used respectively. This opens up the possibility of using mixed ILs as solvents, thereby achieving greater control over solvent properties.

In these complexes mentioned above, $[Emim]^+$ is highly disordered and plays a charge compensation role. At the same time, the water content in ILs can be controlled through IL anions. [Emim]Br is hydrophilic, and the presence of trace amounts of water can act as mineralizers in the formation of complexes, which is essential for the crystallization of metal coordination polymers. While synthesizing MOFs, the more hydrophobic the ILs used, the less likely the ILs cations are to be blocked. Of course, as the chemical properties of the system change (such as the preparation of different materials), the balance between solvent and templating effect will also change. However, IL anions themselves are generally not enclosed in the structure, so this is an induced effect rather than a templating oriented effect. This example also fully demonstrates another particularly interesting feature of the synthesis of metal organic complexes using ILs, which is the preparation of new solvents with specific properties by mixing two different miscible ILs. The use of mixed ILs as solvents can make the solvents suitable for specific reaction chemistry.

6.4.3 The role of ionic liquid solvents in the synthesis of rare earth complexes

Besides being used in the synthesis of transition metal complexes, ILs are widely used in the preparation of rare earth metal complexes. In 2010, Farida Himeur et al. [124]

synthesized three complexes containing lanthanide metals using an IL choline chloride/1,3-dimethylurea mixture as a solvent. In addition to 1,3-dimethylurea acting as a solvent, the ammonium salt or alkyl cation generated by its decomposition could serve as an SDA in the reaction, occupying the coordination site of Ln^{3+} and participating in the structure of the complex. The most important feature of ion participation in the metal organic complexes is the change of reaction solvent from molecular to ionic type, and the changes in solvent properties usually lead to changes in products. In the synthesis of complexes, the strong hydrogen bonding effect of water molecules can make them less reactive than other organic solvents with similar content. Using special ILs as a solvent can reduce the participation of water in the final product even in very low concentrations of water. This type of ionic solvent has a very strong solvation effect, so that some readily hydrolyzable compounds such as PCl_3 can exist in ILs for a considerable period of time without being prone to violent hydrolysis reactions. The special ionic properties are the reason why they can obtain complexes with special properties. In the complex $Ln(TMA)(DMU)_2(Ln(C_9O_6H_3)$ $(CH_3NH)_2CO)_2$; Ln = La, Nd, Eu; TMA = pyromellitic acid, DMU = dimethyl urea [124], ILs have a similar effect, resulting in the urea component remaining intact (or at least partially decomposed). Therefore, dimethyl urea exists in the final skeleton by coordination bonds with metal ions. Another one-dimensional rare earth coordination polymer, $Gd[(SO_4)](NO_3)(C_2H_6SO)_2$, was obtained using Brønsted acidic ionic liquid [HBIM][HSO4] as a solvent, and the IL only acts as a solvent in this synthesis process [126].

Obviously, ILs could act as solvents and SDAs in the synthesis of complexes, such as imidazole, and its derivative ILs, which serve as fillers for pores or holes in the structures, indicating that the volume and geometric structure of the IL determine the obtained complex structure. Usually, larger size imidazole ILs and their derivatives should lead to the formation of complexes with larger pores or holes, increasing the thermodynamic stability of metal organic materials. Changing the size of ILs cation does have a certain impact on the structure, as larger cations form larger open frameworks that require additional space to accommodate larger templates. However, in this approach, the role of SDAs, namely, templates, is not very specific and standardized in guiding or precisely controlling the interaction between templates and structures. More often, it is simply "space filling" [124,126].

6.4.4 Synthesis of metal organic complexes using carboxyl-functionalized ionic liquids

With the increasing demand for ILs and the expansion of their applications, ILs with special structures and properties have received widespread attention. Most ILs based on dialkylated imidazole cations do not have coordination sites and do not coordinate with metals in a similar manner to water molecules. By modifying the cations and anions of conventional ILs with functional groups, functionalized ILs that meet the specific requirements can be obtained, and then more functions could be introduced into the final metal organic complexes. Generally, carboxyl-functionalized ILs are obtained by hydrolysis of ionic liquids with ester groups, which have the acidity and coordination functions.

The synthesis of carboxyl-functionalized ILs has been reported in many literature [127], using these Brønsted acidic ILs as bridging ligands to synthesize a series of novel metal organic coordination polymers. Pierre Farger [128] obtained a two-dimensional complex with 1,3-di(carboxymethyl)imidazole chloride ($[mim(COOH)_2Cl]$) by the solvo-ionothermal approach, where the carboxyl oxygen atom on ILs serves as a coordination atom and forms a coordination bond with the metal. As a result, $[(mimCO_2)_2]_2Co$ shows remarkable axial anisotropy of Co^{2+} ($|D| = 17.7(2)$ cm^{-1}), which is magnetic, and $[(mimCO_2)_2]_2Zn$ shows certain luminescent characteristics. In 2006, Zhaofu Fei et al. [129] synthesized a series of halogenated carboxyl-functionalized imidazole ILs and introduced the carboxyl-functionalized ILs, 1,3-di (carboxymethyl) imidazole bromide $[mim(COOH)_2Br](H_2A)$, into transition metal coordination polymers. Utilizing the acidity of the ligand, $\{[ZnCl(H_2O)A](H_2O)\}_\infty$ and $\{[Co(H_2O)_4A]Br(H_2O)\}_\infty$ were obtained and the effect of changing the anions of ILs was studied. Comparing carboxyl-functionalized imidazole ILs with anions of F^-, Cl^-, Br^-, and ClO_4^-, it can be found that these IL anions with different sizes could form complexes with different sizes and pore structures, which are important factors determining the structure of metal organic complexes.

In $(C_7H_7N_2O_4)_2Sr_4H_2O$, IL cation 1,3-di(carboxymethyl) imidazole ($[mim(COOH)_2]^+$) [130] was introduced into the complexes as a ligand. Using the same method, a series of coordination polymers Ba[ABr], Ca$_2$[A$_3$Br], and Sr[ABr] (A = 1,3-di(carboxymethyl) imidazole ion) [131] were fabricated.

In 2007, Wang Xian-wen et al. synthesized the first chiral complex [PbCl $(C_7H_7N_2O_4)$] ($C_7H_7N_2O_4$ = 1,3-dicarboxymethyl imidazole ion) using 1,3-dicarboxymethyl imidazole IL as ligand [132]. This compound has a right-handed helical rectangular channel and is the first double interpenetrating three-dimensional complex composed solely of symmetric flexible IL ligands without the involvement of any chiral solvents. This fully demonstrates the enormous potential of using IL precursors or functional ILs to synthesize new functional materials under the premise of reasonable design. In 2007, Morris et al. [133] obtained chiral ILs by combining the 1-methyl-3-butyl imidazole (Bmim) cation with L-aspartic acid as an anion and synthesized chiral structural complexes $[(Bmim)_2][Ni(TMA-H)_2(H_2O)_2]$ and $[(Bmim)_2]$ $[Ni_3(TMA-H)_4(H_2O)_2]$ (TMA = pyromellitic acid).

Although anions of ILs have not been seen in the final products, the structural induction potential of anions is very remarkable. Researchers hope to further explore and utilize the characteristics of ionic liquids in the future, introducing some special property of IL anions into inorganic organic hybrid materials. Similarly, Wang Xuan et al. [134] also synthesized coordination polymers containing five Mn (II) with imidazole carboxylic acid zwitterionic ligands, 1-alkyl-3-acetic acid imidazole and 1,3-dicarboxymethyl imidazole ILs, and elucidated the impact of the chiral characteristics of ILs on the structure and properties of chiral metal complexes.

Brendan F. Abrahams et al. [135] synthesized five different structural metal coordination polymers containing metal Cu (II) using chlorinated 1,3-diacetic acid imidazole ([H_2imdc]Cl) ILs. When solvent methanol molecules participate in coordination, the complex forms a two-dimensional bidirectional network structure formed by coordination bonds, and adjacent layers are connected by the anion [BF_4]$^-$. However, when the coordinated and solvent methanol in the lattice were removed, the structure and composition of these complex undergo interesting changes, with [BF_4]$^-$ participating in coordination and F^- forming hydrogen bonds with the H atom on the ligand. It indicates that small molecule solvents in crystals cultivated in solvothermal environments often participate in coordination and occupy pores, and ILs can simultaneously play multiple roles such as solvents and ligands.

In 2005, Zhaofu Fei [136] obtained a zinc complex with 1,3-dicarboxymethyl imidazole IL. After Zn coordinating with 1,3-dicarboxymethyl imidazole, one-dimensional helical channels were formed through the stacking of imidazole rings, where water molecules were filled into the channels via hydrogen bonding. Each chain with one-dimensional spiral channels forms a three-dimensional structure through hydrogen bonding. Nicolas P. Martin et al. [137] synthesized a complex with UO_2 as the unit using 1,3-diacetic acid imidazole cation as a ligand by the hydrothermal method. In the low pH range (0.8–3.1), $(UO_2)_2$(imdc)$_2$(ox) $\cdot$ $3H_2O$ (imdc- = 1,3-diacetic acid imidazole, ox = oxalic acid) was obtained. The oxalate in the reaction system is produced by the decomposition of the imidazole cation under acidic conditions.

Using 1,3-diacetic acid imidazole salt as a bridging ligand, Zhang Suojiang's group [138] fabricated the first three-dimensional lanthanide metal complex [Er_4(μ_3-OH)$_4$(μ_2-O)$_{0.5}$OL_4(H_2O)$_3$] $\cdot$ $Br_{2.90}$ $\cdot$ $Cl_{1.10}$ $\cdot$ $2H_2O$ (Fig. 6.1) and a two-dimensional complex [PrL(H_2O)$_4$Cl] $\cdot$ Br $\cdot$ H_2O (Fig. 6.2). Using the same 1,3-diacetic acid imidazole salt ligand, Xiaochuan Cai [139] synthesized a series of Ln^{3+} complexes and measured their fluorescence properties and thermal stability.

Using Brønsted acidic IL, for example, 1,3-bis(carboxymethyl) imidazolium salt and 1,3-bis(carboxylatoethyl) imidazolium salt, Wang's group fabricated many Ln^{3+} complexes through optimized the synthesis conditions. These Ln^{3+} carboxyl-functionalized IL complexes exhibit excellent physicochemical properties, such as thermal stability, hydrophilicity, and solubility in polar solvents. Moreover, these complexes show promising applications in luminescent probe, fluorescent devices, and magnetic materials [140−145]. Dual-emitting ratiometric luminescent thermometers based on lanthanide metal organic complexes with Brønsted acidic ILs {[Ln(imdc)(CH_3OH)(H_2O)$_3$]Cl_2}$_n$ (Ln(imdc), Ln = Eu and Tb,

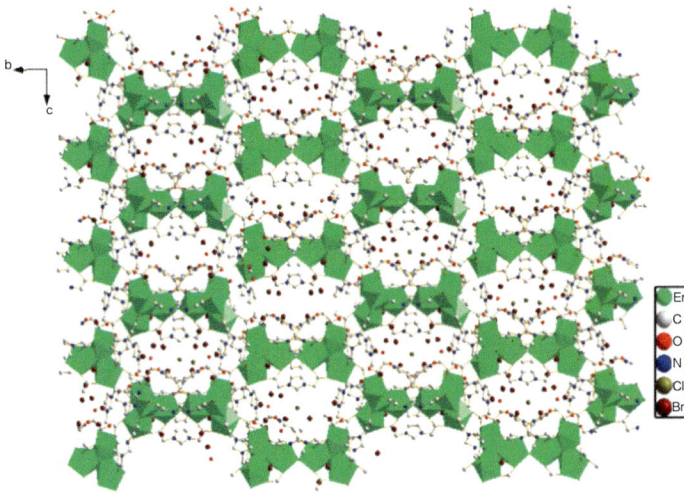

Figure 6.1 View of $[Er_4(\mu_3\text{-}OH)_4(\mu_2\text{-}O)_{0.5}OL_4(H_2O)_3] \cdot Br_{2.90} \cdot Cl_{1.10} \cdot 2H_2O$ down the a axis. Note: Polyhedra represent the ErO_8 (H atoms and hydrogen bonds are omitted for clarity).

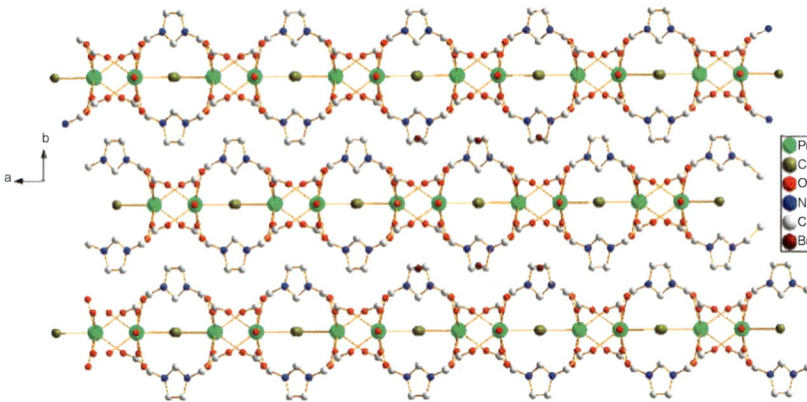

Figure 6.2 1D chains of $[PrL(H_2O)_4Cl] \cdot Br \cdot H_2O$ along the a axis. Note: H atoms and hydrogen bonds are omitted for clarity.

imdc = 1,3-bis(carboxymethyl)imidazolium ion) were obtained for the first time [141]. Two ratiometric luminescent thermometers show good temperature-dependent emission behaviors and linear relationships with temperatures from 353 to 403 K for Eu(imdc) and 323 to 373 K for Tb(imdc), respectively. Furthermore, these luminescent thermometers have high sensitivity in the high temperature range and successfully bring about luminescence color change visible to the naked eye (Fig. 6.3). Compared to other thermometers, they offer better thermal stability and sensitivity for practical applications in the biological and scientific fields by providing accurate and noninvasive monitoring. Besides that, the fluorescence

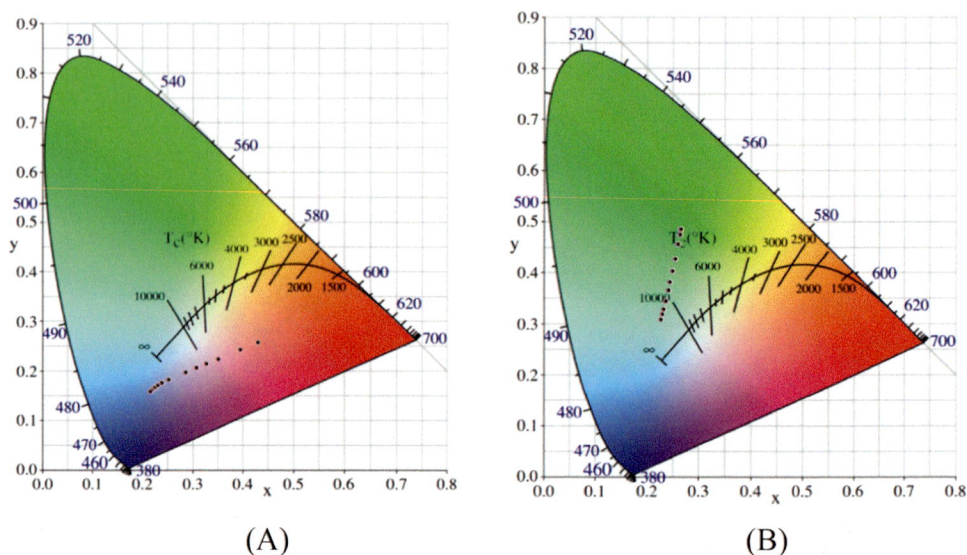

Figure 6.3 Corresponding CIE pictures of Eu(imdc) (A) and Tb(imdc) (B).

performances of [H₂imdc]Cl reveal its quite different characteristics of the red–edge effects and the features of the concentration and excitation wavelength-dependent two component emission. Owing to the distinctive fluorescence properties of [H₂imdc]Cl, this work undoubtedly provides great possibilities for ILs as fluorescent chromophores to fabricate new metal complexes and then opens up new perspectives for the application of ILs and IL-based metal organic complexes.

6.5 Synthesis and application of polymerized ionic liquids

PILs or polymerized ILs or polymeric ILs are a class of IL polymers that feature an IL species in each monomeric repeat unit linked by a polymeric backbone to form a macromolecular architecture. Thus, PILs exhibit some of the special properties of both ILs and polymers, such as thermal stability, high ionic conductivity, low glass transition temperature, and mechanical properties and processability of polymers that effectively overcome the mobility of ILs [146–148]. Hiroyuki Ohno and coworkers first reported the preparation of PIL as a matrix for fast ion conduction in 1998 [149]. After that, the number of papers on PILs increased year by year. PILs have been investigated more extensively and are mainly divided into the following categories: (1) polycationic ILs; (2) polyanionic ILs; (3) polyzwitterion-type ILs; and (4) copolymer-type ILs. Due to these advanced points, the research of PILs in recent years has focused on structurally controllable properties to design and synthesize stable, functional materials with specific properties, which have been widely used in the fields of polymer chemistry, physics, electrochemistry, materials science, catalysis, separation, analytical chemistry, and energy science [150].

6.5.1 Synthesis of poly(ionic liquid)s

Like ILs, various PILs with different structures and functions can be designed and synthesized by modulating monomers' anionic and cationic structures. To date, radical polymerization remains the most commonly used method to synthesize PILs, which has the advantage of a wide range of polymeric monomers, mild and easily controlled reaction conditions, the possibility of using water as a medium, and ease of industrial production [147,151−153]. With the further development of research, other more novel polymerization methods have been initially introduced for the synthesis of PILs, such as controlled reactive atom transfer radical polymerization (ATRP) and reversible addition-fragmentation chain transfer (RAFT) polymerization, ring-opening metathesis polymerization, in situ polymerization, dispersion polymerization, cyclopolymerization, and dehydrogenative coupling polymerization.

6.5.1.1 Polycationic ionic liquids

Polycationic ILs are of great interest to researchers due to their ease of synthesis and versatility, where the cations are covalently bonded to the polymer backbone, and the anions are connected to the cations by ionic bonds. Most efforts in the synthesis of polycationic ionic liquids have focused on imidazole cations, pyridine cations, quaternary ammonium cations, and quaternary phosphonium cations. Here, the imidazole monomer with one or more polymerizable units on the cation like vinyl, styrene, (meth)acrylic, and (meth)acrylamide is the most popular structure that can be incorporated into polymer backbones. Mecerreyes D. and coworkers used 1-vinyl imidazole as a starting material to synthesize the poly(1-vinyl-3-ethyl imidazolium) bromide and poly(1-vinyl-3-butyl imidazolium) chloride and then through an easy anion-exchange reaction to tune the solubility of the poly(1-vinyl-3-alkyl-imidazolium) [154]. It has been reported in the literature that conventional synthetic methods are not favorable to the molecular weight of PILs, which would be disadvantageous to the properties of PILs [155,156]. To further improve the properties of PILs, it is important to exploit new synthetic routes to develop PILs. Yang and his coworkers successfully use a novel three-step process involving radical polymerization of the 1-vinylimidazole monomer, followed by quaternization and anion exchange, to synthesize poly(1-ethyl-3-vinylimidazolium bis(trifluoromethanesulphonylimide)), which is used as a matrix for polymer electrolytes. The results show that the lithium-ion batteries with the PIL using the new synthetic process exhibited more excellent performance than that of the conventional route [157].

6.5.1.2 Polyanionic ionic liquids

In comparison to polycationic ILs and polyanionic ILs, a considerably wider variety of anion structures have the potential to form anionic monomers that can be covalently attached to the polymer backbone, examples of which include PILs with sulfonic acid anions, phosphate anions, and imide anions [158,159]. For instance, Shaplov and coworkers designed three IL monomers with highly conductive bis(trifluoromethylsulfonyl)imide, tricyanomethanide, dicyanamide anions, and mobile aprotic pyrrolidinium cations to synthesize the polyanionic ILs by a radical polymerization procedure. As the results showed, the properties of synthesized polyanionic ILs depended on the size of the attached anion, which mainly affected the solubility, viscosity, thermal stability, glass transition temperature, and ionic conductivity [160]. Yan et al. prepared an ionic liquid monomer of 1-methylimidazolium 2-acrylamido-2-methylpropanesulfonate ([mim][AMPS]) to overcome the immiscibility of AMPS with hydrophobic monomer oils, which can be miscible with monomer oils and cross-linked polymerized with styrene, acrylonitrile, and divinylbenzene to form PAMPS-based polymer membranes used in the proton exchange membranes for fuel cells. It can be concluded that the membranes with PAMPS are transparent, flexible, and easily fabricated and have good and tunable mechanical properties [161].

6.5.1.3 Polyzwitterion-type ionic liquids

Polyzwitterion-type ILs are the polymer backbone of the PILs containing both anionic and cationic groups, with the cations and anions being covalently attached to the polymer backbone [162]. Polyzwitterion-type ILs have been studied to a lesser extent than polycationic and polyanionic ionic liquids. The more classical synthetic pathways are related to Ohno's group. Ohno and his coworkers used dehydrocoupling polymerization of imidazolium-type IL and lithium 9-borabicyclo [3,3,1] nonane hydride, followed by an exchange reaction to obtain an ion-conducting matrix. The resulting polymer electrolyte has an ionic conductivity of more than 10^{-6} S/cm at 50°C [163].

6.5.1.4 Copolymer-type ionic liquids

A copolymerized IL is defined as the copolymerization of two or more IL monomers. These IL monomers typically consist of cationic and anionic monomers linked to each other by covalent or ionic bonds to form copolymers [164]. ATRP and RAFT polymerization are two of the most efficient methods for the synthesis of bulk copolymerized PILs. They allow for better control of the molecular weight of the polymer and the synthesis of a greater variety of structures [165–168]. Li et al. used the RAFT reaction to obtain the copolymer mPEG-b-PS-b-PVB, which was then quaternized with methylimidazole and ion-exchange reaction with TFSI⁻ to give the triblock ionic liquid copolymer mPEG-b-PS-b-PVBmimTFSI. As demonstrated in this study, the introduction of polymeric IL chain segments can effectively enhance the ionic conductivity of copolymers [169]. To obtain the tunable and charged surface of the membranes, Du et al. prepared a series of P(MMA-b-MEBim-Br) (PMEBim-Br) block copolymers by RAFT polymerization in the PVDF porous membranes. The study shows that the chain length of the PMEBim-Br block copolymers has a significant effect on the hydrophilicity and charge properties of the blend membranes and that increasing the chain length gives the membrane better antifouling and separation properties against Rhodamine 6 G [170].

6.5.2 Applications of poly(ionic liquid)s

6.5.2.1 Lithium batteries and capacitors

ILs are a new class of soft-functional dielectric materials that stand out among many materials for electrochemical energy storage applications, mainly due to the properties of high viscosity, ultra-low volatility, excellent thermal stability, and inherent high ionic conductivity [146,171]. As a component of electrolytes, it has a wide range of applications in the field of electrochemical energy storage. However, ILs are particularly suitable for liquid electrolytes, which are struggling to meet the demand for high-safety energy storage batteries. PILs are ILs containing unsaturated chemical bonds derived from polymerization under certain conditions and have excellent properties of both ILs and polymers, which can inhibit the mobility of ionic liquids [150]. In recent years, PILs have attracted more interest in electrochemical devices. Fu et al. constructed a polymer electrolyte based on a polymerized IL poly(diallyldimethylammonium) bis(fluorosulfonyl) imide (PDADMAFSI) as a polymer matrix and $PYR_{13}FSI$ as a plasticizer with high oxidative stability and ionic conductivity for all-solid-state lithium metal batteries. It concludes that the polymer electrolytes with PIL depict excellent compatibility with a lithium metal anode, which achieves stable cycling of Li/NCM811 and Li/LNMO cells. The reason is that the positively charged $PDADMA^+$ chains reduce the coordination of lithium ions to the polymer compared to typical lithium-ion coordination polymer matrices, thus promoting high lithium-ion mobility. LiFSI has low binding energy between Li^+ and FSI^- and the ability to form a stable interfacial phase upon contact with lithium metal [172].

Kang et al. developed a nesting doll-like multilayer solid electrolyte based on PIL (HPILSE) through the in-situ polymerization method for high-safety Li-ion and Na-ion batteries. The HPILSE membrane is fabricated by embedding a radical polymerized C1−4TFSI framework in an EmimTFSI-based electrolyte filled with a network of PDDATFSI porous membranes, which effectively improves the interfacial contact between the electrolyte and electrodes so that enhance the electrochemical performances of $LiFePO_4$/Li-HPILSE/Li and $Na_{0.9}[Cu_{0.22}Fe_{0.30}Mn_{0.48}]O_2$/Na-HPILSE/Na cells [173].

Zhang et al. successfully designed and fabricated a novel "ionic gel in ceramics" hybrid electrolyte with the ability to inhibit dendrites and achieve excellent compatibility with lithium metal anodes. The key point of the ionogel electrolyte is a solid electrolyte consisting of a "polyimide based PIL in salt" ionic gel as an ionic bridge and $Li_{1.3}Al_{0.3}Ti_{1.7}(PO_4)_3$ as a rigid backbone. Volume changes of the electrodes during the charge−discharge process are effectively cushioned by the elasticity of the "poly ionic liquid in salt," which contains precursors in a soft ionic gel state. As a result, optimized lithium/lithium symmetric batteries can last for more than 4000 hours [174].

More importantly, PILs feature molecular tailoring design, allowing unique structures to be synthesized to provide new functions and meet different requirements. This provides diversity and flexibility for high performance electrolytes. For example, polymer electrolytes based on PILs are capable of imparting self-healing properties due to interactions such as hydrogen bonding interactions, ion–ion interaction, and so forth in IL structures.

Ma and Shi et al. reported a novel self-healing PIL-based polymer electrolyte by grafting the Emim$^+$ cation into the PMMA polymer main chain. The self-healing ability of the electrolyte can spontaneously repair the holes and defects at the lithium anode/electrolyte interface caused by the growth of lithium dendrites, thus inhibiting the growth of lithium dendrites. This is mainly due to the hydrogen in the imidazole ring and the oxygen-containing groups in PMMA acting as donors and acceptors of hydrogen bonds, and the intermolecular hydrogen bonds giving the polymer rapid self-healing. The Li|LFP cells assembled with the PIL-based electrolyte can achieve a specific capacity of 134.7 mAh/g on the first cycle discharge at room temperature and a capacity retention rate of 91.2% after 206 cycles. When the operating temperature was increased to 48°C, the battery could be cycled 560 cycles with 74.5% capacity retention. The results indicate that self-healing PIL-based polymer electrolytes have significant competitive advantages in achieving interfacial stability and high safety in all-solid-state lithium-metal batteries [175]. Fu and Yang et al. synthesized a six-armed and dicationic polymeric IL and used it to prepare high-performance self-healing solid electrolytes (DPIL-6-SPE), which can improve the interfacial compatibility with lithium metal anode. The DPIL-6-SPE not only improves the stability and safety of the corresponding battery system but also enhances the self-healing properties of the material through ion–ion interactions. However, the material's self-healing depends on applying an external stimulus [176].

In addition to their applications in batteries, PILs are also involved in the role of capacitors. For example, Qi and Zhao et al. first took advantage of the one-step ionizing radiation technology with clean, simple, and no initiators to prepare a poly(ionic liquid)/MXene gel polymer electrolyte (PIL/MXene GPE) for supercapacitor electrolytes. Supercapacitors based on PIL/MXene GPE have been shown to have excellent electrochemical properties and high stability, with capacitance retention up to approximately 93.02% and even beyond 300 charge–discharge cycles. This provides a new route to fabricate a PIL/MXene-based GPE with superior properties for use in electronic devices [177]. Kwang S. Suh and Rodney S. Ruoff et al. demonstrated that the poly(1-vinyl-3-ethylimidazolium) salt-modified reduced graphene oxide (PIL: RG-O) electrode can effectively improve the interfacial compatibility with the Emim-NTf$_2$ electrolyte, thereby increasing the effective electrode surface area accessible to electrolyte ions. The supercapacitor assembled with the PIL-modified RG-O electrode and Emim-NTf$_2$ electrolyte exhibits a maximum energy density of 6.5 W · h/kg with a maximum power density of 2.4 kW/kg [178].

6.5.2.2 Dye-sensitized solar cells

The structure of dye-sensitized solar cells is simple and consists of five main components: conductive substrate, nanoporous semiconductor film, dye sensitizer, counter electrode, and electrolyte. Among them, PILs are mainly used in the electrolyte part of dye-sensitized solar cells due to their good electrical conductivity [179]. Thomas et al. prepared PVA/PA/PIL/C nanofibrous electrospun membranes consisting of phthaloyl agarose, poly(3-butyl-1-vinyl imidazolium iodide), and conductive carbon (C) incorporated in poly(vinyl alcohol) (PVA) and used them as electrolytes for dye-sensitized solar cells. It shows excellent performances of dye-sensitized solar cells with 5.9×10^{-3} S/cm of ionic conductivity, 4.7×10^{-7} cm^{2}s^{-1} tri-iodide diffusion coefficient, improved charge transport properties, 6.05% power conversion efficiency, and 82% of its initial cell efficiency after 500 h. One of the factors is that the presence of conductive carbon and imidazolium segment in PIL reduces unfavorable recombination reaction, which increases V_{oc}, J_{sc}, and η, leading to superior charge transport properties [180]. Yan's team designed and synthesized bis-imidazolium-based PILs (poly[BVim][Him][TFSI]) employed to construct quasi-solid-state electrolytes without organic solvent for dye-sensitized solar cells. The large π-π structure of the poly [BVim][Him][TFSI] facilitates charge transport and exhibits a conductivity close to that of liquid ionic liquids, and the formation of nano-channels of the PIL facilitates the diffusion of I_3^-/I^-, which enhances the power conversion efficiency of 5.92% of dye-sensitized solar cells [181].

6.5.2.3 Proton exchange membrane fuel cells

Polymer electrolyte membrane fuel cells (PEMFCs) have extensively attracted attention as an effective and promising power source that directly converts the chemical energy in fuel into electrical energy with high efficiency due to their high conversion efficiency, high power density, and fast load response. This dedication to PEMFC has a broad range of application potential in portable, electric vehicles, and communication power supplies. The PEMFC has an all-solid structure and employs a special polymer electrolyte membrane to conduct protons [182]. The commercial membrane in PEMFC is Nafion, which is not feasible for fuel cells to work at temperatures over 80°C due to water evaporation and low water affinity limiting the application [183,184]. Many researchers make efforts to improve the properties under different conditions. It has been demonstrated that PILs, as charged polymers incorporated into PEMs, can keep some unique features of ILs and improve the electrochemical properties of PEMFCs [151,185,186]. Mecerreyes and Zhu were the first to present the protic phosphonium counter-cations PIL, used in proton-conducting membranes, which show a good ability to form membranes and high ionic conductivities in the range of 10^{-8}–10^{-3} S/cm from 30 to 90°C [185]. Wang et al. successfully prepared novel high-temperature proton exchange membranes consisting of fluorinated polybenzimidazole

(6FPBI), cross-linkable PILs [ViBuim][TFSI] (cPIL), and allyl glycidyl ether copolymer with excellent stability and mechanical properties. It is predicted that the cPIL monomer [ViBuim][TFSI] provide [ViBuim][TFSI]. When the content of cPIL in the polymer membrane is 20 wt.%, the phosphate doping was up to 27.8. The proton conductivity of the membrane at 170°C was 0.106 S/cm [187].

It has been demonstrated that membranes based on PILs can achieve high proton conductivity in dry conditions and at high temperatures. However, plasticization of poly-IL chains and IL leakage are some of the issues that need to be addressed for PIL-based membranes. Researchers have taken a lot of effort to construct ion transport network channels by adding appropriate amounts of inorganic fillers to improve the proton conductivity of PEMs to solve the problem of the slow loss of PIL in polymeric membranes. For example, Jana et al. grafted poly(vinylimidazolium)bromide (PVImBr) PIL brush polymers onto the surface of silica NPs (SiNPs), forming the polymers of different molecular weights [PVImBr(L)-g-SiNP] and [PVImBr(H)-g-SiNP] by RATF technology for the construction of the polymer electrolyte membrane with OPBI. The research shows that the membrane has excellent tensile strength and storage modulus, which is related to the good miscibility between OPBI and PVImBr(H)-g-SiNP.

In addition, there are stronger hydrogen and ionic interactions between the PIL chains on SiNPs and OPBI matrix, which increases the capacity of the H_3PO_4 dopant level and reduces its leakage [188].

6.5.2.4 Catalysts and catalyst supported

With their higher mechanical properties, PILs are more favorable for catalytic applications to overcome the difficulty in recovering catalysts from homogeneous ionic liquids. Depending on their role in catalysis, PILs can act as catalysts, catalytic supports, and precatalysts, and different catalytic effects can be achieved through the selection of suitable cations, the design of different molecular structures, and the use of different polymer processing techniques. At the same time, PILs can also serve as good carriers for metallic nanoparticles, allowing the nanoparticles to be uniformly dispersed in the polymer backbone to produce highly efficient heterostructures via anion exchange reactions [189−191]. Gao et al. designed and synthesized a series of swelling PILs with a cross-linked network structure by free radical polymerization of N-vinylimidazolium ILs, sodium 4-vinylbenzoate, and cross-linker [(EG)3-DVim]Br$_2$, which showed excellent homogenous catalytic performance in transesterification between EC and MeOH. The poly[VBim-VBA-DVim]-2.5% with the highest swelling ratio (Q = 12.4 (g/g)) in the mixed solvent MeOH/EC obtained a DMC yield of 78.6% and a selectivity of 98.1%. This shows the achievement of homogeneous catalysis, equivalent to the same activity of the homogeneous IL [Bmim]BA (77% DMC yield, 96.3% selectivity). In addition, poly[VBim-VBA-DVim]-2.5% can be recycled 7 times without significant loss of activity and the structure remains essentially unchanged, showing good cyclic stability. The swollen nature of the PIL in the reaction substrate can increase the mass transfer rate and active site exposure and thus the catalytic efficiency. This provides a new strategy for the design of highly active heterogeneous catalysts [192].

Zhang's group has done a lot of research work on the catalysis of PILs, such as high-value utilization of CO_2 and the recycling of waste [193−196]. For example, a PIL-based Cu^0-Cu^I tandem electrocatalyst (Cu0@PIL@CuI) is successfully constructed to improve the reaction rate and C_{2+} selectivity of CO_2RR. The hybrid Cu (0)@PIL functional PIL layer can introduce a highly dispersed Cu_2O phase and a single Cu site through reaction and coordination with additional copper salts. As a result, it provides sufficient Cu(0)-PIL-Cu(I) interfaces and a surface-dense electrostatic network to stabilize and enrich critical intermediates. In particular, the internal Cu(0) PIL consumes most of the electrons and CO_2, producing modest amounts of FE_{C2+} and abundant CO. The released CO diffuses and accumulates in the PIL layer, enriching the local concentration of *CO. This is followed by dimerization in the PIL-Cu(I) region of the outer layer, which increases the selectivity for C_{2+}. In addition, the weak intermolecular interaction between the PIL units and the critical CO_2RR intermediates facilitates the selectivity of C_{2+} products by reducing the barrier to C-C coupling. These findings provide a novel design concept for achieving efficient and C_{2+} selective CO_2RR catalysts [194].

In addition, carbon dioxide (CO_2) as one of the greenhouse gases, is an important C1 resource, and the cycloaddition reaction of CO_2 with epoxide can realize the high-value utilization of CO_2, which is a green route for CO_2 recycling. Cheng et al. prepared hydrogen bond donor-functionalized nonhomogeneous polymerization IL catalysts (HPILs) by cross-linking copolymerization of hydrogen bond donors and ILs and used them in the synthesis of propylene carbonate (PC) from CO_2 and propylene oxide. The yield of propylene carbonate (PC) was found to be 94%. This was far superior to that of s-PIL (PC yield of 72%) and even close to that of bulk ILs (PC yield of 95%) (shown in Table 6.7) [195].

The chemical recycling of polyethylene terephthalate (PET) waste into valuable chemicals has received a lot of attention from the perspective of a "sustainable society" and "green chemistry." Lv et al. prepared a series of PIL catalysts containing different metal ions for the catalytic methanolysis of PET. The results show that the catalysts are well-suited for the methanolysis of different PET feedstocks. More importantly, the PIL-Zn^{2+} catalyst can be recovered by simple post-reaction filtration and reused up to six times without significantly reducing PET conversion and DMT yield. The mechanism of the methanolysis of PET catalyzed by PIL-Zn^{2+} is proposed. This approach provides for the design of efficient, stable, and recyclable catalysts for PET methanolysis [196].

Table 6.7 CO_2 cycloaddition with propylene oxide[a] [195].

Entry	Catalysts	Sel.[b] (%)	Yield[b] (%)
1	VHEimBr	100	92
2	VHPimBr	100	95
3	VCMimBr	100	36
4	VCEimBr	100	86
5	s-PIL-1	>99	60
6	s-PIL-2	>99	72
7	HPIL-1	>99	84
8	HPIL-2	>99	87
9	HPIL-3	>99	90
10	HPIL-4	>99	94
11	HPIL-5	>99	91
12	HPIL-6	>99	93
13	HPIL-7	>99	94
14	HPIL-8	>99	95
15	HPIL-9	>99	92
16	HPIL-10	>99	90
17[c]	HPIL-7	>99	99
18 [c,d]	HPIL-7	>99	97

Reaction conditions.
[a]PO (0.83 g, 14.3 mmol), catalysts (1.5 mol% of PO, the dosage was according to the EA analysis, see in ESI), 105°C, 2 MPa CO_2, 3 h.
[b]Based on GC analysis.
[c]Temperature = 120°C, 1.5 MPa, t = 3 h.
[d]The catalyst were used for 5 times.

6.5.2.5 Separations

ILs have been successfully used as adsorbents in the field of separation. However, the high viscosity of IL, IL regeneration, and other problems remain to be addressed [197]. The PILs offer the opportunity to solve these problems by combining the properties of ILs and polymers and can be used as a matrix for polymer membranes. For example, Tomé and Marrucho et al. prepared three PILs based on the poly(diallyldimethylammonium) cation with [TFSAM], [TSAC], and [FSI] anions to construct the composite membranes with IL for CO_2/H_2 separation. It is found that the PIL TFSAM-40 IL TFSAM and PIL FSI-40 IL FSI membranes with 40 wt.% IL exhibit excellent CO_2/H_2 permselectivities compared to the conventional membrane containing [TFSI]$^-$ [198]. Husson et al. grafted the PIL (poly [2-(methacryloxy) ethyl] trimethyl ammonium chloride (P[(META)$^+$Cl$^-$])) onto cellulose utilizing the surface-initiated ATRP technology, followed by the ion exchange reaction to obtain three PIL-modified celluloses (P [(META)$^+$X$^-$]-cellulose X = BF$_4^-$, CF$_3$SO$_3^-$, and CH$_3$SO$_3^-$). PIL-modified cellulose has been successfully used for adsorption and CO_2/N_2 separation. It has been shown that the adsorption capacity of modified cellulose on CO_2 is significantly enhanced. For example, at 25°C and 0.078 MPa, the adsorption capacity of P[(META)$^+$CF$_3$SO$_3$$^-$]-cellulose reaches 2.0 mmol CO$_2$/g IL, but has little adsorption on N$_2$, which shows good selectivity for CO$_2$ separation [199].

6.5.2.6 Poly (ionic liquid)-based antibacterial materials

PILs are derived from ionic liquids and contain ionic liquid structural units that contain positively charged groups that can interact with negatively charged bacterial surfaces, causing the hydrophobic chain segments to penetrate the bacterial lipid membrane, leading to bacterial collapse, rupture, and death and exhibit significant bactericidal effects [200,201]. Yan's group conducted extensive research in the area of PIL-based antimicrobial materials [7,200,202−206]. For example, Yan et al. fabricated a PIL-based multilayer nanofiber skin with double gradients by designing the chemical structure of PIL and tuning the parameters of the electrostatic spinning process. It was found that the PIL skin achieves directional sweat transport while ensuring breathability due to the wetting gradient and pore size gradient. At the same time, the carboxylated carbon nanotubes sprayed on the nanofiber membrane provide a stable electrical conductivity due to the electrostatic interaction with the PIL, enabling the collection and monitoring of bioelectrical signals. In addition, the intrinsic broad-spectrum antimicrobial properties of PIL effectively inhibit the growth of microorganisms and prevent bacterial infections, providing users with a comfortable and safe environment to use the device, showing promising applications in wearable devices, health monitoring, and human−computer interaction [200]. In addition, the team also synthesized imidazolium-type PIL membranes with high antimicrobial activity by in situ photo-cross-linking of ionic liquid monomers and anionic exchange reaction with amino acids (L-proline or L-tryptophan) to investigate the effect of anions on antibacterial properties. The polymeric membrane showed a high level of antimicrobial activity against both gram-negative *Escherichia coli* and gram-positive *Staphylococcus aureus* and did not exhibit any significant hemolysis or cytotoxicity against human erythrocytes or skin fibroblasts [202].

References

[1] He Z, Alexandridis P. Ionic liquid and nanoparticle hybrid systems: emerging applications. Advances in Colloid and Interface Science 2017;244:54−70.

[2] Kataria J, Devi P, Rani P. Importance of structures and interactions in ionic liquid-nanomaterial composite systems as a novel approach for their utilization in safe lithium metal batteries: a review. Journal of Molecular Liquids 2021;339:116736.

[3] Hejazifar M, Lanaridi O, Bica-Schröder K. Ionic liquid based microemulsions: a review. Journal of Molecular Liquids 2020;303:112264.

[4] Ueno K, et al. Effect of cation and anion sizes of additive ionic liquid on the crystal structure of poly (vinylidene fluoride) nanofiber. RSC Advances 2023;13(18):12000−8.

[5] Llaver M, Coronado EA, Wuilloud RG. High performance preconcentration of inorganic Se species by dispersive micro-solid phase extraction with a nanosilica-ionic liquid hybrid material. Spectrochimica Acta Part B: Atomic Spectroscopy 2017;138:23−30.

[6] Sabbaghan M, Nadafan M. Annealing in ionic liquid for synthesis of ZnO nanostructures and its effect on linear-nonlinear optical properties. Optical Materials 2023;139:113758.

[7] Zhang T, Guo J, Ding Y, et al. Redox-responsive ferrocene-containing poly (ionic liquid) s for antibacterial applications. Science China Chemistry 2019;62:95−104.

[8] Moumene T, Kadari M, Belarbi EH, et al. Effect of dicationic ionic liquid: Trimethylene bis-methylimidazolium bromide ($[M(CH_2)_3IM^{2+}][2Br^-]$) on the structural, optical and morphological properties of ZnO nanoparticles. Journal of Molecular Liquids 2023;382:122007.

[9] Maneewattanapinyo P, Pichayakorn W, Monton C, et al. Effect of ionic liquid on silver-nanoparticle-complexed Ganoderma applanatum and its topical film formulation. Pharmaceutics 2023;15(4):1098.

[10] Pang C, Liu J, Peng R, et al. Liquid-phase exfoliation of titanium disulfide nanosheets in aqueous ionic liquid solutions for highly efficient CO_2 electroreduction. Journal of Molecular Liquids 2023;381:121814.

[11] Zhao HL, Yao KS, Wang N, et al. Poly (ionic liquid)-mediated green synthesis of 3D AuPt flower-like nanoballs with composition-dependent SERS sensitivity and catalytic activity. Journal of Molecular Liquids 2023;381:121823.

[12] Baker GA, Rachford AA, Castellano FN, et al. Ranking solvent interactions and dielectric constants with [Pt (mesBIAN)(tda)]: a cautionary tale for polarity determinations in ionic liquids. Chemphyschem: a European Journal of Chemical Physics and Physical Chemistry 2013;14(5):1025−30.

[13] Krishnan D, Schill L, Axet MR, et al. Ruthenium nanoparticles stabilized with methoxy-functionalized ionic liquids: synthesis and structure−performance relations in styrene hydrogenation. Nanomaterials 2023;13(9):1459.

[14] Wang J, Ju Y. Detection of Dichlorvos based on Ni/Cu-MOF@Au/Ionic liquid a novel enzyme sensor. Food Science and Technology 2023;48.

[15] Guo X, Peng Z, Traitangwong A, et al. Ru nanoparticles stabilized by ionic liquids supported onto silica: highly active catalysts for low-temperature CO_2 methanation. Green Chemistry 2018;20 (21):4932−45.

[16] Mondal A, Das A, Adhikary B, et al. Palladium nanoparticles in ionic liquids: reusable catalysts for aerobic oxidation of alcohols. Journal of Nanoparticle Research 2014;16:1−10.

[17] Hong GH, Oh JH, Ji D, et al. Activated copper nanoparticles by 1-butyl-3-methyl imidazolium nitrate for CO_2 separation. Chemical Engineering Journal 2014;252:263−6.

[18] Dhar A, Kumar NS, Khimani M, et al. Silica-immobilized ionic liquid Brønsted acids as highly effective heterogeneous catalysts for the isomerization of n-heptane and n-octane. RSC Advances 2020;10(26):15282−92.

[19] Song P, Liu L, Feng JJ, et al. Poly (ionic liquid) assisted synthesis of hierarchical gold-platinum alloy nanodendrites with high electrocatalytic properties for ethylene glycol oxidation and oxygen reduction reactions. International Journal of Hydrogen Energy 2016;41 (32):14058−67.

[20] Shi YC, Chen SS, Feng JJ, et al. Dicationic ionic liquid mediated fabrication of Au@ Pt nanoparticles supported on reduced graphene oxide with highly catalytic activity for oxygen reduction and hydrogen evolution. Applied Surface Science 2018;441:438−47.

[21] Yang H, Dai H, Wan X, et al. Simultaneous determination of multiple mycotoxins in corn and wheat by high efficiency extraction and purification based on polydopamine and ionic liquid bifunctional nanofiber mat. Analytica Chimica Acta 2023;1267:341361.

[22] Khaliq A, Nazir R, Khan M, et al. Co-doped CeO_2/activated C nanocomposite functionalized with ionic liquid for colorimetric biosensing of H_2O_2 via peroxidase mimicking. Molecules (Basel, Switzerland) 2023;28(8):3325.

[23] Tao C, Xu J, Shi S, et al. Environmental-friendly nanocomposite attapulgite modified by β-cyclodextrin and ionic liquid for the adsorption of thiamethoxam. Journal of Water Process Engineering 2023;53:103838.

[24] Liévano JFP, Díaz LAC. Synthesis and characterization of 1-methyl-3-methoxysilyl propyl imidazolium chloride−mesoporous silica composite as adsorbent for dehydration in industrial processes. Materials Research 2016;19:534−41.

[25] Grygiel K, Wicklein B, Zhao Q, et al. Omnidispersible poly (ionic liquid)-functionalized cellulose nanofibrils: surface grafting and polymer membrane reinforcement. Chemical Communications 2014;50(83):12486−9.

[26] Curreri AM, Mitragotri S, Tanner EEL. Recent advances in ionic liquids in biomedicine. Advanced Science 2021;8(17):2004819.

[27] Duczinski R, Polesso BB, Bernard FL, et al. Enhancement of CO_2/N_2 selectivity and CO_2 uptake by tuning concentration and chemical structure of imidazolium-based ILs immobilized in mesoporous silica. Journal of Environmental Chemical Engineering 2020;8(3):103740.

[28] Xie K, Dong Z, Zhai M, et al. Radiation-induced surface modification of silanized silica with n-alkyl-imidazolium ionic liquids and their applications for the removal of ReO_4^- as an analogue for TcO_4^-. Applied Surface Science 2021;551:149406.

[29] Falahati M, Soleimani M, Aflatouni F. Separation and preconcentration of anionic dyes using magnetic nanoparticles with modify polymer ionic liquid. American Journal of Heterocyclic Chemistry 2023;9(1):1−8.

[30] Kou X, Ma Y, Pan C, et al. Effects of the cationic structure on the adsorption performance of ionic polymers toward Au (III): an experimental and DFT study. Langmuir: The ACS Journal of Surfaces and Colloids 2022;38(19):6116−27.

[31] Qiu X, Qin J, Xu M, et al. Organic-inorganic nanocomposites fabricated via functional ionic liquid as the bridging agent for Laccase immobilization and its application in 2,4-dichlorophenol removal. Colloids and Surfaces B: Biointerfaces 2019;179:260−9.

[32] Wang G, Yu N, Peng L, et al. Immobilized chloroferrate ionic liquid: an efficient and reusable catalyst for synthesis of diphenylmethane and its derivatives. Catalysis Letters 2008;123:252−8.

[33] Zhang W, Wang H, Han J, et al. Multifunctional mesoporous materials with acid−base frameworks and ordered channels filled with ionic liquid: synthesis, characterization and catalytic performance of Ti−Zr-SBA-15-IL. Applied Surface Science 2012;258(16):6158−68.

[34] Rafiee E, Shahebrahimi S. Organic-inorganic hybrid polyionic liquid based polyoxometalate as nano porous material for selective oxidation of sulfides. Journal of Molecular Structure 2017;1139:255−63.

[35] Yao BJ, Ding LG, Li F, et al. Chemically cross-linked MOF membrane generated from imidazolium-based ionic liquid-decorated UiO-66 type NMOF and its application toward CO_2 separation and conversion. ACS Applied Materials & Interfaces 2017;9(44):38919−30.

[36] Chen Y, Sadeghzadeh SM. Dendritic fibrous nano-titanium (DFNT) with highly dispersed poly (ionic liquids) as a nanocatalyst for synthesis of dimethyl carbonate from methanol and carbon dioxide. Journal of Molecular Liquids 2023;122201.

[37] Taheri K, Elhamifar D, Kargar S, et al. Graphene oxide supported ionic liquid/Fe complex: a robust and highly stable nanocatalyst. RSC Advances 2023;13(24):16067−77.

[38] Jia H, Chen X, Liu CY, et al. Ultrafine palladium nanoparticles anchoring graphene oxide-ionic liquid grafted chitosan self-assembled materials: the novel organic-inorganic hybrid catalysts for hydrogen generation in hydrolysis of ammonia borane. International Journal of Hydrogen Energy 2018;43(27):12081−90.

[39] Nosov DR, Ronnasi B, Lozinskaya EI, et al. Mechanically robust poly (ionic liquid) block copolymers as self-assembling gating materials for single-walled carbon-nanotube-based thin-film transistors. ACS Applied Polymer Materials 2023;5(4):2639−53.

[40] Galán-Cano F, del Carmen Alcudia-León M, Lucena R, et al. Ionic liquid coated magnetic nanoparticles for the gas chromatography/mass spectrometric determination of polycyclic aromatic hydrocarbons in waters. Journal of Chromatography. A 2013;1300:134−40.

[41] Song X, Li D, Li R, et al. Ionic liquid modified inorganic nanoparticles for gaseous phenol adsorption. Journal of Wuhan University of Technology-Mater. Sci. Ed. 2019;34:787−90.

[42] Ogomi Y, Kato T, Hayase S. Dye sensitized solar cells consisting of ionic liquid and solidification. Journal of Photopolymer Science and Technology 2006;19(3):403−8.

[43] Smarsly B, Kuang D, Antonietti M. Making nanometer thick silica glass scaffolds: an experimental approach to learn about size effects in glasses. Colloid and Polymer Science 2004;282:892−900.

[44] Raucci MG, Fasolino I, Pastore SG, et al. Antimicrobial imidazolium ionic liquids for the development of minimal invasive calcium phosphate-based bionanocomposites. ACS Applied Materials & Interfaces 2018;10(49):42766−76.

[45] Zhu H. Preparation and dye absorption properties of Fe_3O_4/SiO_2 composite. Chinese Journal of Inorganic Chemistry. 2022;34:1649−54.

[46] Ghasemi S, Setayesh SR, Habibi-Yangjeh A, et al. Assembly of $CeO_2−TiO_2$ nanoparticles prepared in room temperature ionic liquid on graphene nanosheets for photocatalytic degradation of pollutants. Journal of Hazardous Materials 2012;199:170−8.

[47] Al Kiey SA, Sery AA, Farag HK. Sol-gel synthesis of nanostructured cobalt oxide in four different ionic liquids. Journal of Sol-Gel Science and Technology 2023;106(1):37−43.

[48] Hara S, Ishizu M, Watanabe S, et al. Improvement of the transparency, mechanical, and shape memory properties of polymethylmethacrylate/titania hybrid films using tetrabutylphosphonium chloride. Polymer Chemistry 2019;10(35):4779−88.

[49] Ikake H, Hara S, Kurebayashi S, et al. Development of a magnetic hybrid material capable of photoinduced phase separation of iron chloride by shape memory and photolithography. Journal of Materials Chemistry C 2022;10(20):7849−56.

[50] Hara S, Kurebayashi S, Sanae G, et al. Polycarbonate/titania hybrid films with localized photoinduced magnetic-phase transition. Nanomaterials 2020;11(1):5.

[51] Wang X, Li X, Aya S, et al. Reversible switching of the magnetic orientation of titanate nanosheets by photochemical reduction and autoxidation. Journal of the American Chemical Society 2018;140 (48):16396−401.

[52] Wang S, Hsia B, Carraro C, et al. High-performance all solid-state micro-supercapacitor based on patterned photoresist-derived porous carbon electrodes and an ionogel electrolyte. Journal of Materials Chemistry A 2014;2(21):7997−8002.

[53] Néouze MA, Le Bideau J, Gaveau P, et al. Ionogels, new materials arising from the confinement of ionic liquids within silica-derived networks. Chemistry of Materials 2006;18(17):3931−6.

[54] He Z, Alexandridis P. Nanoparticles in ionic liquids: interactions and organization. Physical Chemistry Chemical Physics 2015;17(28):18238−61.

[55] Akter M, Faisal MA, Singh AK, et al. Hydrophilic ionic liquid assisted hydrothermal synthesis of ZnO nanostructures with controllable morphology. RSC Advances 2023;13(26):17775−86.

[56] Xiao Y, Chen F, Zhu X, et al. Ionic liquid-assisted formation of lanthanide metal-organic framework nano/microrods for superefficient removal of Congo red. Chemical Research in Chinese Universities 2015;31(6):899−903.

[57] Zhang P, Yuan J, Fellinger TP, et al. Improving hydrothermal carbonization by using poly (ionic liquid)s. Angewandte Chemie (International ed. in English) 2013;52(23):6028−32.

[58] Xia J, Yin S, Li H, et al. Improved visible light photocatalytic activity of sphere-like BiOBr hollow and porous structures synthesized via a reactable ionic liquid. Dalton Transactions 2011;40 (19):5249−58.

[59] Gong J, Yang F, Shao Q, et al. Microwave absorption performance of methylimidazolium ionic liquids: towards novel ultra-wideband metamaterial absorbers. RSC Advances 2017;7(67):41980−8.

[60] Gangaraju D, Shanmugharaj AM, Sridhar V. Graphene oxide facilitates transformation of waste PET into MOF nanorods in ionic liquids. Polymers 2023;15(11):2479.

[61] Du CF, Li JR, Huang XY. Microwave-assisted ionothermal synthesis of SnSex nanodots: a facile precursor approach towards $SnSe_2$ nanodots/graphene nanocomposites. RSC Advances 2016;6(12):9835−42.

[62] Yan H, Ren Y, Zhou G, et al. Aqueous-phase synthesis of heterojunction molecular imprinted photocatalytic nanoreactor via stable interaction forces for improved selectivity and photocatalytic property. Applied Surface Science 2022;579:152174.

[63] Nguyen VH, Shim JJ. Ionic liquid mediated synthesis of graphene−TiO_2 hybrid and its photocatalytic activity. Materials Science and Engineering: B 2014;180:38−45.

[64] Wang J, Cao J, Fang B, et al. Synthesis and characterization of multipod, flower-like, and shuttle-like ZnO frameworks in ionic liquids. Materials Letters 2005;59(11):1405−8.

[65] Zhang GL, Gao XM, Xu XD. Microwave-assisted synthesis of nanocrystalline zirconium dioxide using an ionic liquid. Applied Mechanics and Materials 2013;271:255−8.

[66] Liu YH, Liu PI, Chung LC, et al. Diverse effects of microwave heating on anatase crystallization in ionothermal synthesis of nanostructured TiO$_2$. Journal of Materials Science 2011;46:4826−31.

[67] Chen L, Zhang T, Cheng H, et al. A microwave assisted ionic liquid route to prepare bivalent Mn$_5$O$_8$ nanoplates for 5-hydroxymethylfurfural oxidation. Nanoscale 2020;12(34):17902−14.

[68] Li X, Liu M, Cheng H, et al. Development of ionic liquid assisted-synthesized nano-silver combined with vascular endothelial growth factor as wound healing in the care of femoral fracture in the children after surgery. Journal of Photochemistry and Photobiology B: Biology 2018;183:385−90.

[69] Xiao D, Yuan D, He H, et al. Microwave assisted one-step green synthesis of fluorescent carbon nanoparticles from ionic liquids and their application as novel fluorescence probe for quercetin determination. Journal of Luminescence 2013;140:120−5.

[70] Lorbeer C, Cybinska J, Mudring AV. Europium (III) fluoride nanoparticles from ionic liquids: structural, morphological, and luminescent properties. Crystal Growth & Design 2011;11(4):1040−8.

[71] Lee JH, Cho KK, Lee JR, et al. Manganese fluoride nanoparticles synthesized by microwave irradiation using ionic liquid−ethylene glycol mixtures: room-temperature photoluminescence, crystalline phase, and morphology. Crystal Growth & Design 2021;21(3):1406−12.

[72] Cao J, Fang B, Wang H, et al. Synthesis of needle-like pyramid ZnF(OH) microstructures by microwave-assisted heating in ionic liquids. 231st National Meeting of the American-Chemical-Society 2006;.

[73] Jiang Y, Zhu YJ. Microwave-assisted synthesis of sulfide M$_2$S$_3$ (M = Bi, Sb) nanorods using an ionic liquid. The Journal of Physical Chemistry. B 2005;109(10):4361−4.

[74] Wu Z, Long YF, Lv XY, et al. Microwave heating synthesis of spindle-like LiMnPO$_4$/C in a deep eutectic solvent. Ceramics International 2017;43(8):6089−95.

[75] Li J, Zhang J, Han B, et al. Ionic liquid-in-ionic liquid nanoemulsions. Chemical Communications 2012;48(85):10562−4.

[76] Guo Y, He D, Xia S, et al. Preparation of a novel nanocomposite of polyaniline core decorated with anatase-TiO$_2$ nanoparticles in ionic liquid/water microemulsion. Journal of Nanomaterials 2012;2012:8.

[77] Sun X, Qiang Q, Yin Z, et al. Monodispersed silver-palladium nanoparticles for ethanol oxidation reaction achieved by controllable electrochemical synthesis from ionic liquid microemulsions. Journal of Colloid and Interface Science 2019;557:450−7.

[78] Zhang G, Zhou H, Hu J, et al. Pd nanoparticles catalyzed ligand-free Heck reaction in ionic liquid microemulsion. Green Chemistry 2009;11(9):1428−32.

[79] Wu LG, Shen J, Du CH, et al. Development of AgCl/poly (MMA-co-AM) hybrid pervaporation membranes containing AgCl nanoparticles through synthesis of ionic liquid microemulsions. Separation and Purification Technology 2013;114:117−25.

[80] Wang X, Cheng J, Ji G, et al. Starch nanoparticles prepared in a two ionic liquid based microemulsion system and their drug loading and release properties. RSC Advances 2016;6(6):4751−7.

[81] Xu J, Cui Y, Wang R, et al. Mesoporous La-based nanorods synthesized from a novel IL-SFME for phosphate removal in aquatic systems. Colloids and Surfaces A: Physicochemical and Engineering Aspects 2021;624:126689.

[82] Mirhoseini F, Salabat A. Ionic liquid based microemulsion method for the fabrication of poly (methyl methacrylate)−TiO$_2$ nanocomposite as a highly efficient visible light photocatalyst. RSC Advances 2015;5(17):12536−45.

[83] Demirkurt B, Cakan-Akdogan G, Akdogan Y. Preparation of albumin nanoparticles in water-in-ionic liquid microemulsions. Journal of Molecular Liquids 2019;295:111713.

[84] Zhan T, Zhang Y, Yang Q, et al. Ultrathin layered double hydroxide nanosheets prepared from a water-in-ionic liquid surfactant-free microemulsion for phosphate removal from aquatic systems. Chemical Engineering Journal 2016;302:459−65.

[85] Cejka J. Recent trends in the synthesis of molecular sieves. Studies in Surface Science and Catalysis 2005;157:111−34.

[86] Li Y, Yu J. Emerging applications of zeolites in catalysis, separation and host−guest assembly. Nature Reviews Materials 2021;6(12):1156−74.

[87] Wilson ST, Lok BM, Messina CA, et al. Aluminophosphate molecular sieves: a new class of microporous crystalline inorganic solids. Journal of the American Chemical Society 1982;104(4):1146−7.

[88] Rong Y, Zhang X, Wang H, et al. Imidazolium salts facilitate mechanochemical synthesis of well-dispersed MFI zeolite crystals with c-axis orientation. Microporous and Mesoporous Materials 2022;341:112094.

[89] Rogers RD, Seddon KR. Ionic liquids—solvents of the future? Science (New York, N.Y.) 2003;302(5646):792—3.

[90] Seddon KR. A taste of the future. Nature Materials 2003;2(6):363—5.

[91] Cooper ER, Andrews CD, Wheatley PS, et al. Ionic liquids and eutectic mixtures as solvent and template in synthesis of zeolite analogues. Nature 2004;430(7003):1012—16.

[92] Morris RE. Ionothermal synthesis—ionic liquids as functional solvents in the preparation of crystalline materials. Chemical Communications 2009;21:2990—8.

[93] Lobo RF, Zones SI, Davis ME. Structure-direction in zeolite synthesis. Journal of Inclusion Phenomena and Molecular recognition in Chemistry 1995;21(1—4):47—78.

[94] Parnham ER, Morris RE. Ionothermal synthesis of zeolites, metal—organic frameworks, and inorganic—organic hybrids. Accounts of Chemical Research 2007;40(10):1005—13.

[95] Ma Z, Yu J, Dai S. Preparation of inorganic materials using ionic liquids. Advanced Materials 2010;22(2):261—85.

[96] Li X, Choi J, Ahn WS, et al. Preparation and application of porous materials based on deep eutectic solvents. Critical Reviews in Analytical Chemistry 2018;48(1):73—85.

[97] Zhang T, Doert T, Wang H, et al. Inorganic synthesis based on reactions of ionic liquids and deep eutectic solvents. Angewandte Chemie International Edition 2021;60(41):22148—65.

[98] Azim MM, Stark A. Ionothermal synthesis and characterisation of Mn-, Co-, Fe-and Ni-containing aluminophosphates. Microporous and Mesoporous Materials 2018;272:251—9.

[99] Musa M, Dawson DM, Ashbrook SE, et al. Ionothermal synthesis and characterization of CoAPO-34 molecular sieve. Microporous and Mesoporous Materials 2017;239:336—41.

[100] Benin AI, Hwang SJ, Zones SI, et al. Structural and compositional studies of crystalline MAPO molecular sieves formed in ionothermal synthesis using imidazolium bromides. Microporous and Mesoporous Materials 2019;274:257—65.

[101] Yasong W, Yunpeng X, Zhijian T, et al. Research progress in ionothermal synthesis of molecular sieves. Chinese Journal of Catalysis 2012;33(1):39—50.

[102] Li J, Yu J, Xu R. Progress in heteroatom-containing aluminophosphate molecular. Proceedings of the Royal Society A: Mathematical, Physical and Engineering Sciences 2012;468(2143):1955—67.

[103] Ma H, Tian Z, Xu R, et al. Effect of water on the ionothermal synthesis of molecular sieves. Journal of the American Chemical Society 2008;130(26):8120—1.

[104] Wang L, Xu Y, Wei Y, et al. Structure-directing role of amines in the ionothermal synthesis. Journal of the American Chemical Society 2006;128(23):7432—3.

[105] Fayad EJ, Bats N, Kirschhock CEA, et al. A rational approach to the ionothermal synthesis of an AlPO4 molecular sieve with an LTA-type framework. Angewandte Chemie International Edition 2010;49(27):4585—8.

[106] Drylie EA, Wragg DS, Parnham ER, et al. Ionothermal synthesis of unusual choline-templated cobalt aluminophosphates. Angewandte Chemie 2007;46(41):7839—43.

[107] Harrison WTA. $C_5H_{14}NO \cdot ZnCl (HPO_3)$, an unexpected product from an ionic-liquid synthesis. Inorganic Chemistry Communications 2007;10(7):833—5.

[108] Xu YP, Tian ZJ, Wang SJ, et al. Microwave-enhanced ionothermal synthesis of aluminophosphate molecular sieves. Angewandte Chemie International Edition 2006;45(24):3965—70.

[109] Cai R, Sun M, Chen Z, et al. Ionothermal synthesis of oriented zeolite AEL films and their application as corrosion-resistant coatings. Angewandte Chemie International Edition 2008;47(3):525—8.

[110] Zhao X, Zhang X, Hao Z, et al. Synthesis of FeAPO-5 molecular sieves with high iron contents via improved ionothermal method and their catalytic performances in phenol hydroxylation. Journal of Porous Materials 2018;25:1007—16.

[111] Zhao X, Duan W, Wang Q, et al. Microwave-assisted ionothermal synthesis of Fe-LEV molecular sieve with high iron content in low-dosage of eutectic mixture. Microporous and Mesoporous Materials 2019;275:253—62.

[112] Parnham ER, Morris RE. The ionothermal synthesis of cobalt aluminophosphate zeolite frameworks. Journal of the American Chemical Society 2006;128(7):2204−5.

[113] Lin QF, Gao ZR, Lin C, et al. A stable aluminosilicate zeolite with intersecting three-dimensional extra-large pores. Science (New York, N.Y.) 2021;374(6575):1605−8.

[114] Wei Y, Tian Z, Gies H, et al. Ionothermal synthesis of an aluminophosphate molecular sieve with 20-ring pore openings. Angewandte Chemie 2010;49(31):5367−70.

[115] Wheatley PS, Allan PK, Teat SJ, et al. Task specific ionic liquids for the ionothermal synthesis of siliceous zeolites. Chemical Science 2010;1(4):483−7.

[116] Wu Q, Hong X, Zhu L, et al. Generalized ionothermal synthesis of silica-based zeolites. Microporous and Mesoporous Materials 2019;286:163−8.

[117] Jin K, Huang X, Pang L, et al. [Cu(I)(bpp)][BF₄]: the first extended coordination network prepared solvothermally in an ionic liquid solvent. Chemical Communications 2002;23:2872−3.

[118] Dybtsev DN, Chun H, Kim K. Three-dimensional metal−organic framework with (3,4)- connected net, synthesized from an ionic liquid medium. Chemical communications 2004;14:1594−5.

[119] (a) Parnham ER, Wheatley PS, Morris RE. The ionothermal synthesis of SIZ-6-a layered aluminophosphate. Chemical Communications 2006;4:380−2.

       (b) Taubert A, Li Z. Inorganic materials from ionic liquids. Dalton Transactions 2007;7:723−7.

       (c) Liu H, Wang F, Jia XY, et al. Synthesis, characterization, and 1,3-butadiene polymerization studies of Co (II), Ni (II), and Fe (II) complexes bearing 2-(N-arylcarboximidoylchloride) quinoline ligand. Journal of Molecular Catalysis A: Chemical 2014;391:25−35.

[120] Xu L, Kwon YU, de Castro B, et al. Novel Mn (II)-based metal−organic frameworks isolated in ionic liquids. Crystal Growth & Design 2013;13(3):1260−6.

[121] Liu QY, Li YL, Wang YL, et al. Ionothermal synthesis of a 3D dysprosium−1, 4-benzenedicarboxylate framework based on the 1D rod-shaped dysprosium−carboxylate building blocks exhibiting slow magnetization relaxation. CrystEngComm 2014;16(3):486−91.

[122] Xu L, Choi EY, Kwon YU. Ionothermal synthesis of a 3D Zn−BTC metal-organic framework with distorted tetranuclear [Zn₄ (μ₄-O)] subunits. Inorganic Chemistry Communications 2008;11(10):1190−3.

[123] Liao JH, Huang WC. Ionic liquid as reaction medium for the synthesis and crystallization of a metal-organic framework: (BMIM)₂[Cd₃(BDC)₃Br₂] (BMIM = 1-butyl-3-methylimidazolium, BDC = 1,4-benzenedicarboxylate). Inorganic Chemistry Communications 2006;9(12):1227−31.

[124] Himeur F, Stein I, Wragg DS, et al. The ionothermal synthesis of metal organic frameworks, Ln (C₉O₆H₃)((CH₃NH)₂CO)₂, using deep eutectic solvents. Solid State Sciences 2010;12(4):418−21.

[125] Lin Z, Wragg DS, Warren JE, et al. Anion control in the ionothermal synthesis of coordination polymers. Journal of the American Chemical Society 2007;129(34):10334−5.

[126] Luo Q, Han Y, Lin H, et al. Formation of Gd coordination polymer with 1D chains mediated by Brønsted acidic ionic liquids. Journal of Solid State Chemistry 2017;247:137−41.

[127] (a) Fei Z, Zhao D, Geldbach TJ, et al. Brønsted acidic ionic liquids and their zwitterions: synthesis, characterization and pKa determination. Chemistry—A European Journal 2004;10(19):4886−93.

       (b) Kulkarni PS, Branco LC, Crespo JG, et al. Comparison of physicochemical properties of new ionic liquids based on imidazolium, quaternary ammonium, and guanidinium cations. Chemistry—A European Journal 2007;13(30):8478−88.

[128] Farger P, Guillot R, Leroux F, et al. Imidazolium dicarboxylate based metal-organic frameworks obtained by solvo-ionothermal reaction. European Journal of Inorganic Chemistry 2015;32:5342−50.

[129] Fei Z, Ang WH, Geldbach TJ, et al. Ionic solid-state dimers and polymers derived from imidazolium dicarboxylic acids. Chemistry—A European Journal 2006;12(15):4014−20.

[130] Fei Z, Geldbach TJ, Zhao D, et al. A nearly planar water sheet sandwiched between strontium−imidazolium carboxylate coordination polymers. Inorganic Chemistry 2005;44 (15):5200−2.

[131] Fei Z, Geldbach TJ, Scopelliti R, et al. Metal-Organic frameworks derived from imidazolium dicarboxylates and group I and II salts. Inorganic Chemistry 2006;45(16):6331−7.

[132] Wang XW, Han L, Cai TJ, et al. A novel chiral doubly folded interpenetrating 3D metal-organic framework based on the flexible zwitterionic ligand. Crystal Growth & Design 2007;7 (6):1027−30.

[133] Lin Z, Slawin AMZ, Morris RE. Chiral induction in the ionothermal synthesis of a 3-D coordina-tion polymer. Journal of the American Chemical Society 2007;129(16):4880—1.

[134] Wang X, Li XB, Yan RH, et al. Diverse manganese (II) coordination polymers derived from achi-ral/chiral imidazolium-carboxylate zwitterions and azide: Structure and magnetic properties. Dalton Transactions 2013;42(27):10000—10.

[135] Abrahams BF, Maynard-Casely HE, Robson R, et al. Copper (ii) coordination polymers of imdc$^-$(H$_2$imdc$^+$ = the 1,3-bis (carboxymethyl) imidazolium cation): unusual sheet interpenetration and an unexpected single crystal-to-single crystal transformation. CrystEngComm 2013;15 (45):9729—37.

[136] Fei Z, Zhao D, Geldbach TJ, et al. A synthetic zwitterionic water channel: characterization in the solid state by X-ray crystallography and NMR spectroscopy. Angewandte Chemie International Edition 2005;44(35):5720—5.

[137] Martin NP, Falaise C, Volkringer C, et al. Hydrothermal crystallization of uranyl coordination polymers involving an imidazolium dicarboxylate ligand: effect of pH on the nuclearity of uranyl-centered subunits. Inorganic Chemistry 2016;55(17):8697—705.

[138] Han L, Zhang S, Wang Y, et al. A strategy for synthesis of ionic metal-organic frameworks. Inorganic Chemistry 2009;48(3):786—8.

[139] Chai XC, Sun YQ, Lei R, et al. A series of lanthanide frameworks with a flexible ligand, N, N'-diacetic acid imidazolium, in different coordination modes. Crystal Growth & Design 2010;10 (2):658—68.

[140] Zhong S, Yin Q, Diao Y, et al. Optimization of synthesis conditions, characterization and mag-netic properties of lanthanide metal organic frameworks from Brønsted acidic ionic liquid. Journal of Molecular Structure 2023;1278:134974.

[141] Yin Q, Sun X, Dong K, et al. Dual-emitting ratiometric luminescent thermometers based on lan-thanide metal-organic complexes with brønsted acidic ionic liquids. Inorganic Chemistry 2022;61 (47):18998—9009.

[142] Liu M, Yin Q, Zhong S, et al. Two novel rare earth coordination polymers derived from zwitter-ionic 1,3-Bis (1-carboxylatoethyl) imidazolium bromide: structures and magnetic properties. Journal of Molecular Structure 2022;1250:131665.

[143] Sun H, Liu M, Fu X, et al. Solvothermal synthesis and conformation probe of novel europium complex of brønsted acidic ionic liquid: 1,3-Bis (1-carboxylatoethyl) imidazolium bromide. Zeitschrift Für Anorganische und Allgemeine Chemie 2021;648:97—105.

[144] Wang Y, Fu X, Zhang T, et al. Solvothermal syntheses and properties of europium metal organic framework with 1, 3-bis (carboxymethyl) imidazolium chloride ionic liquid. Journal of Molecular Structure 2020;1200:127081.

[145] Wang Y, Fu X, Liu S, et al. A new gadolinium complex with 1, 3-bis (carboxymethyl) imidazo-lium chloride ionic liquid: solvothermal synthesis, structure and magnetic properties. Journal of Molecular Structure 2020;1217:128340.

[146] Welton T. Room-temperature ionic liquids. Solvents for synthesis and catalysis. Chemical Reviews 1999;99(8):2071—84.

[147] Lu J, Yan F, Texter J. Advanced applications of ionic liquids in polymer science. Progress in Polymer Science 2009;34(5):431—48.

[148] MacFarlane DR, Forsyth M, Howlett PC, et al. Ionic liquids and their solid-state analogues as materials for energy generation and storage. Nature Reviews Materials 2016;1(2):1—15.

[149] Ohno H, Ito K. Room-temperature molten salt polymers as a matrix for fast ion conduction. Chemistry Letters 1998;27(8):751—2.

[150] Qian W, Texter J, Yan F. Frontiers in poly (ionic liquid)s: syntheses and applications. Chemical Society Reviews 2017;46(4):1124—59.

[151] Green O, Grubjesic S, Lee S, et al. The design of polymeric ionic liquids for the preparation of functional materials. Polymer Reviews 2009;49(4):339—60.

[152] Men Y, Drechsler M, Yuan J. Double-stimuli-responsive spherical polymer brushes with a poly (ionic liq-uid) core and a thermoresponsive shell. Macromolecular Rapid Communications 2013;34(21):1721—7.

[153] Yuan J, Mecerreyes D, Antonietti M. Poly (ionic liquid)s: an update. Progress in Polymer Science

2013;38(7):1009−36.

[154] Marcilla R, Alberto Blazquez J, Rodriguez J, et al. Tuning the solubility of polymerized ionic liquids by simple anion-exchange reactions. Journal of Polymer Science Part A: Polymer Chemistry 2004;42(1):208−12.

[155] Yuan J, Antonietti M. Poly (ionic liquid)s: polymers expanding classical property profiles. Polymer 2011;52(7):1469−82.

[156] Sutto TE, Duncan TT. The behavior of Li and Mg ions in a polymerized ionic liquid. Electrochimica Acta 2012;72:23−7.

[157] Yin K, Zhang Z, Yang L, et al. An imidazolium-based polymerized ionic liquid via novel synthetic strategy as polymer electrolytes for lithium ion batteries. Journal of Power Sources 2014;258:150−4.

[158] Mecerreyes D. Polymeric ionic liquids: Broadening the properties and applications of polyelectrolytes. Progress in Polymer Science 2011;36(12):1629−48.

[159] Qian WJ, Guo JN, Yan F. Design, synthesis and application of poly (ionic liquid) based functional materials. Polymer BulletinPolymer Bulletin 2015;10:94−104.

[160] Shaplov AS, Vlasov PS, Armand M, et al. Design and synthesis of new anionic "polymeric ionic liquids" with high charge delocalization. Polymer Chemistry 2011;2(11):2609−18.

[161] Diao H, Yan F, Qiu L, et al. High performance cross-linked poly (2-acrylamido-2-methylpropane-sulfonic acid)-based proton exchange membranes for fuel cells. Macromolecules 2010;43(15):6398−405.

[162] Yoshizawa M, Hirao M, Ito-Akita K, et al. Ion conduction in zwitterionic-type molten salts and their polymers. Journal of Materials Chemistry 2001;11(4):1057−62.

[163] Narita A, Shibayama W, Matsumi N, et al. Novel ion conductive matrix via dehydrocoupling polymerization of imidazolium-type ionic liquid and lithium 9-borabicyclo [3,3,1] nonane hydride. Polymer Bulletin 2006;57:109−14.

[164] Nishimura N, Ohno H. 15th anniversary of polymerised ionic liquids. Polymer 2014;55(16):3289−97.

[165] Mori H, Yahagi M, Endo T. RAFT polymerization of N-vinylimidazolium salts and synthesis of thermoresponsive ionic liquid block copolymers. Macromolecules 2009;42(21):8082−92.

[166] Yuan J, Schlaad H, Giordano C, et al. Double hydrophilic diblock copolymers containing a poly (ionic liquid) segment: controlled synthesis, solution property, and application as carbon precursor. European Polymer Journal 2011;47(4):772−81.

[167] Texter J, Vasantha VA, Crombez R, et al. Triblock copolymer based on poly (propylene oxide) and poly (1-[11-acryloylundecyl]-3-methyl-imidazolium] bromide). Macromolecular Rapid Communications 2012;33(1):69−74.

[168] Vijayakrishna K, Mecerreyes D, Gnanou Y, et al. Polymeric vesicles and micelles obtained by self-assembly of ionic liquid-based block copolymers triggered by anion or solvent exchange. Macromolecules 2009;42(14):5167−74.

[169] Li ZY, Yang Y, Wang J, et al. Synthesis and characterization of mPEG-b-PS-b-PVBMImTFSI triblock ionic liquid copolymer. Engineering Plastics Application Engineering Plastics Applications 2022;50(2):55−60.

[170] Du C, Zhang X, Ma X. The surface tunability and dye separation property of PVDF porous membranes modified by P (MMA-b-MEBIm-Br): effect of poly (ionic liquid) brush lengths. Journal of Polymer Research 2020;27:79.

[171] Wilkes JS. A short history of ionic liquids-from molten salts to neoteric solvents. Green Chemistry 2002;4(2):73−80.

[172] Fu C, Homann G, Grissa R, et al. A polymerized-ionic-liquid-based polymer electrolyte with high oxidative stability for 4 and 5 V class solid-state lithium metal batteries. Advanced Energy Materials 2022;12(27):2200412.

[173] Zhou D, Liu R, Zhang J, et al. In situ synthesis of hierarchical poly (ionic liquid)-based solid electrolytes for high-safety lithium-ion and sodium-ion batteries. Nano Energy 2017;33:45−54.

[174] Song X, Wang C, Chen J, et al. Unraveling the synergistic coupling mechanism of Li^+ transport in an "Ionogel-in-Ceramic" hybrid solid electrolyte for rechargeable lithium metal battery. Advanced Functional Materials 2022;32(10):2108706.

[175] Lin X, Xu S, Tong Y, et al. A self-healing polymerized-ionic-liquid-based polymer electrolyte enables a long lifespan and dendrite-free solid-state Li metal batteries at room temperature. Materials Horizons 2023;10(3):859−68.

[176] Li R, Fang Z, Wang C, et al. Six-armed and dicationic polymeric ionic liquid for highly stretchable, nonflammable and notch-insensitive intrinsic self-healing solid-state polymer electrolyte for flexible and safe lithium batteries. Chemical Engineering Journal 2022;430:132706.

[177] Zhao W, Jiang J, Liu Y, et al. Radiation synthesis strategy of poly (ionic liquid)/MXene gel polymer for supercapacitor electrolyte. Ionics 2023;29(7):2865−75.

[178] Kim TY, Lee HW, Stoller M, et al. High-performance supercapacitors based on poly (ionic liquid)-modified graphene electrodes. ACS Nano 2011;5(1):436−42.

[179] Wang N, Hu J, Gao L, et al. Current progress in solid-state electrolytes for dye-sensitized solar cells: a mini-review. Journal of Electronic Materials 2020;49:7085−97.

[180] Thomas M, Jose S. Electrospun membrane of PVA and functionalized agarose with polymeric ionic liquid and conductive carbon for efficient dye sensitized solar cell. Journal of Photochemistry and Photobiology A: Chemistry 2022;425:113666.

[181] Chen X, Zhao J, Zhang J, et al. Bis-imidazolium based poly (ionic liquid) electrolytes for quasi-solid-state dye-sensitized solar cells. Journal of Materials Chemistry 2012;22(34):18018−24.

[182] Díaz M, Ortiz A, Ortiz I. Progress in the use of ionic liquids as electrolyte membranes in fuel cells. Journal of Membrane Science 2014;469:379−96.

[183] Kim DJ, Jo MJ, Nam SY. A review of polymer−nanocomposite electrolyte membranes for fuel cell application. Journal of Industrial and Engineering Chemistry 2015;21:36−52.

[184] Çelik SÜ, Bozkurt A, Hosseini SS. Alternatives toward proton conductive anhydrous membranes for fuel cells: heterocyclic protogenic solvents comprising polymer electrolytes. Progress in Polymer Science 2012;37(9):1265−91.

[185] Isik M, Porcarelli L, Lago N, et al. Proton Proton conducting membranes based on poly (ionic liquids) having phosphonium counter-cations. Macromolecular Rapid Communications 2018;39(3):1700627.

[186] Chen H, Wang S, Li J, et al. Novel cross-linked membranes based on polybenzimidazole and polymeric ionic liquid with improved proton conductivity for HT-PEMFC applications. Journal of the Taiwan Institute of Chemical Engineers 2019;95:185−94.

[187] Liu F, Wang S, Chen H, et al. Cross-linkable polymeric ionic liquid improve phosphoric acid retention and long-term conductivity stability in polybenzimidazole based PEMs. ACS Sustainable Chemistry & Engineering 2018;6(12):16352−62.

[188] Koyilapu R, Singha S, Kutcherlapati SNR, et al. Grafting of vinylimidazolium-type poly (ionic liquid) on silica nanoparticle through RAFT polymerization for constructing nanocomposite based PEM. Polymer 2020;195:122458.

[189] Pinaud J, Vignolle J, Gnanou Y, et al. Poly (N-heterocyclic-carbene)s and their CO_2 adducts as recyclable polymer-supported organocatalysts for benzoin condensation and transesterification reactions. Macromolecules 2011;44(7):1900−8.

[190] Restrepo J, Lozano P, Burguete MI, et al. Gold nanoparticles immobilized onto supported ionic liquid-like phases for microwave phenylethanol oxidation in water. Catalysis Today 2015;255:97−101.

[191] Liu F, Wang L, Sun Q, et al. Transesterification catalyzed by ionic liquids on superhydrophobic mesoporous polymers: heterogeneous catalysts that are faster than homogeneous catalysts. Journal of the American Chemical Society 2012;134(41):16948−50.

[192] Wang X, Hu H, Chen B, et al. Efficient synthesis of dimethyl carbonate via transesterification of ethylene carbonate catalyzed by swelling poly (ionic liquid) s. Green Chemical Engineering 2021;2(4):423−30.

[193] Duan XQ, Duan GY, Wang YF, et al. Sn-Ag Synergistic effect enhances high-rate electrocatalytic CO_2-to-formate conversion on porous poly (ionic liquid) support. Small (Weinheim an der Bergstrasse, Germany) 2023;19(18):2207219.

[194] Duan GY, Li XQ, Ding GR, et al. Highly efficient electrocatalytic CO_2 reduction to C_{2+} products on a poly (ionic liquid)-based Cu^0-Cu^I tandem catalyst. Angewandte Chemie 2022;61(9): e202110657.

[195] Gou H, Ma X, Su Q, et al. Hydrogen bond donor functionalized poly (ionic liquid)s for efficient synergistic conversion of CO_2 to cyclic carbonates. Physical Chemistry Chemical Physics 2021;23 (3):2005−14.

[196] Jiang Z, Yan D, Xin J, et al. Poly (ionic liquid) s as efficient and recyclable catalysts for methanolysis of PET. Polymer Degradation and Stability 2022;199:109905.

[197] Zunita M, Hastuti R, Alamsyah A, et al. Ionic liquid membrane for carbon capture and separation. Separation & Purification Reviews 2022;51(2):261−80.

[198] Gouveia ASL, Malcaite E, Lozinskaya EI, et al. Poly (ionic liquid)−ionic liquid membranes with fluorosulfonyl-derived anions: characterization and biohydrogen separation. ACS Sustainable Chemistry & Engineering 2020;8(18):7087−96.

[199] Samadi A, Kemmerlin RK, Husson SM. Polymerized ionic liquid sorbents for CO_2 separation. Energy & Fuels 2010;24(10):5797−804.

[200] Zheng S, Li W, Ren Y, et al. Moisture-wicking, breathable, and intrinsically antibacterial electronic skin based on dual-gradient poly (ionic liquid) nanofiber membranes. Advanced Materials 2022;34(4):2106570.

[201] Zheng L, Li J, Yu M, et al. Molecular sizes and antibacterial performance relationships of flexible ionic liquid derivatives. Journal of the American Chemical Society 2020;142(47):20257−69.

[202] Guo J, Xu Q, Zheng Z, et al. Intrinsically antibacterial poly (ionic liquid) membranes: the synergistic effect of anions. ACS Macro Letters 2015;4(10):1094−8.

[203] Guo J, Qin J, Ren Y, et al. Antibacterial activity of cationic polymers: side-chain or main-chain type? Polymer Chemistry 2018;9(37):4611−16.

[204] Zheng Z, Xu Q, Guo J, et al. Structure−antibacterial activity relationships of imidazolium-type ionic liquid monomers, poly (ionic liquids) and poly (ionic liquid) membranes: effect of alkyl chain length and cations. ACS Applied Materials & Interfaces 2016;8(20):12684−92.

[205] Guo J, Qian Y, Sun B, et al. Antibacterial amino acid-based poly (ionic liquid) membranes: effects of chirality, chemical bonding type, and application for MRSA skin infections. ACS Applied Bio Materials 2019;2(10):4418−26.

[206] Zheng Z, Guo J, Mao H, et al. Metal-containing poly (ionic liquid) membranes for antibacterial applications. ACS Biomaterials Science & Engineering 2017;3(6):922−8.

CHAPTER 7

Future outlook

Contents

7.1 Structural characteristics and dynamic stability mechanism of ionic liquids   259
7.2 Action mechanism and regulation law of ionic liquids in process intensification   260
7.3 The development of the chemical process for ionic liquids   261
References   264

Ionic liquids attracted the attention of academia and industry in the 1990s, during which 20−30 research papers on ionic liquids were published in journals every year. Since 2000, the number of papers on ionic liquids has increased at a high rate per year. In recent years, it has begun to increase explosively. The research of ionic liquids has moved on from preliminary exploration to in-depth research and now to industrial applications.

In 2003, Prof. Kenneth R. Seddon and Prof. Robin D. Rogers who are famous experts on ionic liquids, pointed out that ionic liquids represent the development direction of future solvents [1]. Ionic liquids have many unique physical and chemical properties, such as nonvolatility, thermal stability, low melting point, and designability, which expand the application field of ionic liquids, making them new functional materials and media, with great industrial application potential.

As a new medium, ionic liquids have been applied in different fields of reaction and process, providing a new way for energy conservation and emission reduction, and are expected to be a key to the development of energy-saving technology and clean process industry. In the chemical reaction process, as a new generation of common medium with catalytic and dissolution properties, the ionic liquids will probably bring about the innovation of the traditional catalytic process, the breakthroughs in reaction thermodynamics and dynamics, and achieve large-scale industrial applications in the resources, energy, environment, materials, and other fields.

In the face of a wide range of application fields, many new problems need to be solved in the practical application of ionic liquids. Fundamentally, as a new system different from conventional compounds, ionic liquids are not well understood in terms of nature and laws. The lack of understanding of the microscopic nature of ionic liquid systems has become a bottleneck problem in the design and industrial application of ionic liquids. The development of a new theoretical system for ionic liquids has become a challenge for researchers around the world.

At present, a certain understanding of ionic liquids at the molecular level has been obtained, and a preliminary research method for macroscopic properties at the system level has been established. However, the scientific nature of ionic liquid systems is not deeply understood, and reaction regulation of ionic liquid can only be based on macroscopic empirical data and phenomenon observation. The existing theories of ionic liquids cannot explain deeply the difference in chemical reaction behavior between ionic liquid medium and conventional solvent, the nonlinear change of catalytic reaction performance of ionic liquid, the scale effect of reaction and transfer performance, and the control factors of reaction/transference processes. The fundamental cause of the above problems is the lack of an in-depth understanding of the special microstructure of the ionic liquids, as well as the structure—property relationship and regulatory mechanism.

There are some difficult issues that we need to face for a long time in the era of rapid development of science and technology, such as how to accelerate the application of ionic liquids, deeply understand the relationship between microstructure and properties, and efficiently design ionic liquids with new structures and high performance. Theoretically, the microstructure of ionic liquids can be freely constructed to obtain unique properties for system and interface and can be regulated and functional through changes in basic units such as internal molecules/ions and external conditions, so as to obtain coupling properties that cannot be achieved by a single molecule/ion. The possibility could create opportunities for the development of new reactions and processes and has a good application prospect.

Some exploratory research is being conducted on the structure of ionic liquids [2—6]. Professor Dupont and Bica et al. [6,7] also researched the regulation of material synthesis and reaction process in ionic liquid media and found that the structure of ionic liquids has an important impact on material morphology and reaction selectivity. However, in general, the research in this area is scattered, and the scientific significance and application value for special properties of ionic liquid have not been recognized from the perspective of mesoscale and system theory.

The research on the structure—property relationship and the regularity is the long-term theme of scientific research in chemistry, while the breakthrough of theory and technology often breeds in it, and it may also bring about the development and reform of chemistry and chemical engineering. At present, the main part of a study on the structure—property relationship of ionic liquids is experimental work, while the simulation study of ionic liquids mostly focuses on the small and single microstructure. The research in the area has just started, which limits the further research and application of ionic liquids greatly.

The interionic forces of ionic liquids, including hydrogen bonds, electrostatic interactions, van der Waals forces, and so forth needed to be analyzed in depth. The crucial issue of the study on the structure—activity relationship is not only the relationship between positive and negative ion structures and physicochemical properties, but also the special manifestation of hydrogen bonds and electrostatic forces in the microstructure, as well as the relationship between performance in chemical reactions and physicochemical properties. In future work, it is an inevitable development trend to combine experiment and simulation methods, systematically study the ionic microstructure and interaction, and reveal the microscopic nature, macroscopic appearance, and structure—activity relationship of ionic liquids in multiple levels.

The ultimate purpose of studying the structure and properties of ionic liquid systems is to provide a universal and multiscale scientific basis for the design of industrial reaction devices in the practical application of ionic liquids. It is necessary to reveal the nanoscale and mesoscale mechanism, so as to form the theoretical understanding and design basis of ionic liquid reaction processes. However, the application of ionic liquids is in the stage of popularization from laboratory to industrial process, and it is mostly limited to routine apparent work, and systematic research is lacking.

In recent years, the ionic liquid research team in the Institute of Process Engineering of the Chinese Academy of Sciences has carried out preliminary research on the nanomicrostructure and mesoscopic phenomenon of ionic liquids at the systematic level, as well as the related gas—liquid two-phase transfer/reaction processes and scientific issues in process system integration for some important chemical reactions in ionic liquids, such as the cycloaddition reaction process of glycol/carbonate. The nanostructures of ion clusters in ionic liquids and their mesoscale mechanisms and properties were investigated. It was gradually realized that there was difficulty in explanation of the ionic liquid reaction phenomenon and breaking through the efficiency limit according to the traditional thinking and mode. It is also necessary to go deep into the mesoscale level for revealing the scientific nature and then lay a scientific foundation for industrial applications of ionic liquids.

7.1 Structural characteristics and dynamic stability mechanism of ionic liquids

The special properties of ionic liquids are the external manifestation of the complex microstructures and interactions of ionic liquids. The traditional view is that the reason for the special properties mainly lies in the electrostatic interactions of cations and anions. With the development of scientific research, the upgrading of detection methods, and the introduction of simulation methods, in-depth researches were carried out on the structure and properties of ionic liquids, and more attention has been paid to hydrogen bonds in ionic liquids. It was found that the interaction of cations and anions in ionic liquids is the combination of several kinds of forces, such as electrostatic force, hydrogen bond, van der Waals force, and so on. Currently, the hydrogen bond is considered to be one of the most important interactions in ionic liquids [7—10]. In addition to the characteristic hydrogen bond, intermolecular forces such as van der Waals forces and electrostatic forces also play important roles in the structure and properties of ionic liquids.

The ionic liquids present a homogeneous phase at the macro level, with complex dynamic changes of structures in a variety of scales at the micro level under the balance of various forces, including nanostructures such as single ion, ion pair, and ion cluster [7,8]. These structures of different scales and their interactions play an important role in the system and process. It is very difficult to obtain the distribution of dynamic change for ionic structures during the separation and reaction processes. Therefore, the core issues in the study of the structure—property relationship of ionic liquids include not only the relationship between the structure and property of ions or ion pairs, but also the relationship of the ion clusters, nanostructures, their dynamic changes, and macroscopic properties, as well as the systematically understanding of the formation mechanism, dynamic stability, and influence law of different structures [11,12].

It will be more difficult and arduous to study the mixed system of ionic liquids [6,13,14]. The properties of ionic liquids in the system with solvents (such as water) are closely related to the molecular structure, characteristic force, and aggregation state of the solvent. Therefore, in addition to the existing analysis methods, such as in situ experiments, spectral analysis, and molecular simulation methods, it is also necessary to develop new detection methods that can accurately observe hydrogen bonds and other interactions and specific algorithms and simulation software for unique interactions of ionic liquids. In order to reveal the nature and influence law of ionic liquids at different scales, a scientific theoretical basis should be provided for further promoting the application of ionic liquids in reaction and separation.

7.2 Action mechanism and regulation law of ionic liquids in process intensification

In recent years, ionic liquids have been widely used as media/catalysts for chemical reactions, solvents for separation processes, and so forth. They show excellent properties in various reaction/separation processes, such as condensation, oxidation, carbonylation, cycloaddition, and other reaction processes, as well as absorption and extraction processes. It is the basis for further understanding of multiscale structure—property relationships of ionic liquids to design functional ionic liquids and new reactors and develop new processes. The static and dynamic structures of ionic liquid at molecular, nanoscale, and flow field scales may affect the macroscopic physical and chemical properties and its performance in catalysis and separation processes [15,16].

In the course of application, the properties and functions of ionic liquids would change along with the difference in external conditions. The introduction of ionic liquid has a great impact on the microstructure of the reaction system and the distribution of ions and molecules, which will not only change the macroscopic physical and

chemical properties of the system, but also change the chemical action and performance, for example, the mixed catalytic system of ionic liquid with Brønsted acid.
The nanostructures formed by hydrogen bonds and other forces change the structure
of the chemical active center and the acidity of the system [16,17]. Due to various
changes in microstructure, forces, and properties, it is possible to bring about changes
and innovations in the principles of separation and catalysis.

It is of great significance to find the mechanism and regulation of ionic liquid in
absorption, separation, and catalytic reaction, and it is one of the key problems to be
solved in the application of ionic liquids. So far, it is still unclear how the cation and
anion control the reaction and separation performance of ionic liquids. Earlier, it was
believed that the cation mainly affects the physical and chemical properties of ionic
liquids, such as viscosity, density, and so forth, while the anion plays a dominant role
in the chemical and reaction properties of ionic liquids. However, ionic liquids have a
complex microstructure, so researchers will need to solve more complex situations and
problems, such as how the various structural units affect the performance of ionic
liquids and how to control and coordinate the separation and reaction processes.

For example, the kinetic behavior of ions is a key factor affecting the reaction and
transfer properties such as the viscosity of the system. However, there are few relevant
reports on what laws their movements follow. Systematic studies are needed on the movements of ions and of the atoms within ions to explore the universal laws of ionic structure
movements, so as to achieve the purpose of regulating the reaction and transfer properties.

Structural regulation of ionic liquids is an indispensable basis for the reaction process
intensification. The current researches are aimed at the static structures, rarely at the formation mechanism and change rule of ionic dynamic structures. The qualitative and quantitative regulation for the related processes of ionic liquids would be realized by exploring the
action law of internal and external factors on the distribution of ions, understanding the
action mechanism and change law in the ion systems, and realizing the regulation of ion
orientation and nonuniform distribution of nanostructures. We need to exert the advantages of ionic liquids as much as possible, reduce their disadvantages, and think about how
to intensify the reaction process on the basis of an in-depth understanding of the mechanism, so as to open up a broader application field for ionic liquids.

7.3 The development of the chemical process for ionic liquids

In the contemporary era of rapid scientific development and technological advancements, it is difficult to achieve breakthroughs in the process and technological innovation based on traditional chemical principles. It is a way to further explore chemical
reaction processes, form mesoscale control mechanisms, and construct new methods of
process engineering on the basis of multiscale theory and then to support the innovation
and sustainable development of the chemical process. The unique physicochemical

properties of ionic liquids provide a broad development space for the innovation of existing chemical processes. However, the applications of most ionic liquids have been limited to the laboratory or pilot stage until now.

At present, there are few industrial applications of ionic liquids. The representative new technology of coproducing glycol with dimethyl carbonate catalyzed by ionic liquids has been developed and industrialized by the Institute of Process Engineering, Chinese Academy of Sciences. The new technology of coproduction using the supported ionic liquid catalysis solved the problem of large-scale preparation of ionic liquid industrial catalysts, low conversion of raw materials, difficult separation of catalysts, and high energy consumption in transesterification production of dimethyl carbonate; it also reduced energy consumption by more than 30%. The simulation parameters of key reactors and equipment parameters was obtained, and the industrial equipment of 30000 t/a output was built.

The technology of functional ionic liquid battery electrolytes also has a good industrial application prospect. Compared to conventional electrolytes, ionic liquids offer better electrochemical and thermal stability, extended voltage range and energy density, and better safety. In addition, the technology of ionic liquid electrolytes can also be used in supercapacitors to achieve high-voltage and supercapacitance. At present, a 5000 t/a production line has been built in the Henan province with the technology of the Institute of Process Engineering, Chinese Academy of Sciences. The economic, social, and environmental benefits of ionic liquids are obvious, and the market potential is huge.

As mentioned above, in recent years, a great deal of research has been carried out on ionic liquids as unconventional media and their related processes. At present, there is a certain understanding of ionic liquids at the molecular level, and the studies on the system level and macro properties have been preliminarily established. However, the understanding at the mesoscale level between molecules and systems is just in the initial stage. The nature and laws of nanostructures and related phenomena are not well understood. The reaction and process control of ionic liquids still mainly rely on macroscopic empirical data and phenomenon observation, which has become one of the bottleneck problems to be solved in large-scale industrial applications of ionic liquids.

In order to obtain the industrial application of ionic liquids, we must solve the problem of engineering scale-up, of which the core is to find the reaction/transfer law of ionic liquid systems. If we do not understand the nature and laws of the reaction and separation and transfer processes in ionic liquid media, it is difficult to establish a truly innovative technology.

At present, such research cannot be found in this area. There is a lack of research on the reaction-transfer coupling law in ionic liquid media, and there are few related reports. Obtaining the transfer law of ionic liquid complex systems is the necessary condition and cornerstone of process intensification and industrial application and is also one of the difficult problems that scientific research and engineering personnel face. The description of processes related to ionic liquids still relies on traditional simple fluid experience or semiempirical models (such as Whitman's film theory [18,19]). These laws being applied to conventional liquids and with a history of more than half a century are far behind the pace of innovation in new ionic liquids media and processes.

At present, it is found that traditional models, such as the Rodrigue model and Volume of Fluid model for gas—liquid interface tracking, which are applicable to high-viscosity organic solvents, are not applicable to the gas—liquid two-phase system of ionic liquids and produce obvious errors [20]. It is necessary to consider the electrostatic and associated forces of ionic liquids and establish a new gas—liquid mass transfer model to obtain reasonable results that are consistent with the experiment. On the basis of investigation of the structure units of ionic liquid systems and their population distribution, interaction, and coupling mechanism, the amplifying law of industrial process can be obtained suitable for complex ionic liquid systems by in-depth analysis of different factors.

The regulation of traditional chemical processes mainly depends on temperature, pressure, internal components, and so on, which is limited to the regulation mechanism based on the existing theory of transfer principles. However, for ionic liquids, it is difficult to achieve quantitative and qualitative regulation by traditional means. It is more complicated that the phase density of ionic liquids can be adjusted by the change of electric field, magnetic field, and other external fields due to the static electricity or magnetism of ionic liquids, so as to realize the controllable distribution and immobilization of ionic liquids.

Intensifying the external field will make the ionic liquid systems more advantageous and is one of the development directions for future reaction and transfer process regulation. However, the amplification law under the action of electric field, magnetic field, and other external fields will be more special and complex. It is necessary to systematically study the regulation laws of ionic liquid reaction, phase transition, and transfer in reactors and industrial equipment under the influence of many external factors, including external fields. These researches will provide a scientific basis for engineering amplification of ionic liquid-related processes.

Today, with the rapid development of science and technology, the chemical industry is facing new challenges and changes. The main thrust of chemical science is no longer confined to the old concepts and established scopes. In the field of mesoscale research, scientists have walked in the forefront of the world and built another bridge between the microstructure and macroscopic properties. In the foreseeable future, a new theory and method of the reaction and process intensification based on the structural regulation of ionic liquid will be conceived and formed. It will promote the formation of the chain which is from the foundation to the application research and development, including the structure—activity relationship, engineering amplification, and industrial application. The scientific foundation for the innovation of ionic liquid cleaning process will be established and will promote the upgrading and innovation in the industrial technology based on the unconventional media ionic liquids.

References

[1] Rogers RD, Seddon KR. Ionic liquids—solvents of the future. Science 2003;302:792—3.

[2] McCrary PD, Beasley PA, Cojocaru OA, et al. Hypergolic ionic liquids to mill, suspend, and ignite boron nanoparticles. Chemical Communications 2012;48:4311—13.

[3] Foreiter MB, Gunaratne HQN, Nockemann P, et al. Novel chiral ionic liquids: physicochemical properties and investigation of the internal rotameric behaviour in the neat system. Physical Chemistry Chemical Physics 2014;16:1208—26.

[4] Ueno K, Tokuda H, Watanabe M. Ionicity in ionic liquids: correlation with ionic structure and physicochemical properties. Physical Chemistry Chemical Physics 2010;12:1649—58.

[5] Apperley DC, Hardacre C, Licence P, et al. Speciation of chloroindate(iii) ionic liquids. Dalton Transactions 2010;39:8679—87.

[6] Dupont J. On the solid, liquid and solution structural organization of imidazolium ionic liquids. Journal of the Brazilian Chemical Society 2004;15:341—50.

[7] Stoimenovski J, MacFarlane DR, Bica K, et al. Crystalline vs. ionic liquid salt forms of active pharmaceutical ingredients: a position paper. Pharmaceutical Research 2010;27:521—6.

[8] Dong K, Zhang S, Wang D, et al. Hydrogen bonds in imidazolium ionic liquids. The Journal of Physical Chemistry A 2006;110:9775—82.

[9] Na L, Fang L, Haoxi W, et al. One-step ionic-liquid-assisted electrochemical synthesis of ionic-liquid-functionalized graphene sheets directly from graphite. Advanced Functional Materials 2008;18:1518—25.

[10] Deetlefs M, Hardacre C, Nieuwenhuyzen M, et al. Liquid structure of the oonic liquid 1,3-dimethylimidazolium bis{(trifluoromethyl)sulfonyl}amide. The Journal of Physical Chemistry B 2006;110:12055—61.

[11] Shi L, Li N, Yan H, et al. Aggregation behavior of long-chain N-aryl imidazolium cromide in aqueous solution. Langmuir 2011;27:1618—25.

[12] Singh T, Kumar A. Aggregation behavior of ionic liquids in aqueous solutions: effect of alkyl chain length, cations, and anions. The Journal of Physical Chemistry B 2007;111:7843—51.

[13] Goodchild I, Collier L, Millar SL, et al. Structural studies of the phase, aggregation and surface behaviour of 1-alkyl-3-methylimidazolium halide + water mixtures. Journal of Colloid and Interface Science 2007;307:455—68.

[14] Wang H, Wang J, Zhang S, et al. Structural effects of anions and cations on the aggregation behavior of ionic liquids in aqueous solutions. The Journal of Physical Chemistry B 2008;112:16682—9.

[15] Rey-Castro C, Vega LF. Transport properties of the ionic liquid 1-ethyl-3-methylimidazolium chloride from equilibrium molecular dynamics simulation. The effect of temperature. The Journal of Physical Chemistry B 2006;110:14426—35.

[16] Borodin O, Smith GD. Structure and dynamics of N-methyl-N-propylpyrrolidinium bis(trifluoromethanesulfonyl)imide ionic liquid from molecular dynamics simulations. The Journal of Physical Chemistry B 2006;110:11481—90.

[17] El Seoud OA, Pires PAR, Abdel-Moghny T, et al. Synthesis and micellar properties of surface-active ionic liquids: 1-alkyl-3-methylimidazolium chlorides. Journal of Colloid and Interface Science 2007;313:296—304.

[18] Lewis WK, WhitmanW G. Principles of gas absorption. Industrial & Engineering Chemistry 1924;16:1215—20.

[19] Danckwerts PV. Significance of liquid-film coefficients in gas absorption. Industrial & Engineering Chemistry 1951;43:1460—7.

[20] Denis R. Drag coefficient—Reynolds number transition for gas bubbles rising steadily in viscous fluids. The Canadian Journal of Chemical Engineering 2001;79:119—23.

Index

Note: Page numbers followed by "*f*" and "*t*" refer to figures and tables, respectively.

A

AA. *See* Ascorbic acid (AA)
AAIL. *See* Amino acid IL (AAIL)
Ab initio calculations, 16−17, 19
Ab initio molecular dynamics simulations (AIMD simulations), 21−25, 27−28
Abnormal IL wetting behaviour, 31−32
Acetic acid, 144−145
Acetone, butanol, and ethanol (ABE), 133
Acetonitrile (CH₃CN), 190, 193−194
ACN. *See* Acetonitrile (CH₃CN)
Active oxygen species, 161, 166
Adjacent ionic layer, 33
Adsorbent, 63−64
Adsorption−desorption process, 44
AIMD simulations. *See* Ab initio molecular dynamics simulations (AIMD simulations)
Alkyl chain, 9, 77, 109−110
   alkyl side chains of cations, 205−206
   of ILs, 214
1-alkyl-3-acetic acid imidazole, 233−234
1-alkyl-3-methylimidazolium chloride ([Emim] Cl), 81−82
1-allyl-3-methylimidazolium chloride ([Amim]Cl), 83
1-allyl-3-vinylimidazole bis (trifluoromethane sulfonyl) imide ([AVIm][TFSI]), 185
AlPO₄−11 molecular sieves, 226
AlPO₄−5 molecular sieves, 226
Aluminophosphate, 228
Aluminosilicate molecular sieves, 225
AMBER. *See* Assisted Model Building with Energy Refinement (AMBER)
Amines, 226
Amino acid IL (AAIL), 8, 133
Amino acids-based bio-ILs, 72
Amino ionic liquids, 105−106
1-aminopropyl-3-methylimidazolium glycinate ([APmim][Gly]), 105−106
1-aminopropyl-3-methylimidazolium lysine ([APmim][Lys]), 105−106
2-aminopyridine bis(trifluoromethylsulfonyl)imide ([2PyH][NTf₂]), 100−101
Ammonium molybdate production, 105
Ammonium-based ILs, 72−73
Amphiphilic ionic liquids, 3−4
Anions, 1, 15, 76, 89, 150−151, 259−260
   of ILs, 230−231
Annealing process, 153−155
Anomalous stepwise melting process, two-dimensional ionic liquids with, 42−46
APPLE&P. *See* Atomistic Polarizable Potential for Liquids, Electrolytes & Polymers (APPLE&P)
Aprotic [Bmim][BF₄], oxygen reduction reaction behavior in, 161−162
Aqueous electrolyte solutions, 5−6
ARE. *See* Artemisitene (ARE)
Armillaria luteo-virens Sacc ZJUQH100−6 cells, 79
Artemisia annua L., 116
Artemisinin, 116
Artemisitene (ARE), 116
Ascorbic acid (AA), 110
Assisted Model Building with Energy Refinement (AMBER), 30
Atom transfer radical polymerization (ATRP), 237
Atomistic Polarizable Potential for Liquids, Electrolytes & Polymers (APPLE&P), 25
ATRP. *See* Atom transfer radical polymerization (ATRP)

B

Bacteriorhodopsin (BR), 120−121
BAIL. *See* Brønsted acid IL (BAIL)
BAPBIL. *See* Brønsted acid-functionalized porphyrin grafted with benzimidazolium-based IL (BAPBIL)
Batteries, 189
Benzimidazole (BenIm), 103
   ionic liquids, 12
Benzmim⁺ ions, 226−227
Benzyl alcohol, 172

1-benzyl-3-methylimidazolium chloride ((Benzmim)Cl), 226—227

BET. *See* Brunauer—Emmett—Teller (BET)

β-1 lignin model compound, 89—90

β-O-4 lignin model compound, 89—90

Biguanide hydrochloride (BGCl), 158

Binding energy, 149

Bio-ILs. *See* Biocompatible ILs (Bio-ILs)

BiOBr porous nanospheres, 214—217

Biocatalysis, 72

  tunable microenvironment for, 71—79

Biocompatible ILs (Bio-ILs), 72

Biodiesel, 78

  application in biodiesel production, 65—66

Bioethanol, 78

Biofuels, 11, 78

Bioionic liquids, 71—79

  ionic liquids and free enzymes, 72—75

    applications of ionic liquids in hydrophilic enzyme-based catalysis system, 73—74

    applications of ionic liquids in hydrophobic enzyme-based catalysis system, 74—75

  ionic liquids and whole cell catalysis, 75—79

    application of whole cell catalysis with ionic liquids, 78—79

    effects of ionic liquids on microbial cells, 77—78

Biomass, 79—80

  ionic liquids

    biomass conversion, 79—90

    cellulose conversion, 81—85

    hemicellulose conversion, 85—86

    lignin conversion, 86—90

Biphase system, 72

Biphasic aqueous system, 111—112

3,5-bis trifluoromethyl acetophenone (3,5-BTAP), 78—79

Bis-(1-pyridinium) butyl ditoluenesulfonate (Py-PTSA), 61—62

Bis-(1-pyridinium) butyl hydrogen sulfate (Py-HSO₄), 61—62

Bis-(3-methyl-1-imidazolium) butyl ditoluenesulfonate (Im-PTSA), 61—62

Bis-(3-methyl-1-imidazolium) butyl hydrogen sulfate (Im-HSO₄), 61—62

1,3-bis(carboxylatoethyl) imidazolium salt, 234—236

1,3-bis(carboxymethyl) imidazolium salt, 234—236

3,5-bism trifluoromethyl phenyl ethanol [(R)-BTPE], 78—79

Boltzmann constant, 40

Boltzmann distribution, 29

Born-Mayer repulsion, 26

Bovine serum albumin (BSA), 117—118

BR. *See* Bacteriorhodopsin (BR)

Brønsted acid IL (BAIL), 84—85, 233

Brønsted acid-functionalized porphyrin grafted with benzimidazolium-based IL (BAPBIL), 148

Brunauer—Emmett—Teller (BET), 66—67

BSA. *See* Bovine serum albumin (BSA)

Butyl acetate, 176

  application in synthesis of, 66

Butyl citrate, application in, 71

1-butyl imidazolium bis (trifluoromethylsulfonyl) imide ([Bim][NTf₂]), 99—100

1-butyl-3-ethylimidazoldiimide (BMI-TFSI), 185

1-butyl-3-methylimidazolium acetate ([Bmin][Ac]), 72—73

1-butyl-3-methylimidazolium bis (trifluoromethylsulfonyl) imide ([Bmim][NTf₂]), 87—90, 99—100

1-butyl-3-methylimidazolium bromide ((Bmim)Br), 205—206

1-butyl-3-methylimidazolium chloride ([Bmim]Cl), 84, 117

1-butyl-3-methylimidazolium dicyanamide ([Bmim][DCA]), 114—115

1-butyl-3-methylimidazolium hexafluorophosphate ((Bmim)(PF₆)), 87, 105, 220—225

1-butyl-3-methylimidazolium hydrogen sulfate ([Bmim][HSO₄]), 84—85

1-butyl-3-methylimidazolium tetrafluoroborate ((Bmim)(BF₄)) system, 61, 153—155, 217—225

  degradation mechanism of *p*-benzyloxyl phenol in, 167—168

  conversion and Faraday efficiency of electrolyzing PBP, 169*f*

  electrochemical behaviors of *p*-benzyloxyl phenol in, 165—166

  electrochemical depolymerization of lignin in, 180—181, 181*t*

    product distribution of electrochemical lignin depolymerisation, 183*t* oxygen reduction reaction behaviors in, 162–164

1-butyl-3-methylimidazolium [Bmim], 72

C

CA. *See* Caffeic acid (CA); Contact angle (CA)

Caffeic acid (CA), 109

Carbon capture, 42

Carbon dioxide (CO_2), 11, 62–63, 243
  capture, 11
    with ionic liquids, 105–107
  cycloaddition with propylene oxide, 244*t*

Carbon footprints, 131

Carbon materials, 192

Carbon nanotubes, 205–206

Carbon NPs (CNPs), 217–219

Carboxyl-functionalized ionic liquids, metal organic complexes synthesis using, 232–236

Carboxylic acids, 59

Catalysts, 63–64, 66–67, 149, 242–243

Catalytic C_4 alkylation reaction, 10

Catalytic system, 89

Catalytic technologies, 79–80

Cathode potential, 177

Cations, 1, 15, 150–151, 259–260

Cell biocatalysis, 72

Cell membrane, 76–77

Cellulose, 11
  cellulose-derived chemicals, 81, 83–84
  ionic liquids intensify cellulose conversion, 81–85
  macromolecule conversion, 81
  monomer, 81, 83

Cellulose nanofibrils (CNFs), 206

Cetylpyridine chloride (CPC), 120–121

CG force fields. *See* Coarse-grained force fields (CG force fields)

Charge transport layers (CTLs), 150

Charging process, 189–190

Chemical process for ionic liquids, development of, 261–264

Chemical reaction process, 257, 261–262

Chemical science, 263–264

Chemisorption, 105

Chiral alcohols, 78

Chiral ILs, 233

Chlorella pyrenoidosa, 118

Chloro-aluminate ionic liquids, 3

Choline, 72

Choline bis(trifluoromethylsulfonyl) imide ([Choline][NTf_2]), 98

Citric acid, 71

Classical force field, 19–25
  atom types in imidazolium cation, 20*f*

CNFs. *See* Cellulose nanofibrils (CNFs)

CNPs. *See* Carbon NPs (CNPs)

Coal, 79

Coarse-grained force fields (CG force fields), 19, 26–27

Collective variables (CVs), 29

Colloidal dispersion of nanostructured materials, 205–206

Commercial resin, 65–66

Coniferyl alcohol, 176–178

Contact angle (CA), 30–31

Conventional electrolytes, 262

Conventional hydrogen bonds, 8

Conventional ionic liquids, 105

Copolymer-type ionic liquids, 236, 238

Copper, 87

Copper chloride ($CuCl_2$), 102

Coulombic fluids, 5–6

Covalent interaction, 203

CPC. *See* Cetylpyridine chloride (CPC)

CTLs. *See* Charge transport layers (CTLs)

CV method. *See* Cyclic voltammetry method (CV method)

CVs. *See* Collective variables (CVs)

Cyclic voltammetry method (CV method), 161

Cyclopolymerization, 237

CYPs. *See* Cytochrome P450 enzymes (CYPs)

Cytochrome P450 enzymes (CYPs), 123

D

Deacidification process, 1–2

Deep eutectic solvents (DESs), 103, 227

Deep neutral networks, 28

Degradation products, 176–177

Dehydrogenative coupling polymerization, 237

Density functional approximation (DFA), 27–28

Density functional theory (DFT), 7–8, 27

Depolymerization of lignin by electrochemical method in ionic liquids, 179–184

DESs. *See* Deep eutectic solvents (DESs)

Detection methods, 259–260

DFA. *See* Density functional approximation (DFA)

DFT. *See* Density functional theory (DFT)

1,3-di(2-aminoethyl)-2-methylimidazolium bromide (DAIL), 105−106

1,3-dialkyl imidazole aluminum chloride salt, 3

Diaminoethimidazole bromide salt (DAIB), 131−133

1,3-dicarboxymethyl imidazole ILs, 233−234

Dicationic polymeric IL and high-performance self-healing solid electrolytes (DPIL-6-SPE), 240

Diester succinate
  application in synthesis of, 61−62
  compounds, 60

Diethylenetriamine acetate ([DETA][OAc]), 146

Diethylenetriamine hexafluorophosphate ([DETA][PF$_6$]), 146

1,3-diethylimidazolium bis(trifluoromethylsulfonyl) imide ([Emim]NTf$_2$), 84−85

Diffraction spectra, 9

Dihedral angle parameters, 21−25

Dihydric alcohol-based DESs, 103

Diisopropyl succinate, 60

1,4-diketo pyrrole and pyrrole (DPP), 60

Dimethyl carbonate (DMC), 61−63, 69, 262
  application in synthesis of, 61−63, 67−69
  electrolytes, 191

Dimethyl sulfate, 69

Dimethyl sulfoxide (DMSO), 153

3-(dimethylamino)-1-propyl formate amine ([DMAPA]FA), 118

Dimethylammonium bis(trifluoromethanesulfonyl) imide (DMATFSI), 157

Dimethylformamide (DMF), 153

2,5-dimethylfuran (DMF), 79−80, 84−85

1,2-dimethylimidazolium nitrate ([Mmim][NO$_3$]), 103−104

1,4-dioxane and 1-butyl-3-methylimidazolium hexafluorophosphate, 112

DIPI. *See* N,N'-diisopropylimidazolium iodide (DIPI)

Dispersion polymerization, 237

DMC. *See* Dimethyl carbonate (DMC)

DMF. *See* Dimethylformamide (DMF)

DMSO. *See* Dimethyl sulfoxide (DMSO)

DNL-1, 228

1-dodecyl-3-methylimidazolium bromide-functionalized nanosilica, 205−206

Donnan−Steric Pore Model, 133

Double layer capacitors, 189

DOXH. *See* Doxycycline hydrochloride (DOXH)

Doxycycline hydrochloride (DOXH), 146

DPIL-6-SPE. *See* Dicationic polymeric IL and high-performance self-healing solid electrolytes (DPIL-6-SPE)

DRD. *See* Drude oscillator (DRD)

Drude oscillator (DRD), 25

Dual-emitting ratiometric luminescent thermometers, 234−236

Dye-sensitized solar cells, 241

Dynamic scanning calorimetry, 64

E

EFM. *See* Electrostatic force microscopy (EFM)

EGFR. *See* Epidermal growth factor receptor (EGFR)

Einstein's equation, 49−50

Electric field, 25−26

Electrical environments, ionic liquid intensification in, 12−13

Electrocatalytic processes, 144

Electrochemical energy
  storage batteries, 192−193
  storage devices, 184

Electrochemical indirect oxidation process, 161

Electrochemical lignin depolymerization, 179−180
  electrochemical behaviors of *p*-benzyloxyl phenol in [Bmim][BF$_4$], 165−166, 165*f*
  electrochemical behaviors of *p*-benzyloxyl phenol in [HNEt$_3$][HSO$_4$], 169−170
  electrochemical lignin depolymerization using guaiacylglycol-β-guaiacyl ether as model compounds, 171*f*
  using *p*-benzyloxyl phenol as model compounds, 164−179

Electrochemical method in ionic liquids, depolymerization of lignin by, 179−184
  electrochemical depolymerization of lignin in [Bmim][BF$_4$], 180−181, 181*t*
  electrochemical depolymerization of lignin in [HNEt$_3$][HSO$_4$], 181−184, 182*t*

Electrochemical processes, 12

Electrochemical reactions, 143−144

Electrochemical research, 179

Electrochemical window, 196−197

Electrochemically active species, 12
Electrode materials, 189
  for supercapacitors, 192
Electrolysis, 167–168, 170–172
Electrolytes, 149–150, 172, 187
  electrolytic liquid system, 186
  ions, 191
Electron transfer
  channel, 46–47
  and friction feature of two-dimensional ionic
    liquids, 46–49, 48f
  typical EFM phase images, 47f
Electron transport layer (ETL), 155
Electronegativity equalization, 25–26
Electrons, 16–17, 19, 179
Electrostatic force microscopy (EFM), 46
Electrostatic forces, 1, 8, 189–190
Electrostatic interactions, 1–2, 15, 203
Electrostatic potential (ESP), 20–21
Emission reduction, 257
EMT. See Epithelial-mesenchymal transition
    (EMT)
Enzymatic hydrolysis, 122–123
Enzymes, 72–73, 75
  barriers, 21–25, 29
  conservation, 257
  energy-saving technology, 257
Epidermal growth factor receptor (EGFR),
    123–124
Epithelial-mesenchymal transition (EMT),
    124–125
Epoxy compounds, 62–63
Escherichia coli, 77–78
ESP. See Electrostatic potential (ESP)
Ester exchange reaction, 79
Esterases, 74–75
Ethanolamine thiocyanate (EtA), 100–101
Ethyl carbonate electrolytes, 191
1-ethyl-3-methylimidazole acetate ([Emim][OAc]),
    146
1-ethyl-3-methylimidazole diimsi ([Emim][TFSI]),
    185
1-ethyl-3-methylimidazole hexafluorophosphate
    ([Emim][PF$_6$]), 146
1-ethyl-3-methylimidazolium bis[(trifluoromethyl)
    sulfonyl]imide, 35–36
1-ethyl-3-methylimidazolium bromide ([Emim]
    Br), 157–158, 230

1-ethyl-3-methylimidazolium tetrafluoroborate
    ([Emim][BF$_4$]), 81–82
1-ethyl-3-methylimidazolium thiocyanate,
    149–150
Ethyl-4-chloro-3-hydroxybutanoate, 77–78
Ethylamine nitrate (EAN), 7
1-ethylpyridine key [EPY], 109–110
ETL. See Electron transport layer (ETL)
Exothermic reaction, 168

F
FA. See Ferulic acid (FA)
Facile blending process, 130
Far-infrared spectra, 7–8
Faradaic efficiency, 148
Faraday capacitors, 189
Faraday efficiency of ORR, 162
FASP. See Filter-assisted sample preparation
    (FASP)
FCM. See Fluctuating charge model (FCM)
Ferulic acid (FA), 109
FF. See Fill factor (FF)
FFPE liver cancer. See Formalin fixed paraffin-
    embedded liver cancer (FFPE liver cancer)
FFs. See Force fields (FFs)
Fill factor (FF), 152–153
Filter-assisted sample preparation (FASP), 119
Flexible energy storage equipment, 188–189
Flow batteries, 192–193
  application of ionic liquids in, 192–197
    flow batteries with different structures,
      193f
    organic flow battery, 195–197
    water-based flow battery, 194–195
Fluctuating charge model (FCM), 25
Fluorinated hydrophobic ILs, 110
FM. See Force matching (FM)
Folate receptor alpha (FOLR1), 124–125
FOLR1. See Folate receptor alpha (FOLR1)
Force fields (FFs), 17–28
  Ab initio calculations, 19
  Ab initio molecular dynamics, 27–28
  basis set selection, 21
  classical force field, 19–25
  coarse-grained force field, 26–27
  experimental data, 19
  molecular conformation, 21
  penalty function factor, 21–25

Force fields (FFs) (*Continued*)
  electrostatic potentials around by different
    fitting strategies, 23*f*
  fitting charges via double ion pairs, 22*f*
  torsion energy profiles of anion, 24*f*
  polarized force field, 25–26
Force matching (FM), 27
Formaldehyde, 65–66
Formalin fixed paraffin-embedded liver cancer
    (FFPE liver cancer), 125–126
Fossil resources, 79
Fourier-transform infrared spectroscopy
    (FT-IR spectroscopy), 60–61, 64,
    103–104
Free enzymes, 72–75
Friction force, 49
FT-IR spectroscopy. *See* Fourier-transform
    infrared spectroscopy (FT-IR spectroscopy)
2,5-furandicarboxylic acid (FDCA), 84
Furfural, 86

G

Gas
  adsorption process, 40–41
  gas–liquid equilibrium data, 98
  separation process, 97–98
Gaussian software, 19
Gel electrolyte, 188
Ginkgo biloba, 113–114
Ginkgolides, 113–114
GK integral. *See* Green–Kubo integral (GK
    integral)
Glycerol (Gly), 104
  application in synthesis of glycerol monolaurate,
    66–67
Glycoproteins, 123
GO. *See* Graphene oxide (GO)
Graphene (Gra), 33–35, 205–206
  graphene-based polyporous carbon materials,
    192
Graphene oxide (GO), 33–35
Graphite, 170–172
Green additives, 151
Green chemistry principles, 147
Green electrolytes, 184
Green reaction solvent, 10
Greenhouse gases, 11
Green–Kubo integral (GK integral), 49–50
GROMACS. *See* GROningen MAchine for
    Chemical Simulations (GROMACS)

GROningen MAchine for Chemical Simulations
    (GROMACS), 30
Guaiacol, 176–177
Guaiacylglycol-β-guaiacyl ether (GGE), 174
  electrochemical lignin depolymerization using,
    174–179
  HPLC of butyl acetate extractive of reaction
    liquid, 179*f*
  molecular structure of GGE, 175*f*
Guanidine acetate (GAAc), 155–156
Guanidine chloride (GACl), 155–156
Guanidinium hydrochloride (GuHCl), 152–153
Guanidinium sulfate (GASO$_4$), 155–156
Gunidinium thiocyanate (GASCN), 155–156

H

1H-benzotriazole 1 (BenTriz), 103
1H-tetrazole (Tetz), 103
Harmonic functions, 20
HBs. *See* Hydrogen bonds (HBs)
HDO. *See* Hydrodeoxygenation (HDO)
Head-to-head structure model, 47–48
Heavy metal removal, 133
Hemicellulose conversion, ionic liquids
    intensifying, 85–86
Henry's law coefficient, 98
Heteroatoms, 192
Heterogeneous photocatalyst, 148
Heteropoly acid catalytic system, 84
1,6-hexanediamine (HDA), 228
1-hexyl-1-methylpyrrolidine [HMPL], 109–110
1-hexyl-3-methylimidazolium tetracyanoborate
    (Im$_{6,1}$ tcb), 133
High-performance liquid chromatography
    (HPLC), 116
Highly oriented pyrolytic graphite (HOPG), 46
[HNEt$_3$][HSO$_4$]
  electrochemical behaviors of *p*-benzyloxyl
    phenol in, 169–170
  electrochemical depolymerization of lignin in,
    181–184, 182*t*
Hole transport layer (HTL), 155
Homogeneous catalytic reactions, 57–58
  application in synthesis of
    diethyl succinate, 61–62
    dimethyl carbonate, 62–63
    dimethyl succinate, 61
    n-amyl alcohol acetate, 59–60
    succinic acid diisopropyl ester, 60–61
  ionic liquids, 58–63

Homogeneous polymerization IL catalysts (HPILs), 243
HOPG. *See* Highly oriented pyrolytic graphite (HOPG)
HPILs. *See* Homogeneous polymerization IL catalysts (HPILs)
HPLC. *See* High-performance liquid chromatography (HPLC)
HTL. *See* Hole transport layer (HTL)
Hybrid nanomaterials, 217
Hybrid solvents, 97−98
Hydrodeoxygenation (HDO), 86, 88
Hydrogen atoms, 21−26
Hydrogen bonds (HBs), 35, 259−261
    network, 3−4, 7−8, 203
        structure of quasi-liquids formed by ionic liquids, 7*f*
Hydrogen peroxide (H₂O₂), 148
Hydrolase-catalyzed reactions, 73−74
Hydrophilic enzyme-based catalysis system, applications of ionic liquids in, 73−74
Hydrophilic ionic liquids, 77, 111
Hydrophobic alkyl chain, 15
Hydrophobic enzyme-based catalysis system, applications of ionic liquids in, 74−75
Hydrophobic ionic liquids
    distribution ratios of phenol between ionic liquids and water, 109*f*
    liquid−liquid extraction with, 109−110
Hydrothermal method, 214−217
1−2(-hydroxyethyl)-3-methylimadazolium bis (trifluoromethylsulfonyl) imide ([EtOHmim] [NTf₂]), 98−99
1−2(-hydroxyethyl)-3-methylimadazolium tetrafluoroborate ([EtOHmim][BF₄]), 98
1-(2-hydroxyethyl)-3-methylimidazolium dicyanamide ([EtOHmim][DCA]), 98−99
Hydroxyl condensation reaction, application in, 70
Hydroxyl ionic liquids, 98−99
5-hydroxymethylfurfural (HMF), 79−80

I

IBI. *See* Iterative Boltzmann inversion (IBI)
IL-based ABS. *See* Ionic liquid-based aqueous biphasic systems (IL-based ABS)
ILGMs. *See* Ionic liquid gel membranes (ILGMs)
ILMs. *See* Ionic liquid-based membranes (ILMs)
ILPIMs. *See* Ionic liquid polymer inclusion membranes (ILPIMs)

ILPMs. *See* Ionic liquid−polymer membranes (ILPMs)
ILs. *See* Ionic liquids (ILs)
Imidazole (Im), 103
    carboxylic acid zwitterionic ligands, 233−234
    imidazole-based ILs, 72−73
    ring, 35
Imidazolium, 27
    cations, 19−20, 145
    imidazolium-based electrolytes, 149
    imidazolium-based ILs, 205−206
    ring, 19−20
Imidazolium bis(trifluoromethylsulfonyl) imide ([Im][NTf₂]), 101
Imidazolium nitrate([Im][NO₃]), 103−104
Impregnation method, 129
IMPs. *See* Integrated membrane proteins (IMPs)
In situ polymerization, 237
Induced point dipoles (IPD), 25
Industrial applications, 259
Industrial process, 259, 263
Infrared spectroscopy data, 6
Inorganic acids, 85
Integrated membrane proteins (IMPs), 119
Interatomic interactions, 17
Interfacial electron transfer, 46
Interfacial energy, 30−31
Intermolecular forces, 259−260
Intermolecular noncovalent interactions, 18
Intramolecular interaction, 20
Ion clusters, 259
Ion pairs, 5−6
    Z-bonds and types in ionic liquids, 6*f*
Ionic clusters, 8−9
Ionic gel in ceramics, 239
Ionic liquid gel membranes (ILGMs), 129−130
Ionic liquid polymer inclusion membranes (ILPIMs), 128
Ionic liquid-based aqueous biphasic systems (IL-based ABS), 111
Ionic liquid-based membranes (ILMs), 97
Ionic liquid−polymer membranes (ILPMs), 128
Ionic liquids (ILs), 1−3, 15, 37, 57−58, 72−79, 97, 143, 203, 258−262. *See also* Bioionic liquids
    action mechanism and regulation law of ionic liquids in process intensification, 260−261
    applications
        hydrophilic enzyme-based catalysis system, 73−74

Ionic liquids (ILs) (*Continued*)
    hydrophobic enzyme-based catalysis system,
        74—75
    liquid—liquid extraction, 108—117
    of whole cell catalysis with, 78—79
catalytic homogeneous reactions, 10
cellulose conversion, 81—85
as charge compensation and structure guiding
        agent action, 230
cleaning process, 263—264
clusters, 9
    aggregation, 11
CO₂ capture with, 105—107
    amino ionic liquids, 105—106
    conventional ionic liquids, 105
    ionic liquid hybrid solvents, 107
    non-amino ionic liquids, 106—107
complex systems, 262—263
development of chemical process for, 261—264
effects of ionic liquids on microbial cells,
        77—78
electrochemistry process, 159—184
    depolymerization of lignin by electrochemical
        method in ionic liquids, 179—184
    electrochemical lignin depolymerization using
        p-benzyloxyl phenol as model compounds,
        164—179
    molecular orbitals of O₂ and reactive oxygen
        species, 159*f*
    oxygen reduction reaction behaviors in
        different ionic liquids electrolyte systems,
        161—164
electrolytes, 262
extraction separation with similar structure
        compound, 112—117
gas separation, 98—107
hemicellulose conversion, 85—86
IL-strengthened nanomaterials fabricated by sol-
        gel method, 215*t*
ionic liquid-based aqueous biphasic systems
        extraction, 111—112
ionic liquid-grafted nanomaterials, 210*t*
ionic liquid—molecular solvent complex
        liquid—liquid extraction, 110—111
islands, 37—42
lignin conversion, 86—90
membrane separation process with, 128—135
    application in liquid separation, 131—134

ionic liquid membranes and preparation
        strategy, 129—131
    ionic liquid polymer inclusion membranes,
        131
    ionic liquid—polymer membranes, 129—130
    poly(ionic liquid) membranes, 130—131
    supported ionic liquid membranes, 129
metal organic complexes synthesis using
        carboxyl-functionalized ionic liquids,
        232—236
methods for simulating ionic liquid structures,
        16—30
    force field, 18—28
    sampling method, 28—29
    software for molecular simulation, 29—30
mixed ionic liquids as solvents for synthesis of
        porous metal organic complexes, 231
nanomaterials
    prepared via microemulsion method, 223*t*
    prepared via microwave-assisted IL method,
        221*t*
    synthesized by hydro- and solvothermal
        method, 218*t*
nanostructure regulation and process
        intensification of, 9—13
    ionic liquid intensification in electrical and
        magnetic environments, 12—13
    ionic liquid nanostructures intensifying
        reaction process, 10
    ionic liquid structures intensifying separation
        process, 11—12
in new energy batteries, 184—197
    application of ionic liquids in flow batteries,
        192—197
    application of ionic liquids in lithium-ion
        batteries, 184—189
    application of ionic liquids in supercapacitors,
        189—192
    ionic liquids as supercapacitor support
        electrolyte, 189—191
    preparation of electrode materials for
        supercapacitors, 192
NH₃ separation with, 98—105
    hydroxyl ionic liquids, 98—99
    ionic liquid-based hybrid solvents, 103—104
    ionic liquid-based NH₃ separation technology
        and applications, 104—105
    metal ionic liquids, 101—102

protic ionic liquids, 99−101
oxygen reduction reaction behavior in effect of, 161−162
photocatalytic and photoelectrocatalytic process with, 144−150
  application of ionic liquids in photocatalytic systems, 144−148, 146f
  application of ionic liquids in photoelectrocatalytic systems, 148−150
prediction and control of ionic liquid structures, 49−51, 51f
preparation of nanomaterials with, 204−225
for protein complex extraction, 126−128
for protein extraction, 117−126
  ionic liquids for single protein extraction, 117−118
for protein mixture extraction, 118−126
  ionic liquids for membrane proteome extraction, 119−121
  ionic liquids for whole proteome extraction, 122−126
reaction process
  bioionic liquids, 71−79
  ionic liquids intensified biomass conversion, 79−90
  ionic liquids regulate homogeneous catalysis reaction, 58−63
  multiphase reaction based on ionic liquids, 63−71
simulation, 28
  and regulation of two-dimensional ionic liquids, 37−49
  study of ionic liquid structures in interface, 30−37
in solar cells, 150−158
  as additives in perovskite precursor solutions, 151−153
  for interface modification, 157−158
  ionic liquid-modified charge transport layers for perovskite solar cells, 155−157
  as solvent for perovskite solar cells, 153−155
solvents in synthesis of rare earth complexes, 231−232
structural characteristics and dynamic stability mechanism of, 259−260
synthesis
  and application of polymerized ionic liquids, 236−245

of ionic liquid-based metal organic complexes, 229−236
of molecular sieve materials, 225−229
systems, 259, 263
understanding of nanomicrostructures, 1−9
  nanostructure of ionic liquids at different scales, 5−9
  understanding history of ionic liquid structures, 3−5, 4f
Ionic porous organic polymers catalyst, 70
Ionic structures, 260
Ionic thermal method, 192
Ionized electrolytes, 5
Ionothermal synthesis, 225−226
IPD. *See* Induced point dipoles (IPD)
Iron, 87
Irreversible electrode process, 174−176
Isoenzymes, 123
Iterative Boltzmann inversion (IBI), 27

K
Kelvin probe force microscopy (KPFM), 46
KPFM. *See* Kelvin probe force microscopy (KPFM)

L
L-neneneba tryptophan, 115
LAMMPS. *See* Large-scale Atomic/Molecular Massively Parallel Simulator (LAMMPS)
Large-scale Atomic/Molecular Massively Parallel Simulator (LAMMPS), 29
Lauric acid, 66−67
LC-MS. *See* Liquid chromatography-mass spectrometry (LC-MS)
Lennard-Jones potential (LJ potential), 18
Lewis acid, 60−61
Lignin, 159−160, 179−180
  conversion, 86−90
  depolymerization of lignin by electrochemical method in ionic liquids, 179−184
  depolymerization process, 179−180, 182−184
  transformation technologies, 86
Lignocellulose, 79−81
Linear response theory, 49−50
Lipases, 72−75
Liquid chromatography-mass spectrometry (LC-MS), 119
Liquid molten salts, 150−151

Liquid separation
  application in, 131—134
  metal separation, 131—133
  organic separation, 133
  transport mechanism, 135
  water desalination, 133—134
Liquid—liquid extraction, 108
  application of ionic liquids in, 108—117
    ionic liquid extraction separation with similar
      structure compound, 112—117
    ionic liquid-based aqueous biphasic systems
      extraction, 111—112
    ionic liquid—molecular solvent complex
      liquid—liquid extraction, 110—111
    liquid—liquid extraction with hydrophobic
      ionic liquids, 109—110
    principle of liquid—liquid extraction, 108f
  separation process, 97—98
Lithium (Li), 102
  batteries and capacitors, 239—240
  lithium-chelated ILs, 102
  lithium-ion flow battery, 195—196
Lithium bis(trifluoromethanesulfonyl)imide
  (LiTFSI), 157
Lithium-ion batteries, 184
  application of ionic liquids in, 184—189
  liquid electrolyte, 184—187
    common ionic liquid cations, 185f
  solid electrolyte, 187—189
LJ potential. See Lennard-Jones potential (LJ
  potential)
Lysozyme, 111—112

M
Machine learning methods, 28
Macroscopic systems, 17
MAFa. See Methylammonium formate (MAFa)
Magnetic attraction, 12—13
Magnetic environments, ionic liquid intensification
  in, 12—13
Magnetic field, 12—13
Manganese, 87
Manganese oxide (MnOx), 149
Mathias—Klotz—Prausnitz mixing rule, 114—115
MC. See Monte Carlo (MC)
MCCCS. See Monte Carlo for Complex Chemical
  Systems (MCCCS)
MD. See Molecular dynamics (MD)

Mean-squared displacement (MSD), 50
Melamine tail gas, 97—98
Membrane protein complexes (MPCs), 126
Membrane proteins, 119
Membrane proteome extraction, 122
  conformation of bacteriorhodopsin and [C_{12}im]
    Cl simulation system, 120f
  ionic liquids for, 119—121
Membrane separation process with ionic liquids,
  128—135
Membrane-free redox flow batteries, 194—195
Membraneless redox flow batteries, 194—195
Metal, 205—206
  metal-based catalysts, 87—88
  metal-free Brønsted acid-functionalized
    porphyrin grafted with benzimidazolium-
    based IL, 148
  separation, 131—133
Metal chlorides, 101—102
Metal fluorides, 217
Metal ionic liquids (MILs), 101—102
Metal organic complexes synthesis using carboxyl-
  functionalized ionic liquids, 232—236
Metal organic frameworks (MOFs), 229—230
Metal oxides, 217, 219—220
  nanoparticles, 205—206
Methacrylic acid, 111
Methanol (MeOH), 62—63, 67
1-methoxyethyl-3-methylimidazole bromo
  chloroaluminate ionic liquid ([MOEmim]
  Br/AlCl$_3$), 10
Methyl chloride, 69
Methyl methacrylate (MMA), 64
  application in, 64
Methyl orange (MO), 147
Methyl pyrrolidonium salts, 77
1-methyl-3-(3-sulfopropyl)-imidazolium chloride
  ([CPmim][NTf$_2$]), 87—88
1-methyl-3-ethylimidazoldiimide (EMI-TFSI), 185
Methylammonium formate (MAFa), 152—153
Methylation, 51
1-methylimidazolium 2-acrylamido-2-
  methylpropanesulfonate ((mim)(AMPS)),
  237—238
2-methylimidazolium bis(trifluoromethylsulfonyl)
  imide ([2-mim][NTf$_2$]), 101
1-methylimidazolium nitrate ([mim][NO$_3$]),
  103—104

Methyltrioctylammonium trifluoromethanesulfonate (MATS), 157–158
Microbial cells, effects of ionic liquids on, 77–78
Microemulsion, 220, 223t
Microorganisms, 79
Microwave heating, 217
Microwave-assisted ionothermal process, 227–228
Microwave-enhanced ionothermal synthesis, 227
MILs. See Metal ionic liquids (MILs)
Mixed ionic liquids as solvents for synthesis of porous metal organic complexes, 231
MM. See Molecular mechanics (MM)
MMA. See Methyl methacrylate (MMA)
MO. See Methyl orange (MO)
MOFs. See Metal organic frameworks (MOFs)
Molecular chromophores, 149
Molecular dipole moment, 21
Molecular docking, 128
Molecular dynamics (MD), 28–29, 99–100, 162
Molecular FF-based simulations, 30
Molecular mechanics (MM), 16–17
Molecular sieve materials, ionic liquids intensify synthesis of, 225–229
Molecular simulation, 16
    software for, 29–30
Monobenzene ring free radicals, 179–180
Monoethanolamide, 12
Monomer, 66–67
Monte Carlo (MC), 27–29
    molecular simulation program, 29
Monte Carlo for Complex Chemical Systems (MCCCS), 29
Monuclear anions, 76
MPCs. See Membrane protein complexes (MPCs)
MSD. See Mean-squared displacement (MSD)
Multifunctional calcium phosphate-IL bionanocomposites, 214
Multinuclear anions, 76
Multiphase reaction based on ionic liquids, 63–71
    application in
        biodiesel production, 65–66
        butyl citrate, 71
        hydroxyl condensation reaction, 70
        methyl methacrylate, 64
        synthesis of butyl acetate, 66
        synthesis of dimethyl carbonate, 67–69
        synthesis of glycerol monolaurate, 66–67
Multistage melting processes, 42

N

N,N,N',N'-tetramethyl-1,3-propanediamine dihydrochloride ([TMPDA]Cl$_2$), 103–104
N,N-dimethyl-3-aminophenol, 65–66
N,N'-diisopropylimidazolium iodide (DIPI), 229
n-amyl alcohol acetate, application in synthesis of, 59–60
n-butylammonium nitrate (BAN), 149
N-butylpyridine bisulphate([Bpy][HSO$_4$]), 61
N-butylpyridine phosphate([Bpy][H$_2$PO$_4$]), 61
N-ethylpyridine bisulfate ([Epy][HSO$_4$]), 61
N-heterocyclic cations, 101
N-methyl imidazolium bisulfate ([Hmim][HSO$_4$]), 61
N-methyl-2-pyrrolidone-nickel chloride (NMP-NiCl$_2$), 217
Nano-EuF$_3$, 217–219
Nano-micro clusters, 5
Nanocarbon materials, 192
Nanochannel, 35
Nanocomposites, 205–206
Nanoconfined system, 33–37
Nanomaterials
    dispersions in ionic liquids, 207t
    preparation with ionic liquids, 204–225
    prepared via microemulsion method, 223t
    prepared via microwave-assisted IL method, 221t
    synthesized by hydro-and solvothermal method, 218t
Nanomicrostructures, 3–4
    ionic liquids and understanding of, 1–9
    nanostructure of ionic liquids at different scales, 5–9
    understanding history of ionic liquid structures, 3–5
Nanoparticles (NPs), 204, 217
Nanostructures, 260
    of ionic liquids at different scales, 5–9
        hydrogen bond network, 7–8
        ion pairs, 5–6
        ionic clusters, 8–9
    regulation and process intensification of ionic liquids, 9–13
Natural gas, 79
Nesting doll-like multilayer solid electrolyte, 239
New energy batteries, application of ionic liquids in, 184–197

NH$_3$ separation with ionic liquids, 98−105
NILCs. *See* Nonaqueous solvent−induced ionic
    liquid crystals (NILCs)
NMR. *See* Nuclear magnetic resonance (NMR)
Non-amino ionic liquids, 106−107
Nonaqueous solvent−induced ionic liquid crystals
    (NILCs), 112
Nonpolar methyl/methylene groups, 26
Nonpolar/polar behavior of ILs, 214
NPs. *See* Nanoparticles (NPs)
Nuclear magnetic resonance (NMR), 16, 116
Nucleophilic reagent, 179

O

1-octyl quinoline keys [OQU], 109−110
1-octyl-3-methylimidazolium bromide ([Omim]
    Br), 116−117
1-octyl-3-methylimidazolium chloride ([Omim]
    Cl), 116−117
1-octyl-3-methylimidazolium tetrafluoroborate
    ([Omim][BF$_4$]), 116−117
Open-circuit voltage (VOC), 152−153
Organic flow battery, 195−197
    function of ionic liquid nanoparticles in lithium-
    sulfur flow battery, 196f
Organic separation, 133
Organic solid electrolyte, 188−189
Organic solvents, 10, 74−75, 89, 115−116, 187
Organic-based lithium-sulfur flow battery, 195−196
ORR process. *See* Oxygen reduction reaction
    process (ORR process)
Oxidation, 86−87
    oxidation−reduction flow batteries, 192−193
    process, 159, 165−166
Oxidative chemical processes, 159
Oxidative lignin depolymerisation, 160
Oxidoreductases, 72−73
Oxygen atom, 106−107
Oxygen evolution reaction, 163
Oxygen reduction reaction process (ORR
    process), 143−144
    in [Bmim][BF$_4$] containing small amount of
    water, 162−164
    in aprotic [Bmim][BF$_4$] and effect of ionic liquid
    viscosity, 161−162
    in different ionic liquids electrolyte systems,
    161−164
    in protonic ionic liquid [HNEt$_3$][HSO$_4$], 164

P

p-benzoquinone, 172
p-benzyloxyl phenol (PBP), 164
    degradation mechanism of *p*-benzyloxyl phenol
    in [Bmim][BF$_4$], 167−168, 167t
    degradation mechanism of *p*-benzyloxyl phenol
    in [HNEt$_3$][HSO$_4$], 170−174, 171f
    electrochemical behaviors of *p*-benzyloxyl
    phenol in [Bmim][BF$_4$], 165−166
    electrochemical behaviors of *p*-benzyloxyl
    phenol in [HNEt$_3$][HSO$_4$], 169−170
    electrochemical lignin depolymerization using,
    164−179, 165f
    electrochemical lignin depolymerization using
    guaiacylglycol-β-guaiacyl ether as model
    compounds, 174−179, 178f
p-hydroxybenzenesulfonic acid, 65−66
p-xylene (PX), 84−85
PAHs. *See* Polycyclic aromatic hydrocarbons
    (PAHs)
PC. *See* Propylene carbonate (PC)
PCE. *See* Power conversion efficiency (PCE)
PCET process. *See* Proton-coupled electron
    transfer process (PCET process)
PDADMAFSI. *See* Poly
    (diallyldimethylammonium) bis
    (fluorosulfonyl) imide (PDADMAFSI)
PEC cells. *See* Photoelectrochemical cells (PEC
    cells)
PEMFCs. *See* Polymer electrolyte membrane fuel
    cells (PEMFCs)
Peng−Robinson cubic equation, 114−115
Perovskite
    ionic liquids as additives in perovskite precursor
    solutions, 151−153
    materials, 150
Perovskite solar cells (PSCs), 143
    ionic liquid-modified charge transport layers for,
    155−157
    ionic liquids as solvent for, 153−155
Peroxidases, 72−73
PET. *See* Polyethylene terephthalate (PET)
Petroleum, 79
Phenol (Ph), 103
Phenolic hydroxyl groups (Ph-OH), 174−176
Phenolic radicals, 172
Phosphoric acid, 88−89
Photocatalysis, 86, 89

Photocatalysts, 144
Photocatalytic processes, 144
Photocatalytic reactions, 147–148
Photocatalytic systems, application of ionic liquids
        in, 144–148
Photochemical reactions, 143
Photocurrents, 149
Photoelectrocatalytic systems application of ionic
        liquids in, 148–150
Photoelectrochemical cells (PEC cells), 148
Photoelectrochemical process, 144
Photoelectrochemical reduction, 149
Photoredox catalyst, 148
Photosensitizer, 148
Physical vapor deposition process (PVD process),
        38–39
PILMs. *See* Poly(ionic liquid) membranes (PILMs)
PILs. *See* Polymeric ILs (PILs); Protic ionic liquids
        (PILs)
Plasticizer, 131
PO. *See* Propylene oxide (PO)
Polar solvents, 47–48
Polarization, 21–25
    catastrophe, 26
    models, 26
Polarized force field, 25–26
Poly(diallyldimethylammonium) bis(fluorosulfonyl)
        imide (PDADMAFSI), 239
Poly(ionic liquid) membranes (PILMs), 128,
        130–131
Poly(ionic liquid)s (PILs), 147, 203
    applications, 239–245
        catalysts and catalyst supported, 242–243
        dye-sensitized solar cells, 241
        lithium batteries and capacitors, 239–240
        poly (ionic liquid)-based antibacterial
            materials, 245
        proton exchange membrane fuel cells,
            241–242
        separations, 244–245
    synthesis of, 237–238
        copolymer-type ionic liquids, 238
        polyanionic ionic liquids, 237–238
        polycationic ionic liquids, 237
        polyzwitterion-type ionic liquids, 238
Poly(vinyl alcohol) (PVA), 241
Poly(vinylimidazolium)bromide (PVImBr), 242
Polyamide chains, 130

Polyamide membranes, 130
Polyamide/polyacrylonitrile membrane (PA/PAN
        membrane), 131–133
Polyaniline core decorated with TiO$_2$ (PANI-
        TiO$_2$), 220–225
Polyanionic ILs, 236–238
Polycationic ILs, 236–237
Polycyclic aromatic hydrocarbons (PAHs), 206
Polydimethylsiloxane (PDMS), 133
Polyethylene terephthalate (PET), 243
"Polyimide based PIL in salt" ionic gel, 239
Polyionic liquid catalyst, 66
Polymer electrolyte, 187
Polymer electrolyte membrane fuel cells
        (PEMFCs), 241–242
Polymer solid electrolyte, 188–189
Polymeric ILs (PILs), 63–64
Polymerized ion, 64
Polymerized ionic liquids
    applications of poly (ionic liquid)s, 239–245
    synthesis of poly(ionic liquid)s, 237–238
Polyoxometalate-IL (POM-IL), 87
Polypropylene glycol 400 (PPG400), 117–118
Polyunsaturated fatty acid (PUFA), 114–115
Polyvinylidene fluoride (PVDF), 205–206
Polyzwitterion-type ionic liquids, 236, 238
POM-IL. *See* Polyoxometalate-IL (POM-IL)
Porous materials, 189–190
Porous polyamino benzenesulfonic acid, 66–67
Potassium hexafluorophosphate (KPF6),
        152–153
Power conversion efficiency (PCE), 150
Pressure-induced method, 129
1,2-propanediol anion, 62–63
1,3-propanesultone, 65–66
3-propyl-1,2-dimethylimidazole diimide salt
        ([DMPim][TFSI]), 185
1-propyl-3-methylimidazolium bis
        (trifluoromethylsulfonyl)imide ([Pmim]
        [NTf$_2$]), 89–90
1-propyl-3-methylimidazolium iodide, 149–150
Propylene carbonate (PC), 243
Propylene oxide (PO), 62–63
Proteases, 72–75
Protein phosphatase 2 A (PP2A), 123–124
Proteins, 117
    extraction
        ionic liquids for, 117–126

Proteins (*Continued*)
  ionic liquids for single protein extraction, 117–118
  system, 126
ionic liquids
  molecular docking simulation results for protein A-IgG and i-TAN system, 127*f*
  for protein and protein complex extraction, 117–128
  for protein complex extraction, 126–128
  for protein mixture extraction, 118–126
Protic hydrogens, 101
Protic ionic liquids (PILs), 99–101, 103–104
Proton exchange membrane fuel cells, 241–242
Proton nuclear magnetic resonance spectroscopy (^{1}H NMR spectroscopy), 60–61
Proton-coupled electron transfer process (PCET process), 159, 163–164
Proton-type ionic liquids, 7–8
Protonic ionic liquids, 8
  oxygen reduction reaction behaviors in, 164
PSCs. *See* Perovskite solar cells (PSCs)
PVA. *See* Poly(vinyl alcohol) (PVA)
PVD process. *See* Physical vapor deposition process (PVD process)
PVDF. *See* Polyvinylidene fluoride (PVDF)
PVImBr. *See* Poly(vinylimidazolium)bromide (PVImBr)
PX. *See* p-xylene (PX)
Pyridine-based IL nanoparticles, 195–196

Q

QAEP. *See* Quaternized diaminoethylpiperzine (QAEP)
QM. *See* Quantum mechanics (QM)
Quantum dots, 147–148
Quantum mechanics (QM), 16–17
Quaternary ammonium IL, 185–186
Quaternized diaminoethylpiperzine (QAEP), 133

R

Radial distribution function (RDF), 35
Radical polymerization, 237
Rare earth complexes, ionic liquid solvents in synthesis of, 231–232
RB. *See* Rose Bengal (RB)
RB-SILLPs. *See* Rose Bengal immobilized on supported IL-like (RB-SILLPs)

RDF. *See* Radial distribution function (RDF)
Reaction process, 66–67, 258, 261
  ionic liquid nanostructures intensifying, 10
Reaction-transfer coupling law, 262–263
Reaction–amplification law, 2–3
Reactive oxygen species, 165–166
Redox reaction, 189
Relative entropy minimization (REM), 27
REM. *See* Relative entropy minimization (REM)
Renewable resources, 79
RESP method. *See* Restrained ESP method (RESP method)
Restrained ESP method (RESP method), 20–21
Reversible addition-fragmentation chain transfer (RAFT) polymerization, 237
Rhizobium, 79
Rhodococcus R312, 75–76
Ring-opening metathesis polymerization, 237
Room temperature ILs (RTIL), 114–115
Rose Bengal (RB), 147–148
Rose Bengal immobilized on supported IL-like (RB-SILLPs), 147–148
Rotating ringdisk electrode method (RRDE method), 161
RRDE method. *See* Rotating ringdisk electrode method (RRDE method)
RTIL. *See* Room temperature ILs (RTIL)

S

Sampling method, 28–29
SAPT. *See* Symmetry-adapted perturbation theory (SAPT)
Scanning electron microscopy techniques (SEM techniques), 64
SDA. *See* Structure-directing agent (SDA)
SDF. *See* Spatial distribution function (SDF)
SEM techniques. *See* Scanning electron microscopy techniques (SEM techniques)
Semiempirical models, 262–263
Semisolid flow batteries, 192–193
Separations, 244–245
  ionic liquid structures intensifying, 11–12
  process, 9–10, 16, 97, 260
Silica-based zeolites, 228–229
SILMs. *See* Supported IL membranes (SILMs)
SILP. *See* Supported IL phase (SILP)
Simulation
  methods, 259–260

and regulation of two-dimensional ionic liquids, 37—49
  study of ionic liquid structures in interface, 30—37
    nanoconfined system, 33—37, 34*f*
    structure and wetting behavior of ionic liquids at solid surface, 30—33
Single protein extraction, ionic liquids for, 117—118
SIZ-7, 228
SIZ-13, 227
Sodium dodecyl sulfate polyacrylamide gel electrophoresis (SDS-PAGE), 118
Sodium nitrate (NaNO₃), 148
Software for molecular simulation, 29—30
Sol-gel method, 214
Solar cells, application of ionic liquids in, 150—158
Solid acids, 85
Solid electrolyte, 187—189, 191
Solid surface, structure and wetting behavior of ionic liquids at, 30—33
Solid-surface-supported IL (SSIL), 37
Solidified liquid layer, 31—32
Solid—liquid extraction methods, 118
Solvation, 203
Solvents, 260
  extraction, 108
  ionic liquids as solvent for perovskite solar cells, 153—155
  solvent-free ionic solutions, 5—6
Sorafenib, 124—125
Sorbitol, 83
Spatial distribution function (SDF), 50
SSIL. *See* Solid-surface-supported IL (SSIL)
Steric hindrance, 203
Structure-directing agent (SDA), 225—226
Structure—property relationship, 258, 260
Succinic acid diisopropyl ester, 60—61
  application in synthesis of, 60—61
Succinimide (Si), 103
Sulfides, 217
Supercapacitors
  application of ionic liquids in, 189—192
  electrode materials for, 192
  ionic liquids as supercapacitor support electrolyte, 189—191
    double electric layer capacitor, 189*f*
Supported IL membranes (SILMs), 128

Supported IL phase (SILP), 147
Symmetry-adapted perturbation theory (SAPT), 26
Systematic studies, 261

T

TCF. *See* Time correlation function (TCF)
TEG. *See* Triethylene glycol (TEG)
Terpenes, 113
4-tert-butylpyridine (t-BP), 157
Tetramethylammonium hydroxide (TMAH), 155—156
TGA. *See* Thermogravimetric analysis (TGA)
Thermodynamic equilibrium state, 17
Thermodynamic first-order response coefficients, 49
Thermogravimetric analysis (TGA), 61—62, 64
Three-dimensional network, 8—9
Three-dimensional space, 3—4
Time correlation function (TCF), 49—50
TON. *See* Turnover number (TON)
Traditional capacitors, 189
Traditional catalytic process, 257
Traditional chemical processes, 263
Traditional membrane separation process, 133
Traditional models, 263
Traditional organic solvents, 188—189
Traditional salts, 1
Traditional surfactant extraction system, 128
Traditional volatile organic solvents, 79—81
Transmembrane proteins, 125—126
Transport mechanism, 135
Trastuzumab, 126
1,2,4-triazole (Triz), 103
Trichoderma asperellum ZJPH0810, 78—79
Triethylamine acetate (TEAA), 126
Triethylammonium dihydrogen phosphate (TEAP), 72—73
Triethylene glycol (TEG), 102
  triethylene glycol-chelated ILs, 102
Trihexyl tetradecane phosphonium bromide ethyl citrate, 115
Trihexyltetradecylphosphinium [THTDP], 109—110
Trimethylammonium nitrate, 7—8
Triple-cation-based perovskite, 151
Tris(2-hydroxyethyl) methyl-ammonium methylsulfate ([MTEOA][MeOSO₃]), 98—99

Trypsin, 111–112
Turnover number (TON), 62–63
Two-dimensional gel electrophoresis (2-DE), 118
Two-dimensional ionic liquids (2DIIs), 37
  with anomalous stepwise melting process and
    ultrahigh CO_2 adsorption capacity, 42–46,
    43f
  electron transfer and friction feature of two-
    dimensional ionic liquids, 46–49, 48f
  on graphite surface via PVD, 38f
  ionic liquid islands, 37–42
  simulation and regulation of, 37–49

U
UA. See United atoms (UA)
UGTs. See Uridine diphosphoglucuronosyl
    transferases (UGTs)
Ultrahigh CO_2 adsorption capacity
  high CO_2 adsorption capacity of 2D ILs, 45f
  two-dimensional ionic liquids with anomalous
    stepwise, 42–46
United atoms (UA), 19–20
Uridine diphosphoglucuronosyl transferases
  (UGTs), 123
UV-visible absorbance spectrum, 145–146

V
Vacuum-induced method, 129
Valence band (VB), 147
Value-added chemicals, 83, 85
Van der Waals (vdW), 33–35, 203
  force, 1–2
  interaction energy, 162
  interactions, 33–35
Vanadium, 87
  extraction, 116–117
  vanadium-water solution system, 194
Vanillin, 176–177, 182

VB. See Valence band (VB)
vdW. See Van der Waals (vdW)
Vibrio sp. Q67 cells, 78
Vinyl carbonate, 67
1-vinyl imidazole, 237
1-vinyl-3-butyl imidazolium tetrafluoroborate
  (VBImBF$_4$), 133
Vitamin E molecules, 112–113
VOC. See Open-circuit voltage (VOC)
Volatile organic solvents, 60
Volatility of organic solvents, 193–194
Voltammograms, 196–197

W
Water (H_2O), 46, 61
  desalination, 133–134
  environment, 7
  oxygen reduction reaction behaviors in[Bmim]
    [BF$_4$] containing small amount of,
    162–164
  water-based flow battery, 194–195
Wetting processes, 32–33
Whole cell catalysis, 75–79
  application of whole cell catalysis with ionic
    liquids, 78–79
Whole cell catalyst, 77–78
Whole proteome extraction, ionic liquids for,
  122–126

X
X-ray diffraction, 64

Z
Z-scheme heterojunction, 145
Zeolite, 225
Zinc dendrites, 195
Zinc-bromine flow battery, 195